D1238633

Dioxins and Dioxin-like Compounds

in the Food Supply

Strategies to Decrease Exposure

Committee on the Implications of Dioxin in the Food Supply

Food and Nutrition Board

INSTITUTE OF MEDICINE
OF THE NATIONAL ACADEMIES

THE NATIONAL ACADEMIES PRESS
Washington, D.C.
www.nap.edu

THE NATIONAL ACADEMIES PRESS   500 Fifth Street, N.W.   Washington, DC   20001

NOTICE: The project that is the subject of this report was approved by the Governing Board of the National Research Council, whose members are drawn from the councils of the National Academy of Sciences, the National Academy of Engineering, and the Institute of Medicine. The members of the committee responsible for the report were chosen for their special competences and with regard for appropriate balance.

Support for this project was provided by the U.S. Food and Drug Administration of the U.S. Department of Health and Human Services and the U.S. Department of Agriculture under contract number 223-99-2321. The views presented in this report are those of the Institute of Medicine Committee on the Implications of Dioxin in the Food Supply and are not necessarily those of the funding agencies.

Library of Congress Cataloging-in-Publication Data

Dioxins and dioxin-like compounds in the food supply : strategies to decrease exposure / Committee on the Implications of Dioxin in the Food Supply, Food and Nutrition Board.
    p. ; cm.
Includes bibliographical references.
    ISBN 0-309-08961-1 (pbk.); 0-309-52548-9 (PDF)
    1. Dioxins—Toxicology. 2. Food contamination.
    [DNLM: 1. Dioxins—adverse effects. 2. Food Contamination—prevention & control. 3. Environmental Monitoring—methods. 4. Food Supply—standards. 5. Risk Management. WA 701 D595 2003] I. National Academy of Sciences (U.S.). Committee on the Implications of Dioxin in the Food Supply.
    RA1242.D55D573 2003
    615.9′512—dc22

                         2003016085

Additional copies of this report are available from the National Academies Press, 500 Fifth Street, N.W., Lockbox 285, Washington, DC 20055; (800) 624-6242 or (202) 334-3313 (in the Washington metropolitan area); Internet, http://www.nap.edu.

For more information about the Institute of Medicine, visit the IOM home page at: **www.iom.edu.**

"Knowing is not enough; we must apply.
Willing is not enough; we must do."
—Goethe

INSTITUTE OF MEDICINE
OF THE NATIONAL ACADEMIES

Shaping the Future for Health

THE NATIONAL ACADEMIES
Advisers to the Nation on Science, Engineering, and Medicine

The **National Academy of Sciences** is a private, nonprofit, self-perpetuating society of distinguished scholars engaged in scientific and engineering research, dedicated to the furtherance of science and technology and to their use for the general welfare. Upon the authority of the charter granted to it by the Congress in 1863, the Academy has a mandate that requires it to advise the federal government on scientific and technical matters. Dr. Bruce M. Alberts is president of the National Academy of Sciences.

The **National Academy of Engineering** was established in 1964, under the charter of the National Academy of Sciences, as a parallel organization of outstanding engineers. It is autonomous in its administration and in the selection of its members, sharing with the National Academy of Sciences the responsibility for advising the federal government. The National Academy of Engineering also sponsors engineering programs aimed at meeting national needs, encourages education and research, and recognizes the superior achievements of engineers. Dr. Wm. A. Wulf is president of the National Academy of Engineering.

The **Institute of Medicine** was established in 1970 by the National Academy of Sciences to secure the services of eminent members of appropriate professions in the examination of policy matters pertaining to the health of the public. The Institute acts under the responsibility given to the National Academy of Sciences by its congressional charter to be an adviser to the federal government and, upon its own initiative, to identify issues of medical care, research, and education. Dr. Harvey V. Fineberg is president of the Institute of Medicine.

The **National Research Council** was organized by the National Academy of Sciences in 1916 to associate the broad community of science and technology with the Academy's purposes of furthering knowledge and advising the federal government. Functioning in accordance with general policies determined by the Academy, the Council has become the principal operating agency of both the National Academy of Sciences and the National Academy of Engineering in providing services to the government, the public, and the scientific and engineering communities. The Council is administered jointly by both Academies and the Institute of Medicine. Dr. Bruce M. Alberts and Dr. Wm. A. Wulf are chair and vice chair, respectively, of the National Research Council.

www.national-academies.org

v

Reviewers

This report has been reviewed in draft form by individuals chosen for their diverse perspectives and technical expertise, in accordance with procedures approved by the NRC's Report Review Committee. The purpose of this independent review is to provide candid and critical comments that will assist the institution in making its published report as sound as possible and to ensure that the report meets institutional standards for objectivity, evidence, and responsiveness to the study charge. The review comments and draft manuscript remain confidential to protect the integrity of the deliberative process. We wish to thank the following individuals for their review of this report:

Ransom L. Baldwin, University of California, Davis
Alfred O. Berg, University of Washington School of Medicine
Tim Byers, University of Colorado Health Sciences Center
Edward M. Cooney, Congressional Hunger Center
Joanna T. Dwyer, Tufts New England Medical Center
George Fries, Silver Spring, Maryland
Helen H. Jensen, Iowa State University
J. Michael McGinnis, Robert Wood Johnson Foundation
Lars Noah, University of Florida College of Law
Timothy D. Phillips, Texas A&M University
Frank E. Speizer, Wolfson Institute of Preventive Medicine
Virginia A. Stallings, Children's Hospital of Philadelphia
Lynn B. Willett, The Ohio State University

Although the reviewers listed above have provided many constructive comments and suggestions, they were not asked to endorse the conclusions or recommendations nor did they see the final draft of the report before its release. The review of this report was overseen by Patricia A. Buffler, University of California, Berkeley, and John C. Bailar, III, University of Chicago. Appointed by the NRC and IOM, they were responsible for making certain that an independent examination of this report was carried out in accordance with institutional procedures and that all review comments were carefully considered. Responsibility for the final content of this report rests entirely with the authoring committee and the institution.

Preface

Dioxin and dioxin-like compounds, or DLCs, are found throughout the environment: in soil, water sediments, and air. Exposure to these unintentional contaminants occurs largely through the food supply, although at low levels. Even though DLC levels in the environment and in foods have been declining over the past several decades, public concern persists because DLCs are widespread, and there remains a great deal of uncertainty about their potential adverse health effects. There is additional concern regarding the potential impact of exposure on the population groups that are particularly sensitive to exposure to toxic compounds (e.g., developing infants) and those that are more highly exposed than the general population (e.g., breastfeeding infants and groups for whom fish and wild game are important subsistence or cultural food sources).

Against the background of increasing knowledge about potential public health effects of dioxins in the food supply, the National Science and Technology Council's Interagency Working Group (IWG) on Dioxin recognized the need for an independent group to examine the scientific evidence of the impact of the presence of dioxins and related compounds in the food supply on food safety and to assess options to reduce exposure to dioxins through foods that would maintain optimal health and nutritional status for the population. IWG, with support from the U.S. Department of Agriculture, the U.S. Department of Health and Human Services, and other agencies and sponsors, asked the National Academies to explore these issues. Thus, the Institute of Medicine, in consultation with the National Research Council, convened a committee with a diverse background and a broad scope of expertise to address the task put to them by the sponsors.

The committee was charged to review the scientific evidence to identify potential ways to reduce the level of DLCs in foods. Importantly, the committee was instructed to consider the need to protect the public and to assess the potential impact of its recommendations to food and nutrition policies, particularly public education and food assistance programs.

The study sponsors recognized that data limitations would allow only qualitative estimates of net risk, descriptions of exposure reduction, and identification of data needs. In response, the committee offered the sponsoring agencies an array of options to reduce DLC exposure through foods while gathering the data needed to determine future action. These options should also be of use to stakeholders in the agricultural and food-processing industries and to public health professionals.

The final recommendations are selected from the array of options and reflect the committee's conclusion that, because uncertainties exist, the level of risk must be determined with greater certainty before regulatory action can be taken, and that it is in the public's best interest for the government to develop a strategic action plan that includes interim steps to reduce exposure as long as the steps do not lead to undesirable consequences to human health and do not impinge unduly upon the cultural norms of the population.

ACKNOWLEDGMENTS

The Committee on the Implications of Dioxin in the Food Supply was aided in its challenging tasks by the invaluable contributions of a number of individuals. First and foremost, many thanks are due to the committee members who volunteered countless hours to the research, deliberations, and preparation of the report. Their dedication to this project and to a stringent timeline was commendable and was the basis of our success.

Many individuals volunteered significant time and effort to address and educate our committee members during the workshops, and the committee thanks them. The committee wishes to acknowledge the invaluable contributions of the FNB study staff, Ann Yaktine, study director; Tazima Davis, research assistant; and Sanait Tesfagiorgis and Sybil Boggis, senior project assistants. The committee is grateful for the contributions of Tina Rouse and Abigail Mitchell, program officers. Special thanks also to Roberta Wedge, senior program officer, and Jamie Jonker, program officer, for their helpful review and suggestions. The committee also thanks Geraldine Kennedo for logistical arrangements, Marcia Lewis for drawing the chemical structures, and Gail Spears for technical editing. This project benefited from the general guidance of Allison Yates, director, and Linda Meyers, deputy director, of the Food and Nutrition Board.

ROBERT S. LAWRENCE, *Chair*
Committee on the Implications of Dioxin in the Food Supply

Contents

Acronyms and Glossary

ORGANIZATIONS, PROGRAMS, STUDIES

AAP	American Academy of Pediatrics
AHA	American Heart Association
ATSDR	Agency for Toxic Substances and Disease Registry
CDC	Centers for Disease Control and Prevention
CERCLA	Comprehensive Environmental Response, Compensation, and Liability Act
CFIA	Canadian Food Inspection Agency
CSFII	Continuing Survey of Food Intakes by Individuals
EC	European Commission
EPA	U.S. Environmental Protection Agency
EU	European Union
FAO	Food and Agriculture Organization of the United Nations
FDA	U.S. Food and Drug Administration
FDCA	Food, Drug and Cosmetics Act
FSIS	Food Safety and Inspection Service
JECFA	Joint FAO/WHO Expert Committee on Food Additives
HACCP	Hazard Analysis and Critical Control Point
HHS	U.S. Department of Health and Human Services
IARC	International Agency for Research on Cancer
NHANES	National Health and Nutrition Examination Survey
TDS	Total Diet Study
UK	United Kingdom

USDA	U.S. Department of Agriculture
WHO	World Health Organization
WIC	Special Supplement Nutrition Program for Women, Infants and Children

TERMS

CDD	chlorinated dibenzo-*p*-dioxin
CDF	chlorodibenzofuran
DLC	dioxin-like compound; in this report, DLCs are the dioxins, furans, and polychlorinated biphenyls that possess dioxin-like activity
dscf	dry standard cubic feet
dscm	dry standard cubic meter
ED_{01}	1 percent effective dose
GAP	Good Agricultural Practice
GMP	Good Manufacturing Practice
HCDD	hexachlorodibenzo-*p*-dioxin
HMIWI	hospital/medical/infectious waste incinerator
HxCDF	hexachlorodibenzo-*p*-furan
I-TEF	international toxicity equivalency factor system
I-TEQ	international toxicity equivalents
LED_{01}	lower 95 percent confidence interval on the 1 percent effective dose
MRL	minimum risk level
N-TEQ	Nordic toxicity equivalents
OCDD	octachlorodibenzo-*p*-dioxin
OCDF	octachlorodibenzofuran
PCB	polychlorinated biphenyl
PCDD	polychlorinated dibenzo-*p*-dioxin
PCDF	polychlorinated dibenzofuran
PCP	pentachlorophenol
PeCDD	pentachlorodibenzo-*p*-dioxin
PeCDF	pentachlorodibenzofuran
POPs	persistent organic pollutants
SMR	standardized mortality rate
TCDD	2,3,7,8-tetrachlorodibenzo-*p*-dioxin
TCDF	2,3,7,8-tetrachlorodibenzofuran
TDI	tolerable daily intake
TEF	toxicity equivalency factor
TEQ	toxicity equivalents
TWI	tolerable weekly intake

GLOSSARY

Aerobic—Growing, living, or occurring in the presence of molecular oxygen; for example, bacteria that require oxygen to survive.

Anaerobic—Growing, living, or occurring in the absence of molecular oxygen.

Beluga whale fat—Fat of the white whale, *Delphinapterus leucas*; the fat is commonly called blubber.

Body burden—The total amount of a chemical, metal, or radioactive substance present at any time after absorption in the body of a human or animal.

Brominated—Combined or saturated with bromine or any of its compounds.

Carcinogen—An agent capable of initiating development of malignant tumors; may be a chemical, a form of electromagnetic radiation, or an inert solid body.

Chlorinated—Any organic chemical that includes chlorine atoms; chlorinated organic compounds, along with other halogenated organics, have been implicated in health risks such as cancer, endocrine system disruption, birth defects, compromised immune systems, and reduced fertility.

Congener—One of two or more compounds of the same kind with respect to classification.

De novo—Anew; often applied to particular biochemical pathways in which metabolites are newly biosynthesized (e.g., de novo purine biosynthesis).

Dose—A quantity to be administered at one time, such as a specified amount of medication.

Epidemiology—The study of the distribution and determinants of health-related states and events in populations and the control of health problems; the study of epidemic disease.

Exposure—The condition of being subjected to the effects of a substance, such as infectious agents, that may have a harmful effect.

Genotoxin—A toxin (poisonous substance) that harms the body by damaging DNA molecules, causing mutations that may lead to tumors or neoplasms.

Halogenated—refers to a chemical compound or mixture that contains halogen atoms; halogen refers to those elements in the seventeenth column of the periodic table: fluorine, chlorine, bromine, iodine, and astatine.

Heterogeneous—Not of uniform composition, quality, or structure.

Hydrolysis—The splitting of a compound into fragments by the addition of water, the hydroxyl group being incorporated in one fragment and the hydrogen atom in the other.

Immunologic—Pertaining to immunology, a subfield of biology that deals with the study of antigens and the immune process and how humans and higher animals fight off disease.

Isotope dilution—A technique using radioactive tracers that can be used to determine the amount of a single substance in a mixture.

Lactation—The period of the secretion of milk.

Lipophilic—An element that has an affinity for lipid.

Muktuk—The skin and underlying fat (blubber) layer of a whale.

Narwhal mattak—The skin and underlying fat (blubber) layer of the whale *Monodon monoceros*. Mattak is commonly used as the dialect representation of muktuk in Baffin Inuktitut language.

Neurobehavior—Neurological status as assessed by observation of behavior.

Neurodevelopment—Development of the central and peripheral nervous systems starting at conception and going through the life span of an organism.

Persistent organic pollutant—Chemical substance that persists in the environment, bioaccumulates through the food web, and poses a risk of causing adverse effects to human health and to the environment.

Photolysis—Light induced cleavage of a chemical bond, as in the process of photosynthesis.

Relative risk—Rate of the outcome of interest in a population compared with the rate in the reference population.

Temporal—Pertaining to time; limited as to time.

Toxicity—The quality of being poisonous, especially the degree of virulence of a toxic microbe or of a poison.

Toxicokinetic modeling—The time course of disposition (absorption, distribution, biotransformation, and excretion) of xenobiotics (foreign chemicals to which organisms are exposed) in the whole organism.

Tropospheric—Pertaining to the lower layer of the earth's atmosphere in which the change of temperature with height is relatively large; it is the region where clouds form, convection is active, and mixing is continuous and more or less complete.

Vapor phase—Phase when substances transition from a liquid state to gaseous state through the breaking of molecular bonds.

Volatile—Readily vaporizable at a low temperature.

Dioxins and Dioxin-like Compounds

in the Food Supply

Strategies to Decrease Exposure

Executive Summary

Dioxins and chemically-related compounds (referred to collectively as DLCs) occur as widespread, low-level contaminants in animal feeds and the human food supply. Because dioxins accumulate in fatty tissues, consumption of animal fats is thought to be the primary pathway for human exposure. In humans, dioxins are metabolized slowly and accumulate in body fat over a lifetime. Dioxin toxicity and its human health impact have been the subjects of recent re-evaluations by the International Agency for Research on Cancer, the U.S. Agency for Toxic Substances and Disease Registry, the National Institute of Environmental Health Sciences, and the U.S. Environmental Protection Agency (EPA). Data indicate declining levels of dioxin in the environment and in human tissues, although the assessments prepared by the agencies differ and have not yet been reconciled.

Notwithstanding the declining overall levels, public concern about food safety issues such as endocrine disruptors in the food supply and the effects of dioxin-like compounds on children's health and development persists. Further, special populations that consume large amounts of fish and wildlife for cultural reasons (American Indian and Alaska Native tribes) and subsistence fishers have eating patterns that place them at higher risk for exposure levels that may be found to be dangerous.

Against this background, the National Science and Technology Council's Interagency Working Group (IWG) on Dioxin anticipated the need to develop policies to reduce dioxin exposure. IWG, with support from the U.S. Department of Agriculture (USDA), the U.S. Department of Health and Human Services, and other agencies and sponsors asked the National Academies to assist them by identifying potential strategies to meet this need.

The sponsors explicitly enjoined the committee from re-examining the question of whether low doses of dioxin are toxic or to what degree in general or specific populations. Separate scientific reviews have been initiated to reassess and reconcile the different exposure analyses. However, the sponsors anticipated that regulatory and other public health policies would likely be necessary to reduce exposures, especially among vulnerable populations, and foresaw the need to be prepared with evidence-based strategies.

The sponsors charged the Committee on the Implications of Dioxin in the Food Supply to review the scientific evidence to identify potential ways to reduce the levels of dioxin in food, taking into account the need to promote good nutrition and health. Specifically, the committee was charged to:

- Take into account the substantial body of data available in the EPA draft reassessment and other reports on the pathways by which DLCs add to the dioxin body burden by concentrating in food from sources such as animal feeds that contain recycled animal fat or sources contaminated by airborne DLCs, and through the intake of specific foods such as seafood, foods of animal origin (e.g., eggs, dairy products, meat), and plant foodstuffs,
- Review the data on food-consumption patterns of various subgroups in the population that appear to be at increased risk due to physiological state, food practices, or geographic location,
- Identify and describe possible risk-management options that could be instituted to decrease the content of dioxins in food animals, seafood, and other food products, and possible changes in food and/or nutrition policies that would decrease exposure, including, where possible, an assessment of the net risk reduction afforded by a risk-management option, including effects on nutrition, and
- Identify and describe efforts in the United States and other countries to decrease dioxin exposures of specific subgroups of the population through public health or risk communication initiatives, and assess the extent to which federal food and nutrition policies contribute to decreasing exposure to dioxins.

The study sponsors recognized that the limited data available would likely allow only qualitative estimates of net risk, descriptions of exposure reduction, and identification of data needs. The committee did not make any judgments about the risks of human exposure of DLCs through food. Rather, it offered options available to the government to reduce this type of exposure and to increase benefits to nutrition and health, while gathering the data needed to determine future action.

The committee concluded that although direct health effects currently cannot be measured, animal and human epidemiological studies support exposure reduction while potential health risks are more fully studied. In its analysis, the committee developed and used a framework to identify and evaluate options to reduce exposure and suggested actions that the government might take as it continues to collect the data needed to devise a more comprehensive, long-term risk-management strategy.

This report is organized around three specific pathways that lead to DLC exposure through the food supply: (1) animal production systems, (2) human foods, and (3) food-consumption patterns.

Animal Production Systems

DLCs enter the food chain when airborne contaminants that have been deposited on plants or in soil and sediment are taken up by food animals or fish (e.g., through grazing or the direct consumption of feeds that contain DLC-contaminated plant- and animal-based ingredients). Due to geographic variation in DLC levels that may result from long-distance air transport and deposition, naturally occurring and unintended contamination events, and different animal husbandry practices, the exposure of food animals to DLCs in forage and feed varies by region, differences in DLC levels in feed ingredients, and the combination of ingredients in different feeds. The resulting concentration of DLCs in food animals is a consequence of accumulated exposure from these various sources as the animals' body fat increases.

Human Foods

DLCs accumulate in human foods through the animal production systems pathway. Levels of exposure through foods may vary regionally, depending on the amount of locally produced food that is consumed. Most foods, except milk, are processed in bulk and distributed widely, so that foods purchased by consumers are less likely to reflect variable DLC levels. Thus, the exposure of the general population through the food supply appears to be relatively uniform. On the other hand, foods that are caught or harvested in the wild have levels of DLCs that reflect those found in the local environment, so some populations that rely on locally caught fish and wildlife for food may be at risk for higher exposure levels.

Food-Consumption Patterns

As a component of its data-gathering process, the committee commissioned an analysis of the population's DLC intake from foods at current levels of consumption. The analysis presented in this report was based on DLC values gathered in the U.S. Food and Drug Administration's (FDA) Total Diet Study (a market basket survey) and linked to the Continuing Survey of Food Intakes by Individuals, which measures actual food intake. Dietary intake scenarios, based on the analysis, were used to predict the potential for reducing exposure to DLCs by decreasing meat and fish intake and by substituting low-fat (1 percent fat) or skim milk for whole (3.5 percent fat) milk. The scenarios indicated that the greatest exposure to DLCs through food was from animal fats found in meats, full-fat dairy products, and fatty fish.

FRAMEWORK FOR POLICY OPTIONS

The committee was charged by its sponsors to identify, evaluate, and recommend policy options to reduce dietary exposure to DLCs, while taking into consideration the need to maintain good nutrition and health. The committee was also charged to evaluate the net exposure reduction afforded by various risk-management options, including nutrition options, in light of efforts in the United States and abroad to decrease the exposure of sensitive or otherwise vulnerable populations.

The framework for risk-management options established a systematic approach to identify, evaluate, and recommend potential interventions to reduce the exposure of humans to DLCs through the food supply. The committee approached the development of options from the perspective of the three pathways discussed above and it considered the potential nutritional consequences of dietary modifications to reduce DLC exposure through risk-relationship analysis. For each option, the committee considered:

1. Alternate or interim actions,
2. Current barriers to implementation,
3. Anticipated DLC exposure reduction achievable through implementation, and
4. Risk relationships that included decreases (ancillary benefits) or increases (countervailing risks) in other risks.

POLICY OPTIONS TO REDUCE EXPOSURE TO DLCs THROUGH THE FOOD SUPPLY

The discussion of options to reduce DLC exposure through the food supply must consider food safety statutes and other regulatory policies and procedures that frame and constrain the adoption of exposure reduction options. Within the federal government, FDA has the primary food safety regulatory jurisdiction over DLCs in food. EPA considers food safety and acceptable levels of DLCs in fish when setting acceptable air and water emission levels for DLCs. FDA is the enforcement agency for DLCs in animal feed and human food, with the exception of meat and poultry, which are under the jurisdiction of USDA's Food Safety and Inspection Service (FSIS). FDA has broad authority to control DLCs in animal feed and human food, constrained by the requirement to make factual showings concerning DLCs and their risk in order to implement regulatory control.

The committee considered a range of possible interventions, but not all options were put forward as recommendations. Options considered to reduce DLC exposure through animal production systems included:

- Require testing for DLC levels in forage, feed, and feed ingredients,
- Establish tolerance levels for DLCs in forage, feed, and feed ingredients,
- Restrict the use of animal products, forage, feed, and feed ingredients that originate from specific areas that are considered to be contaminated, and
- Restrict the use of animal by-products in agriculture, animal husbandry, and manufacturing processes.

Options considered to reduce DLC exposure through the food supply included:

- Require testing and publishing of data on DLC levels in the human food supply, including food products, dietary supplements, and breast milk, to use in establishing tolerance levels in foods,
- Establish enforceable standards for DLC levels in processed foods and in packaging that comes in direct contact with food, and
- Require cleaning or washing practices for all vegetable, fruit, and grain crops that potentially had contact with soil.

Options considered to reduce DLC exposure through food-consumption pathways included:

- Increase the availability of low-fat and skim milk in federal feeding programs targeted to children, which currently favor the provision of whole milk (e.g., the National School Lunch Program),
- Establish a maximum saturated fat content for meals served in schools that participate in federal child nutrition programs, and
- Promote changes in dietary-consumption patterns of the general population that more closely conform to recommendations to reduce consumption of animal fats, such as the recommendations of the Dietary Guidelines for Americans.

Consideration of Consequences of Actions to Reduce Exposure to DLCs Through Dietary Intervention

Because of the persistent nature of DLCs and the uncertainty about their toxic effects, almost any action taken to reduce their concentration in the food supply will have a long-term, rather than an immediate, impact on human health. Such actions will, however, have an immediate impact on the food supply, which could, in turn, have other health and nutritional effects.

With regard to the possible detrimental nutritional effects due to changes in food-consumption patterns, the committee noted that current dietary recommendations for the general population stress the benefits of a reduced intake of saturated fats to decrease the risk of many chronic diseases. Thus, changes in dietary patterns to reduce DLC exposure that involve a reduction of animal fats, the primary source of saturated fats in our diet, would generally have beneficial rather than detrimental nutritional effects.

Apart from the general population, the committee also took into account DLC-sensitive population groups, such as developing fetuses and infants, that are vulnerable due to developmental immaturity, and groups such as breastfeeding infants, subsistence fishers, and American Indian and Alaska Native fish-eating populations that receive higher-than-background levels of DLC exposure. Although not a highly exposed population, preadolescent and teenage girls and young women were of concern to the committee because they accumulate, over time, body burdens of DLCs that can, when they enter their child-bearing years, become a potential source of exposure for their developing infants in utero and while breastfeeding.

Recommendations to reduce exposure were tailored to the particular concerns of these groups, including weighing the benefits of breastfeeding against the risks of DLC exposure for infants, early intervention to reduce lifetime body burdens in children, especially girls, and cultural practices important to American Indian and Alaska Native tribes and other groups.

RISK-MANAGEMENT RECOMMENDATIONS AND RESEARCH PRIORITIES

The committee considered both the scientific uncertainties in risk at current levels of exposure and the concern within the general population about exposure to DLCs. It recognized that there are substantial gaps in the data that have to be filled before many of the identified policy options can be adopted. Based on the analysis of current data and deliberations concerning the strategic options available to the government, the committee recommended several risk-management actions. The committee's recommendations are qualitative rather than quantitative in light of the paucity of data to support specific reduction goals, and they fall into four categories: (1) general strategic recommendations, (2) high-priority risk-management interventions, (3) other risk-management interventions that deserve consideration, and (4) research and technology development to support risk management.

Strategic and High-Priority Risk-Management and Research Recommendations

The following risk-management strategies were determined to be of greatest priority to achieve a reduction in human exposure to DLCs through the food supply:

- **Strengthen interagency coordination for DLC risk management.** The committee recommends that the sponsoring agencies empower an interagency coordination group with the authority and mandate to develop a risk-management strategy to reduce DLCs in the food supply.
- **Interrupt the cycle of DLCs through forage, animal feed, and food-producing animals.** The committee recommends that the development of a risk-management strategy for DLCs give high-priority attention to reducing the contamination of animal forage and feed and interrupting the recycling of DLCs that results from the use of animal fat in animal feed.
- **Reduce DLC intakes in girls and young women.** The committee recommends that the government place a high public health priority on reducing DLC intakes by girls and young women in the years well before pregnancy is likely to occur.

The committee recommends that the government place a priority on the following research and technology development efforts:

- **Develop cost-effective analytical methods and reevaluate the use of toxicity equivalents in assessing DLC exposure.**
- **Increase research efforts aimed at removing DLCs from animal forage and feed.**
- **Expand the National Health and Nutrition Examination Survey's data collection of DLC body burdens.**
- **Increase research efforts on the effects of dietary DLCs on fetuses and breastfeeding infants.**
- **Increase behavioral research on achieving dietary change.**
- **Develop predictive modeling tools and apply them in studies to assess the effects of potential interventions on reducing DLCs in the food supply.**

General Strategic Recommendations

Important progress has been made in reducing new discharges of DLCs into the environment. With respect to DLC exposure through food, most of the effort has focused on assessing the potential risks of DLCs. Given that the risk assessments conducted have raised concerns about the health impacts of DLCs and that there is no benefit, but possible harm, from DLC exposure through foods, the committee considers it appropriate for the federal government to focus its efforts on exposure-reduction strategies. To move effectively toward reducing human exposure to DLCs through food, the federal government should begin by pursuing the following strategic courses of action: (1) establish an integrated risk-management strategy and action plan, (2) foster collaboration between the government and the private sector to reduce DLCs in the food supply, and (3) invest in the data required for effective risk management.

Develop an Integrated Risk-Management Strategy and Action Plan

Considering the large number of federal agencies with responsibility for the safety of food, reduction of DLCs in the food supply will require action across the system. The committee recommends that, as an initial step, federal agencies including FDA, FSIS, and EPA, create an interagency coordination group to develop and implement a single, integrated risk-management strategy and action plan.

Foster Collaboration Between the Government and the Private Sector to Reduce DLCs in the Food Supply

DLC exposure through food is a shared problem that requires shared, collaborative solutions. The committee recommends that as part of the process of developing an integrated risk-management strategy and action plan, the federal government create an atmosphere and program of collaboration with the private sector that involves agriculture, the food processing industry, health organizations, and consumers, which would include ongoing collection and re-evaluation of data.

Invest in the Data Required for Effective Risk Management

There are significant gaps in the data required to devise, implement, and evaluate risk-management interventions to reduce DLC exposure through food. The cost of analyzing DLC congeners is extremely high and often creates an impediment to effective data collection and analysis techniques. Research priority should be given to the development of less costly analytical methods for determining the level of DLCs in animal feed and human food. The committee recommends that the development of a plan and a commitment of resources for data collection and analysis be a central element of the risk-management strategy and action plan for reducing DLC exposure through food.

High-Priority Risk-Management Interventions

Interrupt the Cycle of DLCs Through Forage, Animal Feed, and Food-Producing Animals

Findings. Animal forage and feed are primary pathways for DLC contamination of the human food supply. This typically occurs by airborne deposition of DLCs on forage and plants used for animal feed. When animals consume contaminated forage and feed, DLCs are stored in their fat and subsequently enter the human food supply. In addition to plant material used as animal feed that may contain DLCs, several billion pounds of rendered animal fat are used annually as a feed ingredient, which serves to recycle DLCs and leads to the possibility of increasing levels of DLCs in meat and other animal-derived food products. The committee considers the animal forage, feed, and production stage of the food system to be a key leverage point for reducing DLC exposure through food because it is the primary point of entry of these compounds into the human food supply.

Recommendations. The committee recommends that the government's risk-management strategy for DLCs give high-priority attention to reducing the contamination of animal forage and feed and interrupting the recycling of DLCs that result from the use of animal fat in animal feed.

As an initial step, the government, in collaboration with the animal production and feed industries, should establish a nationwide data-collection effort and a single data repository on the levels of DLCs in animal forage and feed, which should be accessible for both public and private use. Government and industry should also begin collaboration immediately to define voluntary guidelines for good animal feeding and production practices that would reduce DLC levels in forage and feed and minimize other potential sources of DLC exposure during animal production.

The committee further recommends that the government, in collaboration with the animal production industry, identify means to achieve the reduction or elimination of DLC-containing animal fat as a component of animal feed. However, the committee recognizes that doing so could have unintended, negative consequences: (1) increased cost of food, (2) problems of unused animal fat disposal, (3) increased food spoilage, and (4) changes in the taste of food that consumers find unacceptable. The government should consider setting legally binding limits on DLCs in forage and feed only when more complete data are generated and a better understanding is developed of how DLC contamination can be avoided.

Reducing DLC Exposure in Girls and Young Women

Findings. Fetuses and breastfeeding infants may be at particular risk from exposure to DLCs due to their potential to cause adverse neurodevelopmental, neurobehavioral, and immune system effects in developing systems, combined with the potential for exposure of breastfeeding infants to comparitively high levels of DLCs in breast milk. Data suggest that because DLCs accumulate in the body over time, waiting until pregnancy to reduce DLC intake has no significant impact on the mother's level or the baby's exposure in utero or through breastfeeding. Therefore, intervention to reduce DLCs must occur in the years well before pregnancy. Substituting low-fat or skim milk for whole milk, especially when coupled with other substitutions of foods lower in animal fat by girls and young women in the crucial years before pregnancy, could reduce DLC intakes and resulting levels of DLCs during pregnancy.

Recommendations. In order to reduce DLC body burdens in the future for women with child-bearing potential, the committee recommends, as an immediate intervention, that the government take steps to increase the availability of foods low in animal fat in government-sponsored school breakfast and lunch programs and in child- and adult-care food programs. Specifically, the committee recommends that low-fat and skim milk be made readily available in the National School Lunch Program. In addition, the committee recommends that participants in the Special Supplemental Nutrition Program for Women, Infants and Children be encouraged, except for children under 2 years of age, to choose low-fat or skim milk and low-fat versions of other animal-derived foods in their food packages. Further, to reduce other sources of animal fat, the committee recommends that USDA's Economic Research Service undertake a detailed analysis to determine the feasibility of and identify barriers to setting limits on the amount of saturated fat that should be allowed in meals served under the National School Lunch and Breakfast Programs.

Other Risk-Management Interventions That Deserve Consideration

Although more data are needed, there are several other specific interventions that could be considered as part of an integrated risk-management strategy and action plan for reducing DLC exposure through food. These include: (1) reducing DLC-discharge sources in animal production areas, (2) removing DLC residues from foods during processing, particularly by the removal of fat from meat products through trimming, (3) providing advisories and education to highly exposed populations, and (4) educating the general population about strategies for reducing exposure to DLCs.

Research and Technology Development to Support Risk Management

A broader research and technology agenda is needed to support risk-management efforts to reduce exposure to DLCs through food. Among many possible subjects for such efforts, the committee recommends that the government consider placing a priority on:

1. Development of low-cost analytical methods and a review of toxicity equivalents,
2. Research to support the removal of DLCs from animal feed,
3. Expansion of the National Health and Nutrition Examination Survey's data collection on DLC body burdens,
4. Research on the effects of dietary DLCs on fetuses and breastfeeding infants, and
5. Behavioral research on achieving dietary change and, where feasible, predictive modeling studies on DLCs in the food supply.

1

Introduction

BACKGROUND

Dioxins and furans are unintentional contaminants that are released into the environment from combustion processes. The combustion of plant material from forest, brush, and range fires contributed to preindustrial deposition of dioxins into soil, sediment, and clay. Postindustrial sources are varied and include industrial burning (e.g., steel, coke, ceramic, and foundry), landfill fires, structural fires, utility pole and transformer storage yards, crematories, and backyard barrel burning of trash and woody and other plant material.

The geographic distribution of dioxins is a function of source and transport. The reservoir source of dioxins constitutes previous releases from combustion, soil deposition, volatolized and transported particulates, and soil run-off, which were sequestered and are now being rereleased into the environment. Thus, geographic distribution and accumulation of dioxins is not necessarily dependent upon a nearby source.

Dioxins and related compounds, including dioxin-like polychlorinated biphenyls (referred to collectively as DLCs), accumulate, through the food chain, into the lipid component of animal foods. However, levels of dioxins in the environment, and thus exposures to humans, have been declining since the late 1970s. Exposures, based on human tissue samples, decreased by about 75 percent between 1986 and 1996 (Papke, 1998). Even so, public concern persists with regard to the safety of the food supply and potential adverse outcomes to DLC exposure, especially in sensitive and highly exposed population groups. Sensitive groups within the general population include developing fetuses and infants.

Highly exposed groups include breastfeeding infants, subsistence fishers, and American Indian and Alaska Native tribes for whom DLC-containing fish and wild game are important cultural food sources. These populations may be at increased risk not only from exposure to DLCs through certain foods, but also at nutritional risk if the availability of these foods is limited.

Although DLCs have been extensively studied as a contaminant, there is still a great deal of controversy regarding their potential for toxicity and the implications for human health. DLC exposure through foods occurs primarily by consumption of animal fats. However, many foods that are sources of DLCs are also sources of important nutrients, such as calcium and vitamin D in milk and cheese; protein, iron, and niacin in meats; protein, vitamin A, and iron in eggs; and omega-3 fatty acids in fish.

In order to evaluate and recommend risk management strategies to reduce DLC exposure through foods, consideration must be given to the potential impact of changes to food and nutrition policies, particularly those related to public education and food assistance programs, on the nutritional status and health of the population at large and to sensitive and highly exposed groups.

THE COMMITTEE AND ITS CHARGE

Following a request by federal agencies to the National Academies, an expert committee was appointed to review existing reports on the impact of DLCs on the safety of the food supply and to offer options to further reduce exposure to these contaminants, while considering the need to maintain health and optimize nutritional status, particularly with regard to sensitive and highly exposed groups. The Food and Nutrition Board, in consultation with the Board on Agriculture and Natural Resources and the Board on Environmental Studies and Toxicology, brought together an ad hoc committee to study the implications of DLCs in the food supply.

The charge to the committee was to: (1) take into account the substantial body of data available in the U.S. Environmental Protection Agency's draft reassessment (EPA, 2000) and other reports on the pathways by which DLCs add to the dioxin body burden by concentrating in foods from sources such as animal feed, and through intake of specific foods such as seafood, foods of animal origin (e.g., eggs, dairy products, meats) and plant foodstuffs; (2) review the data on food-consumption patterns of various subgroups of the population that appear to be at increased risk due to physiological state, food practices, or geographic location; (3) identify and describe possible risk management options that could be instituted to decrease the content of dioxins in food animals, seafood, and other food products, and possible changes in food and nutrition policies that would decrease exposure, including, where possible, an assessment of the net risk reduction afforded by a risk management option, including effects on nutrition; (4) estimate uncertainty in net risk and identify key data needs; if the uncertainty is too great due to a lack of data, provide a qualitative description of the potential for net risk reduction; (5) use existing estimates of dioxin risk as much as possible; adjust chemical risk estimates derived through upper bound method, if necessary, to allow comparison with nutrition benefits estimated using central tendency methods; and (6) identify and describe efforts in the United States and other countries to decrease dioxin exposures of specific subgroups of the population through public health or risk communication initiatives, and assess the extent to which federal food and nutrition policies contribute to decreasing exposure to dioxins.

The committee approached its charge by gathering information from existing literature and from workshop presentations by recognized experts (see Appendix C for workshop agendas), commissioning an analysis of DLC exposure through foods, deliberating on issues relevant to the task, and formulating an approach to address the scope of work. Reports and other data releases, such as the Centers for Disease Control and Prevention's *Second National Report on Human Exposure to Environmental Chemicals* (NCEH, 2003), occurred subsequent to the committee's deliberations. However, much of the information contained in these reports was provided, in part, to the committee by agency representatives at the open sessions of the committee meetings.

The committee developed an analytical framework to identify, evaluate, and formulate recommendations to reduce DLC exposure to the general population and to sensitive and highly exposed subgroups. This analytical framework, described in Chapter 6, organizes a wide array of policy options in the form of a matrix. Within the matrix, the committee developed a set of general categories that allowed it to array and analyze options and to ask detailed questions about each potential option that would help the committee recommend the most feasible interventions to reduce DLC exposure. This array of options is discussed in Chapter 7.

The committee's recommendations, which flow from the analytical framework, comprise those interventions that the committee determined would be both feasible and effective in reducing DLC exposure through the food supply, while not compromising good nutrition and health.

ORGANIZATION OF THE REPORT

This report is organized into eight chapters that describe what is known about DLCs in agricultural and human food pathways and how DLCs move through the food chain from animal feeds to human food sources. Current evidence is used as a basis for recommended options to reduce exposure through the food supply. Chapters 2 and 3 summarize current evidence for health impacts and environmental sources of exposure to DLCs. Chapters 4 and 5 discuss issues of exposure through forage and feeds, human foods, and human food-consumption patterns. Chapter 6 presents the framework for developing policy options, and

Chapter 7 discusses and summarizes the committee's deliberations on the array of options to reduce exposure. Chapter 8 provides a summary of the committee's findings, recommendations, and needs for future research.

The content of this report reflects the committee's fidelity to its charge. The committee utilized the available evidence as the basis of its deliberations and recommendations to reduce the impact of DLC exposure through the food supply. The separate issue of determining a safe or minimal level of exposure to DLCs, which was not part of the committee's charge, will be a challenge to address and requires much more research.

REFERENCES

EPA (U.S. Environmental Protection Agency). 2000. *Exposure and Human Health Reassessment of 2,3,7,8-Tetrachlorodibenzo-p-Dioxin (TCDD) and Related Compounds.* Draft Final Report. Washington, DC: EPA.

NCEH (National Center for Environmental Health). 2003. *Second National Report on Human Exposure to Environmental Chemicals.* NCEH Publication No. 02-0716, Atlanta, GA: Centers for Disease Control and Prevention.

Papke O. 1998. PCDD/PCDF: Human background data for Germany, a 10-year experience, *Environ Health Perspect* 106:723–731.

2

A Summary of Dioxin Reports, Assessments, and Regulatory Activity

This chapter begins with a brief summary of evaluations from several governmental bodies on the toxicity of chlorinated dibenzo-*p*-dioxins (CDDs) and related compounds, including chlorodibenzofurans (CDFs) and polychlorinated biphenyls (PCBs) with dioxin-like activity, and on the potential human health effects from exposure to these compounds. For brevity, these compounds will be referred to in this report collectively as "dioxin-like compounds" (DLCs), except when there is a need to refer to one of the specific compounds. This chapter also contains information on DLC-related regulations and guidelines that have been established in the United States, a number of European countries, Japan, and Australia for the environment, feeds, and foods. The chapter concludes with discussions on DLC monitoring and research programs and on methods of chemical analysis for DLCs in feeds and foods.

EVALUATIONS BY GOVERNMENTAL BODIES

This section presents information about the toxicity of DLCs. For CDDs and CDFs, dioxin-like activity requires chlorination of the parent compounds at the 2, 3, 7, and 8 positions; for PCBs, this activity requires chlorination at four or more positions (with at most one ortho substitution). While there are 75, 135, and 209 different CDDs, CDFs, and PCBs, respectively, only 7, 10, and 12 of them are considered to have dioxin-like activity (Figure 2-1).

This section begins with a description of toxicity equivalency factors and toxic equivalencies, which are systems that have been developed to compare the potential toxicities of various dioxin congeners (i.e., compounds that have similar

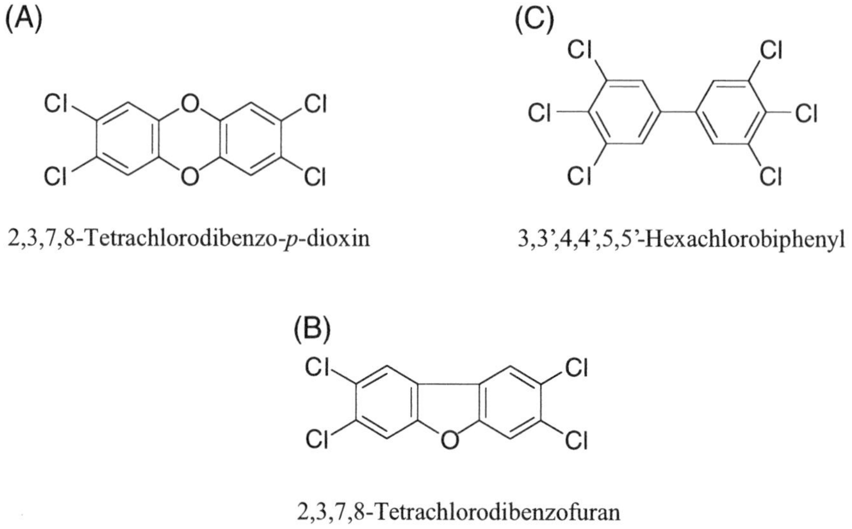

(A)

2,3,7,8-Tetrachlorodibenzo-*p*-dioxin

(C)

3,3',4,4',5,5'-Hexachlorobiphenyl

(B)

2,3,7,8-Tetrachlorodibenzofuran

FIGURE 2-1 Chemical structures of representative congeners of (A) chlorinated diben-zo-*p*-dioxins, (B) chlorodibenzofurans, and (C) polychlorinated biphenyls.

chemical structures or belong to closely related chemical families). Potential adverse human health effects (including cancer and noncancer effects) from exposures to DLCs are then summarized. Because this report focuses on general (i.e., background) exposures to DLCs through feed and food pathways, the discussion on adverse health effects focuses on general exposures, but a brief description of adverse health effects from high exposures to DLCs is also included. The final part of this section consists of a summary of body burden and intake data in the general population and certain vulnerable groups.

Review of Major Reports on DLCs

The body of literature on the toxicity of DLCs is voluminous, and the topic of potential human health effects from exposure to DLCs is highly controversial among scientists. It is beyond the scope of this report to provide a detailed critical analysis and evaluation of the DLC toxic effects data. However, to put some of the information presented in later chapters into context, the committee decided that it is appropriate to include a relatively brief description of the toxicity of DLCs. Several governmental bodies have recently compiled information about DLCs and evaluated their toxicity; the committee has drawn on these evaluations, which are listed below.

- The Agency for Toxic Substances and Disease Registry's (ATSDR) *Toxicological Profile for Chlorinated Dibenzo-p-dioxins* (ATSDR, 1998). This report primarily covers the toxicity of CDDs. It alludes to CDFs and PCBs generally only in discussions of contaminant concentrations in environmental media and foods since these compounds (particularly CDDs and CDFs) are commonly analyzed simultaneously. (ATSDR has published separate toxicological profiles on CDFs and PCBs [ATSDR 1994, 2000].)

- A report prepared for the European Commission DG Environment, *Evaluation of the Occurrence of PCDD/PCDF and POPs in Wastes and Their Potential to Enter the Foodchain* (Fiedler et al., 2000). This report provides a cursory discussion of toxicity, body burdens, and intakes and focuses on the occurrence of persistent organic pollutants (POPs) in environmental media, animal feeds, and pathways of contamination. It addresses only about a dozen POPs, but CDDs and CDFs are discussed thoroughly because of the relatively large databases for these compounds.

- A report prepared for the European Commission DG Environment and the U.K. Department of the Environment and Transport and the Regions, *Compilation of EU Dioxin Exposure and Health Data* (AEA Technology, 1999). This report contains a thorough review of exposure and health effects data.

- An initial report and an update from the Scientific Committee on Food of the European Commission, Health and Consumer Protection Directorate-General, *Opinion of the Scientific Committee on Food on the Risk Assessment of Dioxins and Dioxin-like PCBs in Food* (Scientific Committee on Food, 2000, 2001). The initial report focuses on dietary exposure and toxicity of DLCs; it does not address pathways of contamination. The update, based on scientific information made available after release of the initial report, includes new toxicity and tolerable intake information only.

- The International Agency for Research on Cancer's (IARC) *IARC Monographs on the Evaluation of Carcinogenic Risks to Humans. Volume 69: Polychlorinated Dibenzo-para-dioxins and Polychlorinated Dibenzofurans* (IARC, 1997). The focus of this report is assessment of the potential human carcinogenicity of CDDs and CDFs, but other issues, such as other types of toxicity, environmental occurrence, and human exposure, are also covered.

- The U.S. Environmental Protection Agency's (EPA) draft *Exposure and Human Health Reassessment of 2,3,7,8-Tetrachlorodibenzo-p-Dioxin (TCDD) and Related Compounds* (EPA, 2000). This draft reassessment is the most extensive compilation of data on the environmental occurrence and toxicity of DLCs and the consequent human exposures and risks.

Toxicity Equivalents and Toxicity Equivalency Factors

The biological activities of DLCs vary and, since humans are usually exposed to mixtures of DLCs, the toxicity of an exposure depends on the particular composition of the mixture. It is desirable to express the expected biological activity of mixtures using a common metric. The biological activities of the various dioxin congeners are compared to the activity of 2,3,7,8-tetrachlorodibenzo-p-dioxin (TCDD). TCDD is the most biologically potent of the DLCs, and the greatest amount of toxicity information has been gathered for this dioxin congener. The toxic potency of a mixture of DLCs is therefore expressed in TCDD toxicity equivalents, or TEQs. As an example, exposure to a mixture of DLCs with a potency of 2 ng TEQ/kg means that the total mixture is expected to have the potency of an exposure equal to 2 ng TCDD/kg. The TEQ value for a mixture is calculated by multiplying the mass or concentration of each DLC by a toxicity equivalency factor (TEF) and summing across all DLCs present. TEFs are calculated as a way to express the activity of DLCs in relation to TCDD as determined by various biochemical or toxicological assays. A sample TEQ calculation is shown in Table 2-1. (For further discussion of the derivation and use of TEQs and TEFs see EPA, 2000.)

Any mixture that yields a certain TEQ concentration is assumed to have the same toxic potential as another mixture with the same TEQ. Although the TEF system is useful for determining toxicity in mixtures of DLC congeners, it cannot be used to simplify environmental fate and transport analyses of DLCs because individual congeners differ in their physical and chemical properties, an important consideration in fate modeling. Several TEF schemes have been developed over the years; they differ regarding the inclusion of dioxin-like PCBs and TEFs

TABLE 2-1 Sample Toxicity Equivalents (TEQ) Calculation for a Mixture of Dioxin-like Compounds

Compound	Mass Concentration (ng/kg)	Toxicity Equivalency Factor	TEQ Concentration (ng TEQ/kg)
TCDD	2	1	2
1,2,3,4,7,8-HxCDD	35	0.1	3.5
1,2,3,4,6,7,8,9-OCDF	12	0.001	0.012
3,3′,4,4′,5-penta-CB (PCB-126)	46	0.1	4.6
2,3,4,4′,5-penta-CB (PCB-114)	186	0.0005	0.093
Total	281		10.205

NOTE: TCDD = 2,3,7,8-tetrachlorodibenzo-p-dioxin, HxCDD = hexachlorodibenzo-p-dioxin, OCDF = octachlorodibenzofuran, CB = chlorinated biphenyl, PCB = polychlorinated biphenyl.

for certain compounds. The biological activity of a mixture of DLCs may be estimated differently depending on the TEF system used and the particular compounds analyzed. Appendix Tables A-1 and A-2 present the common TEF systems.

Although the use of TEFs provides a convenient method for assessing the toxicity of mixtures of DLCs, there are limitations to this method. These limitations include uncertainties associated with the TEF values assigned to each DLC congener; whether TEF values are constant across all responses and ranges of dose; whether all effects of DLCs, including TCDD and PCB congeners, are mediated via the arylhydrocarbon receptor; whether a TEQ for a DLC mixture is the sum of the toxicity of each DLC present in mixture; and whether all mixtures with the same TEQ have the same toxicity.

Potential Human Health Effects from Exposure to DLCs

Chemical dose is typically measured as an intake in units of mass per unit of body weight, such as 2 mg of calcium/kg. However, for compounds that are cleared slowly from the body, such as DLCs, intakes are not very useful for understanding toxicity profiles, dose-response modeling, or interspecies comparisons. In the case of DLCs, a given intake can create very different internal concentrations in the species and target organs of interest, depending on the duration of dosing and toxicokinetic and toxicodynamic parameters (e.g., the fat content of the target tissue and the half-life of the specific dioxin congener in an organism). Reviews of DLC toxicity generally refer to an internal exposure metric, such as body burden, or to concentration in a particular tissue when such data are available. Different body burden definitions may be used, such as steady state, lifetime average, or peak concentrations. Because DLCs are preferentially associated with lipids, body burden concentrations are frequently given in units of mass per mass of lipid. Various reports cited in this chapter provide different measurement units. No conversions to other units were made for any of these values to ensure the accuracy of the values as presented by the study author.

Effects Observed in Humans at General Exposures

General exposure is defined here as that received through everyday life. The general population, breastfeeding infants, and consumers of contaminated fish are considered to have general exposures to DLCs. Occupationally exposed workers and victims of unintentional releases are not included. An exception is made for the Times Beach, Missouri investigations, since body burdens of TCDD in exposed subjects were similar to those of the general population, despite widespread soil contamination.

The reports cited earlier (AEA Technology, 1999; ATSDR, 1998; EPA, 2000; Fiedler et al., 2000; IARC, 1997; Scientific Committee on Food, 2000,

2001) consider all human data, regardless of dose, or focus on high-dose populations because of the greater probability of observing an effect. These reports describe certain populations as "poisoned" or "highly exposed," including those in pesticide manufacturing studies, the Ranch Hand studies, the Seveso reports, and the Yusho and Yu-cheng investigations (see below). There are several studies of cancer in pesticide applicators, but they are confounded by lack of specific exposure information. Studies of Swedish (Axelson and Sundell, 1974, as cited in EPA, 2000) and Finnish (Riihimäki et al., 1982, 1983, as cited in EPA, 2000) applicators that used both 2,4-D (2,4-dichlorophenoxyacetic acid) and 2,4,5-T (2,4,5-trichlorophenoxyacetic acid) showed a slight increase in relative risk for cancer, but exposure to TCDD or other DLCs was not quantified.

A study of 1,261 Air Force veterans known as Ranch Hands, who were responsible for aerial herbicide spraying in Vietnam, showed that the men had a median TCDD blood serum level that was 12.4 ppt (range 0–618 ppt) compared with 4.2 ppt in controls (Air Force veterans engaged in cargo transport), with the greatest exposure in the nonflying ground personnel (median value 23.6 ppt) (Wolfe et al., 1990, as cited in EPA, 2000). Follow-up studies through the early 1990s suggested that this group did not have an increased risk of cancer death, although it was noted that in general the Ranch Hands did not have elevated TCDD levels significantly above background and were still a relatively young group, taking into account a 20-year cancer latency period (EPA, 2000).

In 1976, an unintended industrial release in Seveso, Italy, exposed a large population of people to TCDD. At the time of the release, serum levels of TCDD were as high as 56,000 ppt for the most highly exposed children, who developed severe chloracne (Mocarelli et al., 1991, as cited in EPA, 2000). Twenty years after the release, tissue plasma levels of TCDD were measured in randomly selected residents. Those from the area with the greatest initial exposure had geometric mean TCDD levels of 53.2 ppt ($n = 7$; range 1.2–89.9 ppt); those with the next greatest exposure had 11.0 ppt ($n = 51$); and those with the lowest exposure had 4.9 ppt ($n = 52$) (Landi et al., 1996, 1998, as cited in EPA, 2000). In a 15-year follow-up, the overall cancer mortality in the residents did not appear to be increased compared with a control group from outside the exposed area, although significant excess mortality risks occurred in the lowest exposure group for esophageal cancer in males and bone cancer in females (Bertazzi et al., 1997, 1998, as cited in EPA, 2000). EPA reports that the cancer data appear to be contradictory and are difficult to interpret because of the small number of cases, problems with exposure classification, and a 15-year rather than a 20-year follow-up (EPA, 2000).

Two significant occurrences of poisoning of food oils with PCBs and furans have been reported in Japan. The first occurred in 1968 in the Yusho incident, in which 1,900 people unintentionally consumed up to 2 g of rice oil that contained a 1:250 ratio of furans to PCBs. Tissue studies indicated that both the furans and the PCBs were retained for many years. There was a significantly increased risk of liver cancer in males 15 years after the incident, although determination of rice oil as the culprit was problematic (Kuratsune et al., 1988, as cited in EPA, 2000). Noncancer effects included acneform eruptions, hyperpigmentation, and hyperkeratosis and ocular lesions. The second incident (Yu-chen) was a contamination of cooking oil in Taiwan in 1979 that affected 2,000 people and resulted in similar noncancer effects. Six months after the incident, blood PCB levels ranged from 11 to 720 ppb with a mean value of 49 ppb and most values less than 100 ppb (Chen et al., 1980, as cited in EPA, 2000).

There have been few epidemiological investigations of the health effects of DLCs in populations with no extraordinary exposure circumstances. A large population of children from an industrial region in the Netherlands is currently being studied for a variety of health outcomes in relation to exposure to DLCs (Vreugdenhil et al., 2002), and Koopman-Esseboom and colleagues (1994a) examined DLC levels in blood and human milk in two cohorts, one industrial and one rural, in the Netherlands (see below).

Cancer. Epidemiological studies on carcinogenicity from general exposures to DLCs are sparse, but experimental animal studies provide strong support of carcinogenicity. The IARC monographs (1997) include no cancer epidemiology studies regarding general exposures to CDDs and express no opinion about the potential for carcinogenicity from such exposures. All of the investigations described in the monographs are cohort studies of workers with known, inferred, or presumed exposure to CDDs (usually TCDD); cohort studies of the Seveso population with unintentional TCDD exposure; or case-control studies in which subjects had known or expected contact, usually on the basis of occupation, with chlorophenoxy herbicides that likely contained TCDD. The monograph on CDFs describes a few studies of cancer in humans, including investigations that showed a moderate increase in cancer incidence and mortality in Swedish fishermen and consumers of Baltic Sea fish. Stomach cancer incidences in the Swedish consumers were 2.2 (1.3–3.5) and 1.6 (1.0–2.4) per thousand, compared with consumers of Atlantic Ocean fish and regional referents, respectively. Squamous-cell skin cancer rates were 1.9 (1.2–3.1) and 2.3 (1.5–3.5) per thousand compared with Atlantic Ocean fish consumers and the referent group, respectively (IARC, 1997).

EPA (2000) includes no cancer epidemiology studies regarding general exposures to DLCs, but it does provide dose-response data for populations that are highly exposed to DLCs. The relative risk for total cancer for the lowest-exposed stratum in the epidemiological studies cited by EPA ranged from 0.9 to 1.24 (see Appendix Table A-3). However, as reported by EPA (2000), in some exposed populations such as those in Seveso, Italy, the calculated relative risk for specific cancers was considerably higher (e.g., the relative risk for connective and soft tissue sarcoma in males was 2.8). In studies that have identified specific cancers arising after DLC exposure, the evidence is equivocal. However, the cumulative evidence of all studies is consistent with the possibility that DLC exposure is carcinogenic.

None of the evaluations reviewed by the committee (AEA Technology, 1999; ATSDR, 1998; EPA, 2000; Fiedler et al., 2000; IARC, 1997; Scientific Committee on Food, 2000, 2001) derives a conclusion about the carcinogenic potential of DLCs to humans solely from general exposures.

Noncancer Effects. Several potential noncancer health effects from general exposures to DLCs have been reported in recent evaluations of DLC toxicity (ATSDR, 1998; EPA, 2000; Fiedler et al., 2000; IARC, 1997). These evaluations indicate that there are possible adverse neurobehavioral effects and changes in the distribution of thyroid hormone concentrations in breastfed infants compared with formula-fed infants. Koopman-Esseboom and colleagues (1994b) found a negative correlation between extended DLC exposure and thyroid hormone levels in infants and mothers. The Scientific Committee on Food (2000, 2001) also reports neurobehavioral effects and changes in thyroid hormone status in breastfed infants, but stated that these adverse effects were "subtle, within the normal range, and considered without clinical relevance," whereas Weisglas-Kuperus and coworkers (2000) and Patandin and coworkers (1998) concluded that DLCs have a negative effect on neurodevelopment, birth weight, and immunity. EPA (2000) identifies one study that found changes in the distribution of alanine aminotransferase and aspartate aminotransferase concentrations in breastfed infants.

Adverse effects on infant birth weight, neurodevelopment, neurobehavior, thyroid hormone status, and the immune system in children have been reported (AEA Technology, 1999). These observations were made in infants in the general population and in children whose mothers ate DLC-contaminated fish from Lake Michigan. Neurodevelopmental delays in breastfed infants in a Dutch study were also reported (Huisman et al., 1995). A negative correlation was found between neurodevelopment in infants and breast-milk concentrations of PCBs. At 42 months of age, the cognitive decrement remained (Patandin et al., 1999), although the psychomotor deficits seen at younger ages had disappeared (Lanting et al., 1998). A study of early learning in school-age Dutch children found that prenatal exposures to DLCs resulted in impaired cognitive and motor abilities, which could persist when the home environment was less than optimal, but no long-term impairment could be measured in children raised in more optimal environments (Vreugdenhil et al., 2002). One evaluation (AEA Technology, 1999) notes that people with exposure to TCDD from the Times Beach contamination showed reduced immune response and changes in T-lymphocyte differentiation, and people eating relatively large amounts of Baltic Sea fish showed changes in T-cell lymphocytes, but IARC (1997) reports contradictory findings about cell-mediated immunity in the Times Beach subjects. EPA (2000) considers the data about immunological effects from exposures to DLCs to be inconclusive.

Effects Observed in Humans at Higher Exposures

Cancer. The IARC (1997) evaluation discusses the body of epidemiological literature on CDDs, but it focuses on the high-exposure cohorts with documented TCDD exposure (all occupational exposures) in its evaluation of whether CDDs are carcinogenic to humans. It concludes that the overall standardized mortality rate for all cancer types combined was 1.4 per thousand (95 percent confidence interval, 1.2–1.6) for the most highly exposed subgroups in these occupational cohorts; statistically significant dose-response trends in two of the studies strengthen that opinion.

IARC also concludes, however, that the epidemiological data provide "limited evidence" of a carcinogenic effect of TCDD in humans, although the IARC monographs (1997) upgrade 2,3,7,8-TCDD from classification 2A, Probably Carcinogenic to Humans, to 1, Carcinogenic to Humans. In its monograph on the carcinogenicity of CDFs, IARC notes that male victims of the Yusho incident had a threefold excess of liver cancer mortality, but that no such excess occurred in victims of the Yu-cheng incident, and finds the evidence for the carcinogenicity of CDFs in humans to be inadequate.

The reports of the Scientific Committee on Food (2000, 2001) rely largely on the IARC (1997) evaluation and state that TCDD should be regarded as a human carcinogen, though not a direct-acting genotoxin. The report prepared by Fiedler and colleagues (2000) repeats the IARC (1997) evaluation of the carcinogenicity of CDDs. AEA Technology (1999) concludes that the epidemiological data suggest that TCDD exposure increases the rates of all cancers. On the basis of the epidemiological and animal studies, ATSDR (1998) concludes that TCDD may be a human carcinogen.

EPA (2000) considers TCDD to be a human carcinogen and that other DLCs are likely to be human carcinogens, based on a combination of epidemiological and animal cancer studies and mechanistic information. The epidemiological data alone do not demonstrate a causal association between exposure and cancer, but suggest that the compounds are multisite carcinogens, increasing cancer at all sites, lung cancer, and perhaps other particular cancers. EPA (2000) describes TCDD as a nongenotoxic carcinogen and a potent promoter.

Noncancer Effects. A number of noncancer human health effects have been associated with high exposures to DLCs. These effects are listed below.

- Chloracne and other dermal effects (ATSDR, 1998; EPA, 2000; Fiedler et al., 2000; IARC, 1997).

- Changes (AEA Technology, 1999; ATSDR, 1998; Fiedler et al., 2000; IARC, 1997) or possible changes (EPA, 2000; Scientific Committee on Food, 2000, 2001) in glucose metabolism and in diabetes risk.
- Alterations (ATSDR, 1998) or possible alterations (EPA, 2000) in thyroid function.
- Alterations in growth and development (AEA Technology, 1999; IARC, 1997; Scientific Committee on Food, 2000, 2001), neurodevelopment (AEA Technology, 1999; IARC, 1997), and neurobehavior (AEA Technology, 1999; IARC, 1997; Scientific Committee on Food, 2000, 2001). EPA (2000) considers postnatal developmental effects on neurobehavior to be possibly related to exposures to DLCs. Fiedler and colleagues (2000) note that there was an altered sex ratio among births after the Seveso unintended exposure.
- Increased gamma-glutamyl transferase concentrations (EPA, 2000; Scientific Committee on Food, 2000, 2001).
- Altered concentrations of reproductive hormones in men (EPA, 2000).
- Alterations in liver function (Fiedler et al., 2000; IARC, 1997) and hepatic effects (ATSDR, 1998).
- Possibly altered serum lipid (EPA, 2000; Scientific Committee on Food, 2000, 2001) and cholesterol concentrations (EPA, 2000).
- Possibly altered alanine aminotransferase and aspartate aminotransferase concentrations (EPA, 2000).
- Alterations in the immune system (Fiedler et al., 2000; IARC, 1997).
- Possible ocular changes (ATSDR, 1998; IARC, 1997).
- Increased mortality from cardiovascular disease (AEA Technology, 1999; Scientific Committee on Food, 2000, 2001).

Toxicity Benchmarks

Estimates of Tolerable Intakes

Several governmental bodies have derived or recommended acceptable daily intakes or similar parameters for TCDD or DLCs as a group (AEA Technology, 1999; ATSDR, 1998; EPA, 2000; Fiedler et al., 2000; Scientific Committee on Food, 2000, 2001). These guidance levels are summarized in Table 2-2.

ATSDR (1997) defines a minimal risk level (MRL) for a hazardous substance (e.g., DLCs) as an estimate of daily human exposure that is likely to be without appreciable risk of adverse noncancer health effects over a specified duration and route of exposure. For 2,3,7,8-TCDD, ATSDR (1998) derives MRLs for oral exposure to over three intervals: acute (14 days or less), intermediate (15–364 days), and chronic (1 year or more). An MRL is "an estimate of the daily human exposure to a hazardous substance that is likely to be without appreciable risk of adverse non-cancer health effects over a specified duration of exposure"

TABLE 2-2 Estimates of Tolerable Intakes of Dioxins and Dioxin-like Compounds

Reference	Cancer	Noncancer
ATSDR, 1998	None given	Acute oral MRL: 200 pg/kg/d (TCDD) Intermediate oral MRL: 20 pg/kg/d (TCDD) Chronic oral MRL: 1 pg/kg/d (TCDD)
EPA, 2000	1×10^{-3} pg/kg/d (TCDD) *or* 0.001 pg/kg/d at a 1 in 1 million excess risk level	None given
Scientific Committee on Food, 2000, 2001	14 pg TEQ $_{DFP\text{-}WHO98}$/kg/wk	
AEA Technology, 1999; Fiedler et al., 2000	1–4 pg TEQ/kg/d (CDDs, CDFs, PCBs)	

NOTE: MRL = minimal risk level, TCDD = 2,3,7,8-tetrachlorodibenzo-*p*-dioxin, TEQ = toxicity equivalents, CDD = chlorinated dibenzo-*p*-dioxin, CDF = chlorodibenzofuran, PCB = polychlorinated biphenyl.

(ATSDR, 1998); the MRL does not pertain to cancer. The MRLs for TCDD are, respectively, 0.0002 µg/kg/d (200 pg/kg/d), 0.00002 µg/kg/d (20 pg/kg/d), and 0.000001 µg/kg/d (1 pg/kg/d). The acute MRL is based on immunological effects in female mice, the intermediate MRL on immunological effects in guinea pigs, and the chronic MRL on developmental effects in rhesus monkeys. In each case, uncertainty factors are applied to address animal-to-human extrapolation, interindividual variation in response, and (if necessary) a low-effect to no-effect dose extrapolation.

The report by Fiedler and colleagues (2000) reiterates the World Health Organization (WHO)-recommended tolerable daily intake (TDI) of DLCs of 1 to 4 pg TEQ/kg/d and also WHO's recommendation that exposures should be reduced as much as possible. Similarly, the AEA Technology (1999) report encourages member states of the European Union to adopt the recent WHO recommendation, and notes that PCBs contribute about half of dietary TEQ exposure. The Scientific Committee on Food (2000, 2001) derived a tolerable weekly intake (TWI) of DLCs based on TCDD body burdens associated with sensitive effects in experimental animals: developmental and reproductive effects in rats and monkeys and endometriosis in monkeys. Body burdens associated with the lowest-observed-adverse-effect levels in the relevant studies ranged from about 30 to 100 ng TCDD/kg. The estimated human dietary intakes needed to produce these body burdens were then calculated, and safety factors (ranging from 3 to 10) were applied. The lowest TDI thus calculated was 2 pg TCDD/kg, corresponding to a

TWI of 14 pg TCDD/kg. The TWI is intended to apply to all DLCs expressed as TEQs using the WHO 1998 TEF system. Cancer is not considered one of the more important adverse effects of background DLC exposure; the Scientific Committee on Food (2000, 2001) believes that TCDD is a nongenotoxic carcinogen best evaluated with a threshold model, and that TCDD body burdens associated with cancer in laboratory animal bioassays and high-exposure human populations are "several orders of magnitude higher" than body burdens in the general population.

The IARC (1997) and EPA (2000) reports make no quantitative recommendations about intake or exposure to DLCs. EPA (2000) notes that if it were to develop reference doses or concentrations (life-long tolerable intakes for noncancer health effects), the values would be less than current exposures and, therefore, not useful for risk management. (Reference doses and concentrations are normally used to characterize the health risk posed by incremental exposures to compounds in the absence of significant background exposures.) IARC (1997) reports intake guidelines from several countries:

- Canada, 10 pg I-TEQ/kg/d (CDDs and CDFs)
- Nordic countries, 0 to 35 pg TCDD/kg/wk
- Netherlands, 1 pg dioxin/kg/d
- Sweden, 5 pg dioxin/kg/d
- Japan, 10 pg/kg/d (CDDs and CDFs)

EPA (2000) derives quantitative upper-bound estimates of the excess lifetime cancer risk posed by exposure to DLCs. Cancer potencies (or slope factors) are derived from three occupational epidemiology investigations and range from 9×10^{-3} pg/kg/d TCDD (for all cancers) to 3×10^{-4} pg/kg/d TCDD (for lung cancer specifically).[1] A meta-analysis of these studies yields a cancer potency of 1×10^{-3} pg/kg/d TCDD, which is EPA's preferred value based on human data. Potencies based on dose-response data from laboratory animal bioassays of TCDD range from 3×10^{-3} pg/kg/d to 1×10^{-4} pg/kg/d. EPA's preferred value based on animal data is 1×10^{-3} pg/kg/d TCDD. Overall, EPA (2000) believes a slope of 1×10^{-3} pg/kg/d (for all cancers) is most appropriate for TCDD. (This slope implies that a daily dose of 0.001 pg TCDD/kg poses an upper-bound excess lifetime cancer risk of 1 in 1 million.) In developing cancer potencies, EPA (2000) fitted linear models to study datasets to derive estimates of the 1 percent effective dose (ED_{01}) and the lower 95 percent confidence limit on the ED_{01} (LED_{01}) either for all cancers combined or for particular cancers (e.g., lung, liver). In that analysis,

---

[1] A potency of 8.6×10^{-3} pg/kg/d TCDD implies an additional 8.6 extra cancer cases per 1,000 exposed people for every 1 pg/kg/d increase in TCDD exposure.

dose meant average excess body burden, and intakes corresponding to the ED_{01} and LED_{01} values were computed. A linear model was used because, in both the human and rodent datasets, there were too few dose levels to support more complicated models. The other reports that were examined (AEA Technology, 1999; ATSDR, 1998; Fiedler et al., 2000; IARC, 1997; Scientific Committee on Food, 2000, 2001) do not contain slope factors for TCDD or other DLCs.

Risk Estimates for Current Exposure

ATSDR (1998) does not comment on the degree of health risk posed by current exposures to DLCs. IARC (1997) does not comment on the issue with respect to CDFs and states only that its evaluation "[does] not permit conclusions to be drawn on the human health risks from background exposures to 2,3,7,8-TCDD." Fiedler and colleagues (2000) repeat IARC's (1997) statement verbatim, and also states that "subtle effects" of DLC exposure may be occurring in the general population; however, no quantitative risk assessment is performed. AEA Technology (1999) also concludes that subtle health effects may be occurring, without giving quantitative risk estimates. The reports of the Scientific Committee on Food (2000, 2001) do not perform a quantitative risk assessment, but they note that (1) current body burdens of TCDD are much lower than those associated with cancer in experimental and epidemiology reports, and (2) "a considerable proportion of the European population [exceeds] the TWI derived by the Committee."

Based on its estimate of the cancer potency of DLCs, 1×10^{-3} pg TEQ/kg/d, EPA (2000) concludes that intakes of about 3 pg TEQ/kg/d in the general population (three times the current intake level), pose an upper-bound, excess cancer risk greater than 1 in 1,000. If a small part of the population were to receive intakes two to three times higher than the upper-bound mean of 3 pg TEQ/kg/d, as EPA (2000) concludes, the excess cancer risk for those individuals would be proportionally higher as well.

EPA (2000) does not generate a specific number describing the risk of noncancer health effects posed by current exposures to, or body burdens of, DLCs. Rather, it calculates the margin of exposure between current intakes or body burdens and toxicological "points of departure." A point of departure is a value near the low end of the dose-response curve for a sensitive health endpoint, such as a no-observed-adverse-effect level, a lowest-observed-adverse-effect level, or an ED_{01}, based on human or laboratory animal data. If the margins of exposure are 100 to 1,000 or more, EPA considers it unlikely that significant human health effects would occur at background exposures or at incremental exposures plus background. EPA's (2000) calculated margins of exposure range from 1.2 to 238, with some values less than 10, based on several points of departure and a human body burden of 5 ng TEQ/kg. Most points of departure appear to be ED_{01} or LED_{01} estimated from laboratory animal experiments, derived by fitting certain models (which allowed for nonlinear forms) to the dose-response data.

Body Burdens and Intakes of DLCs

Data on body burdens of DLCs for the general population and for susceptible groups are presented below. There are several issues associated with the estimation and use of body burdens, including the validity of assumptions about the percent body fat (typically assumed to be 25 percent) for lipid-adjusted tissue concentrations, the uncertainty associated with intake rates and half-lives, and species extrapolations from short-term animal studies to steady-state human exposures (EPA, 2000).

General Population

Body Burdens. Recent body burden data (DLCs were measured in blood or adipose tissue) for persons living in Europe and North America are presented in several reports (AEA Technology, 1999; ATSDR, 1998; EPA, 2000; IARC, 1997). These data are summarized below.

- The range of mean CDD/CDF body burdens for the general population in the United States is 8.5 pg I-TEQ/g of lipid to 50.0 pg $TEQ_{DF-WHO98}$/g of lipid.
- EPA's (2000) preferred estimate of the mean body burden of CDD/CDF in the U. S. general population is 21.1 pg $TEQ_{DF-WHO98}$/g of lipid.
- Dioxin-like PCBs have not been routinely assayed for in human tissues.
- EPA's (2000) best estimate of the mean body burden of CDD, CDF, and PCBs in the U.S. general population is 25.4 pg $TEQ_{DFP-WHO98}$/g of lipid. PCBs contribute 5.3 pg TEQ/g of this body burden.
- The range of mean CDD/CDF body burdens in the general populations of European and Scandinavian countries is 2.1 pg Nordic-TEQ/g of lipid to 57 pg I-TEQ/g of lipid.

Intake of DLCs. Dietary intake is widely believed to contribute up to 90 percent of human exposure to DLCs (see Chapter 5). Only the EPA (2000) document contains original estimates, and many of the other documents rely on the same published reports of intakes. The majority of intake estimates are based on a combination of contaminant concentrations in foods and food-consumption rates, but EPA (2000) also uses toxicokinetic modeling to estimate the average intakes that produce current body burdens of DLCs. The data are summarized below.

- The best estimate of current CDD/CDF intake for adults from all sources in the United States is 41 pg $TEQ_{DF-WHO98}$/d, or 0.59 pg $TEQ_{DF-WHO98}$/kg/d.
- The best estimate of current dioxin-like PCB intake for adults from all sources in the United States is 25.2 pg $TEQ_{P-WHO98}$/d, or 0.36 pg $TEQ_{P-WHO98}$/kg/d.
- Based on the two previous estimates, the best estimate of current CDD, CDF, and PCB intake for adults from all sources in the United States is about 65 pg $TEQ_{DFP-WHO98}$/d, or about 1 pg $TEQ_{DFP-WHO98}$/kg/d.
- The preceding estimates are based on analysis of diet. Toxicokinetic modeling of the average past intake that would produce the current average adult body burden of 25.4 pg $TEQ_{DFP-WHO98}$/g of lipid gives 146 pg $TEQ_{DFP-WHO98}$/d. This suggests that (1) past intakes, on average, were considerably higher than current intakes, and (2) current body burdens are not at steady state given current intakes.
- European estimates of mean adult dietary intakes of CDD/CDF range from about 0.4 to 1.5 pg I-TEQ/kg/d.
- Intakes of CDD/CDF and PCBs among children are estimated to be higher than in adults on a body-weight basis, but decrease with age, both in the United States and the United Kingdom. In the United States, combined intakes are estimated to be 3.6 pg $TEQ_{DF-WHO98}$/kg/d (54 pg $TEQ_{DF-WHO98}$/d) at 1 to 5 years of age, 1.96 pg $TEQ_{DF-WHO98}$/kg/d (59 pg $TEQ_{DF-WHO98}$/d) at 6 to 11 years of age, and 1.00 pg $TEQ_{DF-WHO98}$/kg/d (64 pg $TEQ_{DF-WHO98}$/d) at 12 to 19 years of age; similar declines were seen in the United Kingdom.
- EPA estimates that 70 percent of dietary TEQ comes from five compounds: TCDD, 14 pentachlorodibenzo-*p*-dioxins (1-PeCDD), 14 pentachlorodibenzo-*p*-furans (1-PeCDF), 10 hexachlorodibenzo-*p*-furans (1-HxCDF), and PCB 126; of these, PCB 126 contributes the most TEQ.

EPA estimates that the greatest contributors to adult dietary penta-*p*-chlorodioxin and pentachlorofuran exposure from foods are (in order of highest to lowest TEQ): beef, fish and shellfish, and dairy products and milk, and the greatest estimated contributors to adult PCB exposure are: fish and shellfish, beef, and dairy products and milk (EPA, 2000).

EPA (2000) estimates that intakes at the "upper end" (mean plus 3 standard deviations) of the general population are two to three times higher than mean intakes, based on measured variability in both fat consumption (assumed to be closely linked to DLC intake) and in consumption rates of specific foods. Estimates of intakes at the upper end of the distribution in a few European countries are also two to three times larger than mean intakes.

Contribution of PCBs to Intakes. The contribution of PCBs to total TEQ intakes is estimated as:

- 37 percent for the U.K. mean total dietary exposure (AEA Technology, 1999; Fiedler et al., 2000),
- 52 percent for the Netherlands median total dietary exposure (AEA Technology, 1999; Fiedler et al., 2000),
- 49 to 57 percent for the Swedish mean total dietary exposure (AEA Technology, 1999; Fiedler et al., 2000),
- 48 to 62 percent for dietary intake in Spain (AEA Technology, 1999),
- approximately 50 percent in Finland, the Netherlands, Sweden, and the United Kingdom (Scientific Committee on Food, 2000),
- approximately 80 percent in Norway (Scientific Committee on Food, 2000), and
- approximately 37 percent for the mean and upper-percentile U.S. total dietary exposure (EPA, 2000).

Time Trends. Several of the reviewed reports describe data suggesting that a decrease in DLC body burdens and intakes has occurred over recent decades. Specifically, EPA (2000) applied nonsteady-state toxicokinetic modeling to a database of TCDD concentrations in body lipids collected since the 1970s, the results of which suggested that TCDD intakes increased from the 1940s through the 1970s and then began to drop. During the period of peak intake, daily exposure to TCDD may have reached 1.5 to 2.0 pg/kg/d, whereas EPA's best current estimate of TCDD intake is 5.6 pg/d, or 0.08 pg/kg/d for adults. EPA (2000) also argues, on the basis of body burden data from the 1980s through the present, that average DLC body burdens in the 1980s to the early 1990s were approximately 55 pg $TEQ_{DFP-WHO98}$/g of lipid, whereas the current average is about 25 $TEQ_{DFP-WHO98}$/g of lipid. EPA further reports that:

- The estimate of dietary intake of DLCs in U.K. foods decreased from 240 pg $I-TEQ_{DF}$/d in 1982 to 69 pg $I-TEQ_{DF}$/d in 1992, with PCB intake dropping from 156 pg $TEQ_{P-WHO94}$ to 46 pg $TEQ_{P-WHO94}$.
- Diet samples collected in the Netherlands in 1978, 1984 to 1985, and 1994 indicate that mean intakes were, respectively, 4.2 pg $I-TEQ_{DF}$/kg/d, 1.8 pg $I-TEQ_{DF}$/kg/d, and 0.5 pg $I-TEQ_{DF}$/kg/d, while mean PCB intakes were 11 pg $TEQ_{P-WHO94}$/kg/d, 4.2 pg $TEQ_{P-WHO94}$/kg/d, and 1.4 pg $TEQ_{P-WHO94}$/kg/d.
- In Germany, the current mean intake of CDD/CDF is estimated at 69.6 pg $I-TEQ_{DF}$/d, compared with 127.3 pg $I-TEQ_{DF}$/d in 1990.
- The mean total concentration of CDD/CDF in blood in Germany decreased from about 720 pg/g in 1991 to about 373 pg/g in 1996.

- AEA Technology (1999) reports three time-trend analyses of DLC intake, from the Netherlands, Germany, and the United Kingdom. The Dutch data on DLCs in foods gathered in 1978, 1984 to 1985, and 1994 indicate a statistically significant decreasing trend in adult intake over time, with a 50 percent decrease in I-TEQ/kg/d over each 5.5-year interval in that period. Dietary studies conducted in Germany in 1989 and 1995 indicate a 45 percent decrease in I-TEQ intake. Dietary studies in the United Kingdom in 1982, 1988, and 1992 suggest a 45 percent decrease in intake at each time point compared with the preceding point. All of these estimates likely pertain to the average consumer. AEA Technology (1999) relates some body burden data indicating a decrease in the mean burden of DLCs: blood sampled in Germany between 1988 and 1996 shows a 64 percent decline in I-TEQ concentrations (presumably CDD/CDF) over the period, or 12 percent per year. Those same data are also cited by Fiedler and colleagues (2000).
- Independent of changes over time in the DLC content of food and in dietary intake, it is generally recognized that body burdens of DLCs increase with age due to the long half-lives of these compounds.
- There is evidence of decreasing concentrations of DLCs in human breast milk:
  — There has been an approximately 25 percent decrease in $I\text{-TEQ}_{DF}$ in cow and human milk fat between 1990 and 1994 in Germany.
  — In Birmingham, England, breast-milk concentrations were 37 pg $I\text{-TEQ}_{DF}$/g of fat in 1987 to 1988 and 18 pg $I\text{-TEQ}_{DF}$/g of fat in 1993 to 1994, while concentrations at similar times in Glasgow, Scotland, were 29 pg $I\text{-TEQ}_{DF}$/g of fat and 15 pg $I\text{-TEQ}_{DF}$/g of fat, respectively.
  — $I\text{-TEQ}_{DF}$ concentrations in human breast milk in rural areas of Finland decreased from 20.1 pg/g of fat in 1987 to 13.6 pg/g in 1992 to 1994, and in urban areas from 26.3 pg/g to 19.9 pg/g.
- A WHO-sponsored study in 11 countries suggested an annual decrease in CDD/CDF of 7.2 percent in human breast milk between 1977 to 1987 and 1992 to 1993.

Persistence of DLCs in the Body. Intakes and body burdens of DLCs are related through the half-lives of these compounds in the body. The Scientific Committee on Food (2000, 2001) notes that half-lives for TCDD range from 5 to 11 years. At least 20 to 30 years of exposure are needed to reach steady state if the half-life of TCDD is 7.5 years (the value the Scientific Committee on Food [2000, 2001] uses in its calculations of body burden or intake); half-lives also increase with age, perhaps due to changes in metabolism and fat burden. IARC (1997) describes several studies of CDD/CDF (mostly TCDD) half-lives in humans experiencing large exposures (e.g., Ranch Hand personnel) and reports that TCDD half-lives ranged from 5.1 to 11.3 years, increasing with increasing body fat, but not with age or changing proportion of body fat. Little information was available on the half-lives of other CDD congeners; CDFs have half-lives of 3.0 to 19.6 years.

EPA (2000) compiled half-life estimates for several CDDs and CDFs. The estimates range from 3.7 to 15.7 years for CDDs and from 3 to 19.6 years for CDFs. EPA estimates the half-life of TCDD in humans to be 7.1 years, based on a 25 percent fat volume for a 70-kg individual. There is no evidence to support that weight loss significantly reduces DLC body burdens.

Sensitive Groups

A subgroup of the general population that is more sensitive to the toxic effects of exposure to DLCs is developing fetuses. Infants and young children may be considered both sensitive (due to developmental immaturity) and more highly exposed (due to intake per body mass) subgroups.

Subgroups of the general population that are exposed to higher levels of DLCs through the food supply include indigenous populations in northern North America, subsistence and sports fishermen, small populations whose food supplies are affected by local contamination, and breastfed infants.

Body Burdens of Fetuses and Infants. Estimates of the total DLC body burden of the developing fetus have been made by comparing levels in maternal and cord plasma samples obtained at birth. This estimate is based on the assumption that the DLC level obtained would be representative of in utero exposure. Koopman-Esseboom and colleagues (1994a) determined that PCB concentrations, collected on a cohort in the Netherlands in 1990, were approximately 20 percent of maternal plasma values, expressed on a concentration basis. In contrast to these findings, Jacobson and colleagues (1984), in the Michigan Maternal/Infant Cohort Study, found that PCB levels expressed on a lipid basis were equivalent between maternal plasma and cord serum samples.

A few measurements of body burdens of DLCs in infants are available in EPA's (2000) report; they are presented in Appendix Table A-4. EPA (2000) also modeled potential body burdens under four feeding scenarios for the first year of life: formula-feeding only and breast milk for 6 weeks, 6 months, or 1 year. After the first year, all models assumed the age-appropriate intake rates for the general population. In the third and fourth scenarios, the maximum body burden of DLCs in the first year, on a lipid basis, is 46 pg $TEQ_{DFP-WHO98}$/g, and all scenarios converge on a body burden of about 12 pg $TEQ_{DFP-WHO98}$/g (lipid) around age 10 years. The intake assumptions used to calculate the body burdens during infancy are described below.

Body Burdens of Fish Consumers. People who eat large quantities of fish or who eat contaminated fish might receive unusually large doses of DLCs. EPA

(2000) notes that people who consume fish with typical CDD/CDF contaminant levels, but at very high rates of intake, could receive two to five times the mean intake for the general population.

EPA (2000) describes fish-eating populations whose body burdens of DLCs (on a TEQ basis) are considerably higher than in comparison populations, but the differences are no more than threefold. Possible exceptions are studies of dioxin-like PCBs in the blood of fishermen on the Gulf of the St. Lawrence River, in the blood of Swedish fishermen, and in breast milk of Inuit women in Arctic Quebec. In the first population, PCB TEQ concentrations were 20-fold higher in the fishermen than in blood donor controls; in the second, mass concentrations of individual PCBs were 1.5 to 8 times higher in men with high fish intake compared to none; and in the third, mass concentrations of PCB congeners were three- to tenfold higher. Finally, a group of sport-fish eaters in the United States had blood concentrations of total PCB (i.e., dioxin-like and nondioxin-like) more than three times higher than controls, but the specific difference in the dioxin-like PCBs was not described. These data are included in Appendix Table A-4.

ATSDR (1998) describes a study of CDD concentrations in the blood of three groups of Swedes: high-fish consumers, moderate-fish consumers, and non-fish consumers. TEQ blood levels were 3.6 times higher in the high-fish consumers (63.5 pg TEQ/g of lipid) than in the non-fish consumers (17.5 pg TEQ/g of lipid). ATSDR (1998) also reports an investigation of DLCs in the blood of Inuits in Canada, which found significantly higher concentrations of CDD, CDF, and PCB. These data are also shown in Appendix Table A-4.

Body Burdens of People Residing Near Local Sources of Contamination. EPA (2000) describes a number of investigations of DLC body burdens in people residing near local sources of contamination, such as wood preservative facilities, reclamation plants, and steel factories. These data are summarized in Appendix Table A-4.

DLC Intake of Infants. ATSDR estimates of DLC intake by breastfed infants are given in Appendix Table A-5. EPA (2000) models the intake of DLCs by infants during 12 months of nursing using a dosing equation that requires, among other inputs, the concentration in milk fat, milk ingestion rate, and infant body weight. The concentration of CDD/CDF/PCB in milk fat is assumed to be 25 pg $TEQ_{DFP-WHO98}$/g initially, EPA's best estimate of the mean body burden in the general population (per gram of lipid). Using this model, EPA estimates that DLC concentrations in breast milk decline by 30 to 50 percent after 1 year of breastfeeding. The infant's intakes at birth, 6 months, and 1 year of age are thus 242 pg $TEQ_{DFP-WHO98}$/kg/d, 55 pg $TEQ_{DFP-WHO98}$/kg/d, and 22 pg $TEQ_{DFP-WHO98}$/kg/d, respectively; the average over the year is 92 pg $TEQ_{DFP-WHO98}$/kg/d (EPA, 2000). This model indicates that 1 year of breastfeeding contributes 12 percent of the lifetime dose of DLCs.

Although approximately 52 percent of women initiate breastfeeding, only about 20 percent are still breastfeeding after 6 months (IOM, 1991). Thus, the greatest number of infants exposed to DLCs through breastfeeding would likely be exposed during the first 6 months of life. In addition to its own model of DLC exposure in infants, the EPA (2000) report also reviews a previously published model for DLC exposure through breast milk over time. This model estimated that TCDD levels in breast milk declined by 20 percent every 3 months from initial exposure (Patandin et al., 1999), suggesting that the greatest exposure would occur in the first 12 weeks of breastfeeding.

Using data on concentrations of CDD/CDF in human milk from European countries, AEA Technology (1999) estimates that the average daily intake by infants in 1993 could have ranged from 106 pg I-TEQ/kg/d (rural areas) to 144 pg I-TEQ/kg/d (industrial areas). Estimates for the United Kingdom are 110 pg I-TEQ/kg/d at 2 months of age and 26 pg I-TEQ/kg/d at 10 months of age. Fiedler and colleagues (2000) and the Scientific Committee on Food (2000, 2001) state that intakes of DLCs by infants are one to two orders of magnitude greater than that for adults.

ATSDR (1998) reports estimates of CDD/CDF intakes in TEQ for breastfed infants of 83.1 pg TEQ/kg/d and 35 to 53 pg TEQ/kg/d.

DLC Intakes of Fish Consumers. AEA Technology (1999) reports Swedish data suggesting that Baltic Sea fishermen had 6.3 times the mean dietary intake of CDD/CDF in Nordic TEQ, at 11.7 to 12.5 pg N-TEQ/kg/d.

DLC Intakes of People Residing Near Local Sources of Contamination. Information on intakes of DLCs by people near local sources of contamination is presented in Appendix Table A-5.

DIOXIN REGULATIONS AND GUIDELINES

As the toxic effects of DLCs were recognized, several countries, including the United States, implemented regulations designed to reduce or control exposure to DLCs. The main focus of DLC regulatory activity has been to reduce the release of DLCs into the environment, particularly through the control of stack emissions from waste incinerators. These efforts have yielded significant reductions in environmental emissions since the 1970s. Regulations have also been implemented that focus on DLC uptake rather than source reduction, limiting total human body burden levels. In recent years, as food has been identified as the primary route to human exposure, several new regulations and programs have been implemented to reduce DLC contamination in food. Within the United States, because of the complex pathway of DLC exposure (see Chapter 3), DLC regulation responsibilities can fall across a number of different federal and state agencies and departments for the environment, agriculture, public health, and food safety.

Tolerable Intakes

Total human exposures to DLCs in the forms of daily, weekly, and monthly tolerable intakes have been established by a number of countries and international organizations. These limits focus on exposure to DLCs and potential health effects, without regard to the source.

United States

ATSDR (1998) has established human MRLs for hazardous waste sites in the United States. These MRLs, which serve as a screening tool, may also be viewed as a mechanism to identify those hazardous waste sites that are not expected to cause adverse health effects. Three temporal exposure scenarios with three different health endpoints of concern were considered (see Table 2-3).

Other Countries and Organizations

WHO, the WHO/Food and Agriculture Organization of the United Nations (FAO) Joint Expert Committee on Food Additives (JECFA), and the European Commission (EC) have established daily, weekly, and/or monthly tolerable intakes for DLCs. Australia and Japan have also adopted tolerable DLC intake levels (see Appendix Table A-6).

JECFA. In 2001, JECFA established a provisional tolerable monthly intake of 70 pg TEQ/kg of bodyweight/mo. Additionally, in 1998 it established a range of DLC TDIs of 1 to 4 pg WHO-TEQ/kg of bodyweight (Tran et al., 2002).

EC. The EC has established a TDI of 2 pg TEQ/kg/d (Tran et al., 2002).

Australia. The Australian Commonwealth Department of Health and Aging has supported a national health standard tolerable monthly intake of 70 pg TEQ/

TABLE 2-3 Human Exposure Limits for Hazardous Waste Site Cleanup in the United States (Agency for Toxic Substances and Disease Registry Minimal Risk Levels)

Source	Type of Endpoint	2,3,7,8-Tetrachlorodibenzo-*p*-dioxin or Toxicity Equivalents (pg/kg/d)	Days of Exposure
Chronic exposure	Immunological	1	365 or more
Intermediate exposure	Lymphatic	20	15–364
Acute exposure	Developmental	200	1–14

SOURCE: ATSDR (1998).

kg of bodyweight for DLCs, which conforms to the JECFA tolerable intake and lies at the mid-point of the WHO guideline. The National Health and Medical Research Council released this proposal for public comment in late January 2002 (Tran et al., 2002).

Japan. Public concern in Japan led to the passing of the Law Concerning Special Measures Against Dioxin in July 1999. This law set the level of TDI of DLCs at 4 pg TEQ/kg/d, in addition to establishing environmental quality standards (Tran et al., 2002).

Environmental Regulations

DLCs are regulated in a number of countries through the establishment of DLC limits for specific environmental exposure media (e.g., air, water, soil/ sediment). In the following sections, more detailed summaries of existing DLC regulations in various countries are provided.

United States

U.S. federal regulations cover two basic areas with regard to DLCs: air emissions and water. There are a number of environmental acts that give EPA jurisdiction to create regulations regarding DLCs in the environment, including the Clean Air Act, the Clean Water Act, the Safe Drinking Water Act, and the Resource Conservation and Recovery Act. Appendix Table A-7 details the current federal regulations for DLCs.

In general, the air emissions regulations are promulgated under the Clean Air Act, and to some extent, the Resource Conservation and Recovery Act. Air emissions regulations are codified in Title 40 of the *Code of Federal Regulations*, Part 60. A variety of incinerators and combustors are regulated in 40 C.F.R. part 60, including municipal waste combustors, hospital/medical/infectious waste incinerators, and hazardous waste incinerators. Regulation of these sources of DLCs is fairly recent, with most of the laws coming into force in the late 1990s. Limits are set for both new and existing units and are usually given as a total amount of dioxins and furans.

DLC levels in water are regulated under the Safe Drinking Water Act, which was implemented in 1994. The maximum contaminant level, which is the legally enforceable limit, was set at 3×10^{-8} mg/L. A maximum contaminant level goal of zero was set as a voluntary health goal, but it is not legally enforceable.

In addition to federal regulations, some states have also established DLC regulations and guidelines. Regulations regarding the presence of DLCs in the air or water from California, Illinois, Massachusetts, Missouri, New Jersey, New York, and Wisconsin are provided in Appendix Table A-8. As with the federal regulations, many of the state laws regarding DLCs are recent, and some do not come into force for several more years. Because waste incinerators have been some of the largest contributors to DLC levels in the air, they have been the main focus of many state regulations.

Other Countries and Organizations

Many countries and multinational organizations also have established emission-reduction regulations. Air and water emissions from waste incineration, water quality, and risk management are the primary focus of these regulations. Many of these countries have programs that are newly established, so no data exist on their effectiveness.

EC. The EC has implemented several directives and regulations that are intended to control or reduce the release of DLCs into the environment, as presented in Appendix Table A-9. Member states are legally required to incorporate EC directives into their national legislation within a specified period of time. Some countries go beyond the requirements of the EC directive by (1) establishing target concentrations that are more stringent than those of the directives, (2) establishing target concentrations where none was set by the directives, or (3) addressing processes or media not regulated by existing directives. Appendix Table A-10 presents a detailed summary of each member state's DLC legislation and guidelines that go beyond the EC directives. Appendix Table A-11 presents existing limits in a summary form.

Australia. In its 2001–2002 Federal Budget, the Australian Commonwealth Government announced funding of $5 million over four years for a national program to reduce DLCs in the environment. The State and Territory Environment Ministers endorsed this National Dioxins Program in June 2001. The program is managed by Environment Australia, cooperatively with the states and other commonwealth agencies. The program consists of three stages: an initial-data gathering phase, assessment of the impact of DLCs on human health and the environment, and reduction or, where feasible, elimination of releases of DLCs in Australia. The first task of the National Dioxins Program will be to gather as much data as possible about DLCs in Australia, including levels occurring in food and people. Additionally, the program will develop an updated, comprehensive inventory of sources and emissions of DLCs.

Japan. The passing of the Law Concerning Special Measures Against Dioxin in July 1999 was in response to public concern about the levels of DLCs in Japan. This law states that "Businesses shall take necessary measures to prevent environmental pollution by DLCs that are generated in the course of conducting their business activities, as well as measures to remove such DLCs, and must also cooperate with measures taken by the federal and local governments for prevention of environmental pollutions by DLCs and for their removal." Additionally, this law set the level of TDI of DLCs, established environmental quality standards for air, water, and soil, and set stricter standards to regulate emissions to air and water. These standards are summarized in Appendix Table A-12.

Feeds and Foods

As DLC exposure has been linked to food, many countries and multinational organizations have begun to consider ways to reduce this risk. They are working to develop or have established regulations and guidelines for acceptable contamination levels and management practices for animal feeds and foods.

United States

Currently, there are no U.S. legal standards (tolerances) for the presence of dioxins, as a class, in feeds or foods (Appendix Table A-13), although there are tolerances for PCBs in animal feeds and human foods (21 C.F.R. §109.30) and for an action level for PCBs in meat (FDA Compliance Policy Guide 565.200). The U.S. Food and Drug Administration (FDA), the U.S. Department of Agriculture (USDA), and EPA all conduct monitoring and studies on the incidence of DLCs in feeds and foods and are actively researching its sources and methods to limit human exposure through food. FDA (2000) has also issued revised guidance entitled "Guidelines for Industry: Dioxin in Anti-caking Agents Used in Animal Feed and Feed Ingredients."

Other Countries and Organizations

Countries or entities other than the United States have established acceptable limits for various specific food types. Other regulations and guidelines are under consideration.

EC. In November 2001, EC adopted new regulations that set legally binding maximum limits for the presence of polychlorinated dibenzo-*p*-dioxins (PCDD) and polychlorinated dibenzofurans (PCDF) in terms of pg WHO-PCDD/F-TEQ/ g of fat or product for feeds and foods. This regulation took effect in July 2002. According to the EC's Community Strategy for Dioxins, Furans, and Polychlorinated Biphenyls, these maximum limits are set at a strict but feasible level in order to discard unacceptably contaminated products (Scientific Committee on Food, 2001). It is intended that these levels will be revised to include dioxin-like PCBs, and that the levels will decrease over time. EC also foresees establishing target levels that will reasonably ensure that a large majority of the European population will be within its TWI for DLCs. These target levels would be the driving force behind measures necessary to further reduce emissions of DLCs into the environment.

Additionally, in March 2002, EC recommended a series of action limits for the DLC content in feeds and foods. These levels are not legally binding, but if they are exceeded, it should trigger an investigation of possible contamination sources by EC member-state authorities. The levels are higher than the legally binding content limits adopted the previous year. EC recommended that authorities monitor food and take measures to reduce or eliminate contamination sources when DLC levels are found above the action threshold. These legally nonbinding action levels and legally binding levels are presented in Appendix Tables A-14, A-15, and A-16.

WHO/FAO. WHO, in collaboration with FAO through the joint WHO/FAO Codex Alimentarius Commission, is considering establishing guideline levels for DLCs in foods. Currently, there is no JECFA standard for DLCs in foods.

Monitoring and Research Programs

In order to determine the extent of and changes in DLC contamination and human exposures, it is necessary to establish actual levels of DLCs in food and the environment in which those foods are produced. In addition, basic and applied research on such topics as establishing pathways, the effectiveness of interventions, and mechanisms for the fate and transport of DLCs are very important to understanding the implication of DLCs in the food supply.

Many countries have established monitoring programs for DLCs or have incorporated DLC analysis into on-going monitoring programs. An understanding of the patterns and levels of DLC residues in foods, feeds, human tissues, and the environment can be used in contamination and exposure reduction efforts.

Monitoring in the United States

Most DLC monitoring programs in the United States are conducted at the federal level, although some are also conducted at the state level. At both levels, the majority of the ongoing programs are environmental sampling (air, soil, and sediment monitoring). It appears that at the present time, monitoring of feeds and foods are only being carried out by federal agencies, such as FDA, EPA, and USDA's Food Safety and Inspection Service. These are discussed below.

Wisconsin, New York, Illinois, and California have some form of environmental monitoring, either broad surveys or more targeted studies of identified locations or issues. A complete assessment, at the state level, of DLC monitoring and regulatory activities would necessitate a comprehensive 50-state survey. No monitoring of food takes place at the state level. The U.S. federal and state environmental and food monitoring programs are summarized in Appendix Tables A-17 and A-18, respectively.

Environmental, Feed, and Food Monitoring. FDA currently uses a subsample of between 200 and 300 food items from the Total Diet Study (TDS) to analyze for DLCs. This analysis is conducted separately from the analysis in the TDS (see Chapter 5). The sampling is conducted once per year and has been completed up to 2001. The food samples typically chosen are those that have not previously been analyzed for DLCs or those that may contain animal fats.

In addition to TDS sampling, FDA conducts a targeted sampling study aimed at foods that are potentially variable in contaminant levels, such as fish, vegetable oils, and dietary supplements. For example, a number of different fish varieties may be sampled rather than just one species in order to understand the sources of DLCs and the variability across species. This sampling is conducted on a yearly basis and generally includes 500 to 1,000 samples.

FDA also follows up on any unusually high values in any of their studies to determine sources of DLCs in the food supply. FDA does not target any specific imported foods, but it tries to create a representative sample of the diet of the general U.S. population, which may include imported foods. When FDA does investigate an imported food, it tends to look at imports from the top three countries for that product.

USDA and EPA conducted a joint program of three surveys for DLCs in beef, pork, and poultry, using 60 to 80 samples in each survey taken from federally inspected slaughterhouses in the United States. These studies were not repeated or continuous studies, but rather one-time events. Sixty-three beef samples were collected in May and June 1994 and examined for CDDs and CDFs. The sampling for the pork survey took place in August and September 1995 and yielded 78 final samples. It was the first survey for CDDs and CDFs in pork in the United States. Sampling was conducted in September and October 1996 for poultry, with a final sample size of 80. This poultry survey was also the first of its kind in the United States to survey for CDDs and CDFs (see later section, "Concentrations of DLCs in Foods").

Human Biomonitoring. The Centers for Disease Control and Prevention's (CDC) National Health and Nutrition Examination Surveys (NHANES) are a series of studies that have collected data on the health and nutritional status of the U.S. population since the early 1960s. Between 1998 and 2001, the dietary component of NHANES and the USDA/Agricultural Research Service Continuing Survey of Food Intakes by Individuals merged; NHANES also became a continuous and annual survey. The sampling plan for each year follows a complex, stratified, multistage, probability cluster design to select a representative sample (approximately 5,000 individuals) of the noninstitutionalized, civilian U.S. population.

The NHANES protocol includes a home interview followed by a standardized physical examination in a mobile examination center. As part of the examination protocol, blood is obtained by venipuncture from participants aged 1 year and older, and urine specimens are collected from people aged 6 years and older. The venipuncture is performed to obtain laboratory results that provide prevalence estimates of disease, risk factors for examination components, and baseline information on the health and nutritional status of the population. Recently included among the NHANES laboratory measures are serum dioxins, furans, and coplanar PCBs.

CDC's National Center for Environmental Health, Division of Laboratory Sciences, performs the environmental chemical analysis of the blood or urine specimens collected in NHANES. The first *National Report on Human Exposure to Environmental Chemicals* (CDC, 2001) did not include DLC measurements. The second national report (NCEH, 2003), using data from the 1999–2000 NHANES survey, included DLC results, along with other environmental chemical analytes; data for people older than 12 years, including major demographic attributes (e.g., race and sex); and approximately 2,500 samples (from approximately 10,000 participants for the 2-year period, 5,000 participants per year). Unfortunately, none of these data were available in time for inclusion in this report. Human exposure data from the 2001–2002 NHANES survey on DLCs (and other environmental chemicals) is estimated to be released in fall 2003. It is anticipated that all four years of the NHANES data (1999–2002) will be combined for a more refined demographic analysis; it is not known when these data will be released.

Monitoring Programs of Other Countries and Organizations

EC. Appendix Table A-19 summarizes the nationally funded monitoring programs' research activities that were underway as of 1999 in each EC member state. Programs that were completed by 1999 are not included in this list.

Currently, there are several additional on-going DLC surveys in the United Kingdom, including cow's milk studies, wild and farmed fish studies, total diet studies, and a baby food study that is about to be launched (Personal communication, M. Gem, U.K. Food Standards Agency, May 3, 2002). Beginning in July 2002, EC member states are required to conduct food surveillance studies. In 2002, the United Kingdom will collect 62 food samples, concentrating on foods containing fat; the number of samples will be doubled in 2003 (Personal communication, M. Gem, U.K. Food Standards Agency, May 3, 2002). These results will be published in Food Safety Information Sheets. Several countries, including Austria, Belgium, Denmark, Finland, Germany, the Netherlands, Spain, Sweden, and the United Kingdom, have also participated in the WHO assessment of DLC concentrations in human breast milk.

WHO/FAO. WHO is involved in several monitoring studies of human breast milk and food. Through its European Center for Environment and Health in Bilthoven, the Netherlands, WHO conducts periodic studies on concentrations of DLCs in human breast milk, predominately in European countries.

Since 1976, WHO has been responsible for the Global Environment Monitoring System's Food Contamination Monitoring and Assessment Program. This program provides information on levels and trends of contaminants in food through its network of participating laboratories in over 70 countries around the world. The main objectives of the program are to collect data on levels of certain priority chemicals (including DLCs) in foods, to provide technical coordination with countries wanting to implement monitoring studies on foods, and to provide information to JECFA on contaminant levels to support its work on international standards on contaminants in foods.

Australia. DLCs are not routinely monitored in Australia, and there are very few data on the levels of DLCs in either the environment or in food. However, a survey of DLC levels in foods is currently being conducted by the Department of Health and Aging, the Australia New Zealand Food Authority, and the Australian Government Analytical Laboratories.

Japan. Japan's Law Concerning Special Measures Against Dioxins requires that businesses conduct surveillances of DLC concentrations in emission gas, effluent, ash, dust, and other compounds at least once a year. These results are to be submitted to prefectural governors. Beyond the regulatory requirements, it appears that there are not any ongoing surveillance programs in Japan (see Appendix Table A-20). However, several studies were conducted in 1998 and 1999 in order to identify DLC concentrations in blood, air/indoor air/soil, dust and soot/water, and food. These studies include:

- The State of Dioxin Accumulation in the Human Body, Blood, Wildlife, and Food: Findings of the Fiscal 1998 Survey. Sponsored by the Ministry of the Environment: Environmental Health and Safety Division, Environmental Health Department, Environment Agency of Japan (cited in Tran et al., 2002).
- Survey on the State of Dioxin Accumulation in Wildlife: Findings of the Fiscal 1999 Survey. Sponsored by the Ministry of the Environment: Environmental Risk Assessment Office, Environmental Health Department, Environment Agency of Japan (cited in Tran et al., 2002).
- Detailed Study of Dioxin Exposure: Findings of the Fiscal 1999 Survey. Sponsored by the Ministry of the Environment: Environmental Risk Assessment Office, Environmental Health Department, Environment Agency of Japan (cited in Tran et al., 2002).

Canada. The Feed Program in the Canadian Food Inspection Agency (CFIA) routinely monitors for contaminants in livestock feeds as part of their National Feed Inspection Program. In a preliminary survey conducted by CFIA in 1998–1999, 24 fishmeals and feeds and 9 fish oils were sampled across Canada and tested for dioxins, furans, and PCBs. The results are summarized in Appendix Table A-21. CFIA is currently utilizing these results to develop a continuing monitoring plan for dioxin, furans, and PCBs and future regulatory approaches.

Research Programs in the United States

While significant academic and industrial research on DLCs exists, many governmental organizations also have an active role in promoting and conducting research on DLCs. Much of this research is in complement to on-going monitoring and surveillance programs and includes a variety of modeling studies to evaluate the behavior of DLCs in the air, how they move through the environment, and how they become part of the food supply. The U.S. government has also been conducting research into the effects of exposure to DLCs on humans and examining historical data to determine how DLC levels change through time. Representative federal research programs are summarized in Appendix Table A-22.

Research Programs in Other Countries and Organizations

European Commission. Appendix Table A-23 summarizes the nationally funded research activities that were underway as of 1999 in each EC member state. Programs that were completed by 1999 are not included in this list.

WHO/FAO. WHO is involved in several research studies. The major research endeavor includes working with the United Nations Environmental Programme to provide risk assessments of POPs, including DLCs.

Chemical Methods for Analysis of DLCs in Feeds and Foods

Not all feeds or food products have been found to be at equal risk for DLC contamination. While commonly associated with feeds and foods containing animal fats, DLCs can, however, also be found in vegetables, fruits, and cereals. The need for detection of DLCs at these low levels makes the current quantitative methods of analysis expensive and challenging to perform, which limits the number of laboratories available to conduct these tests (Hass and Stevens, 2001). In order to efficiently develop a reliable picture of DLCs in the food supply, both screening methods (which can be used to analyze a large number and variety of feed and food samples), and trace analysis (which can quantify low levels of DLCs in follow-up to a positive screening result) can be useful.

Both current screening and trace analysis methods follow a two-part procedure: extraction/separation of the sample, where the compounds of interest are isolated from the matrix; and instrumental analysis, where DLCs are detected. The major challenge with regard to food samples is the extraction/separation of DLCs from other compounds in the food matrix. Techniques for DLC extraction from fruits, hard vegetables, soft vegetables, grains, dairy products, fish, and meats are very different, and composite foods and food additives are especially challenging.

Screening Methods

Because of their speed and cost efficiency, significant efforts have been made in recent years to develop screening assays for determining DLC and PCB contamination. However, screening assays provide speed and cost savings at the expense of specificity and a lower level of detection (Hass and Stevens, 2001). They do provide the sensitivity of conventional assays and, most importantly, minimize false negatives. Although to quantify contamination levels, trace analysis must follow a positive screening result, screening methods can be very useful in detecting a contamination event or identifying critical control points in a potential contamination pathway.

Two cost-effective approaches have been developed for screening purposes: instrumental methods and biotechnology approaches. Both approaches rely on the same basic extraction methods used in trace analysis, while reducing the cost of the analytical measurement.

Of the two screening methods, the development of the CALUX method was supported under a Small Business Innovation Research Grant from the National Institute of Environmental Health Sciences. FDA's Center for Veterinary Medicine, Arkansas Regional Laboratory, has a licensing agreement to use the CALUX method for its DLC research.

Instrumental Methods

Instrumental methods of screening for the presence of DLCs respond to the physical properties of the compounds. Interfering compounds that were not removed during the initial extraction procedure and may cause an overestimate of DLC contamination levels can be identified in an initial analysis. A secondary clean-up of the sample can be performed and the sample can be reanalyzed, reducing the number of false positives that this methodology produces.

Biotechnology Approach

The biotechnology approach is based on the chemical reactivity of compounds and uses immunoassay-type tests and arylhydrocarbon receptor-type tests.

In comparison studies, the biotechnology approach has been found routinely to overestimate DLC content in the presence of interfering compounds, which results in false positives (Hass and Stevens, 2001). While some interfering compounds are removed during the extraction phase of the test, the residual presence of these compounds is not detectable during a biotechnology-based assay. Even more seriously, this approach also can have a problem with false negatives due to analyte loss during the extraction phase (Hass and Stevens, 2001). Assays of duplicate aliquots from a single sample can minimize this problem, but this doubles the cost of analysis.

Analytical Methods for Analysis of DLCs in Feeds and Foods

Trace Analysis

The analytical approach used most frequently to detect DLCs in feeds and foods relies on isotopic dilution. This method is isomer specific, very sensitive, and robust, although expensive and demanding. Following the extraction phase, a known amount of the isotope ^{13}C is added to the sample, which creates a mixture of forms of the compound of interest that are chemically identical, yet distinguishable by mass spectrometry. Using a combination of gas chromatography and high-resolution mass spectrometry allows determination of the ratio between each analyte and its associated isotopically labeled standard, leading to accurate quantification of analyte concentration. The overall accuracy of the assay depends on the ability to spike the sample with the isotope accurately, weigh the sample, and measure the ratio. The effects of interfering compounds and minor sample losses due to handling are detectable and correctable. The analytical cost estimates associated with the standard analytical method for DLCs obtained from a number of sources are summarized in Appendix Table A-24.

EPA-Approved Method for Analysis of Dioxins and Furans in Wastewater

In 1997, to augment less sensitive methods approved earlier, EPA Method 1613: Tetra- Through Octa-Chlorinated Dioxins and Furans by Isotope Dilution High Resolution Gas Chromatography/High Resolution Mass Spectrometry (HRGC/HRMS), EPA 821-B-94-005, was approved. Method 1613 is the most sensitive analytical test procedure approved under the Clean Water Act for the analysis of CDDs and CDFs and was developed to meet the need for more stringent pollutant monitoring and control. Method 1613 also allows determination of the 17 tetra- through octa-chlorinated, 2,3,7,8-substituted CDDs and CDFs.

Method 1613 extends minimum levels of quantitation of CDDs and CDFs into the low parts-per-quadrillion range for aqueous matrices and the low parts-per-trillion range for solid matrices. Furthermore, the use of isotope dilution techniques, internal standard calibration, and the 1600 series method quality control protocol results in improved sensitivity, precision, and accuracy. These improvements have been validated through both intra- and interlaboratory validation studies. Method 1613 is also intended to encourage advances in technology and reductions in the cost of analysis by allowing the use of alternate extraction and clean-up techniques. The analyst is permitted to modify the method to overcome interferences or to lower the cost of measurements, provided that all method equivalency and performance criteria are met.

Concentrations of DLCs in Foods

This section presents recent data (1990 or later) regarding the concentrations of DLCs in European and North American foods, as provided by AEA Technology (1999), ATSDR (1998), EPA (2000), Fiedler and colleagues (2000), IARC (1997), and the Scientific Committee on Food (2000, 2001).

Recent Contamination Levels in Foods

Appendix Table A-25 provides DLC values for foods other than breast milk. The data are very heterogeneous with regard to collection date, the number of samples of a particular food, sampling method (individual versus composite samples), the compounds analyzed, the unit of analysis (e.g., fat, fresh weight, dry weight), and the state of the food (e.g., raw or cooked). Numbers in the table may represent means, ranges of means, or ranges of observations.

Temporal Trends

Evidence of temporal trends in the data on DLC contamination levels in foods is presented in some reviews, if only indirectly, in decreasing estimates of dietary intakes of DLCs. EPA (2000) reports, specifically that:

- Concentrations of dioxins and furans in U.K. cows' milk declined from 1.1 to 3.3 pg I-TEQ$_{DF}$/g of lipid in 1990 to 0.67 to 1.4 pg I-TEQ$_{DF}$/g of lipid in 1995.
- The mean pg I-TEQ$_{DF}$/g of lipid in German milk declined by about 25 percent between 1990 and 1994.
- Examination of U.S. foods preserved over the last several decades suggests that dioxin and furan concentrations were two to three times higher in the 1950s to 1970s than at present, while PCB concentrations were ten times higher.

IARC (1997) states that PCDD/PCDF in milk, dairy products, eggs, poultry, and "fatty food composites" in the United Kingdom decreased markedly during the 1980s. Fiedler and colleagues (2000) cite a decrease in the level of contamination of German foods, most markedly for dairy products, meat, and fish.

In 1998, an EPA study compared the DLC concentration in historic samples to current DLC concentrations derived from post-1993 national food surveys for beef, pork, poultry, and milk (Winters et al., 1998). The surveys' principal objective was to determine the national average concentration of DLCs in the lipids of these animal-fat products. National mean TEQ concentrations from these surveys are shown in Appendix Table A-26.

Appendix Table A-27 presents the PCDD/PCDF and PCB TEQ concentrations of the 14 historical samples, as well as TEQ concentrations normalized and expressed as a percent of current concentrations for the most similar food type. For example, the 1908 beef ration percentage of 38 percent means that the 0.34 pg TEQ/g of lipid PCDD/PCDF (calculated at nondetects = ½ limit of detection) is 38 percent of the current beef concentration of 0.89 pg TEQ/g of lipid (at nondetects = ½ limit of detection), as determined by the recent national EPA beef survey.

Although not necessarily representative of these food types or their respective time period, it should be noted that all 10 samples from 1957 to 1982 were higher in PCDD/PCDF TEQ than the current mean concentrations (at nondetects = ½ limit of detection), and that 12 of the 13 samples taken from 1945 through 1983 were higher for PCB TEQ. If the samples are indicative of past concentrations of DLCs, normalized TEQ suggests a PCDD/PCDF concentration two to three times higher during the period of peak environmental loading, while PCB TEQ may have been over 10 times current concentrations. EPA plans to continue analyzing historic meat and dairy products as additional samples become available.

Contribution of Food Groups to DLC Exposure

According to the Scientific Committee on Food (2000), the major sources of dietary exposure to PCDD/PCDF intakes in Europe are milk and dairy products (16 to 39 percent), meat and meat products (6 to 32 percent), and fish and fish products (11 to 63 percent). Fish was a particularly large contributor in Finland and Sweden, fruits and vegetables in Spain, and cereals in the United Kingdom. In Germany, milk, meat, and fish contributed 31 percent, 23 percent, and 17 percent, respectively, of dietary I-TEQ from PCDD/PCDF (Scientific Committee on Food, 2000).

Temporal Trends

Several of the reviewed reports describe data that suggest a decrease in DLC intakes over recent decades. AEA Technology (1999) reports three time trend-analyses of DLC intake. Dutch data on DLC in foods gathered in 1978, 1984 to 1985, and 1994 indicate a statistically significant decreasing trend in adult intake over time, with a 50 percent decrease in I-TEQ/kg/d over each 5.5-year interval in that period. Dietary studies conducted in Germany in 1989 and 1995 indicate a 45 percent decrease in I-TEQ intake. Diet studies in the United Kingdom in 1982, 1988, and 1992 suggest a 45 percent decrease in intake at each time point compared to the preceding point. All of these estimates likely pertain to the average consumer. Independent of changes over time in the DLC content of food and in dietary intake, it is generally recognized that body burdens of DLCs increase with age due, in part, to the long half-lives of these compounds.

Summary

This chapter summarizes reports on the toxicity and risk of DLCs and on regulatory activity in the United States and other countries that are widely recognized to reflect current status and knowledge of these compounds (AEA Technology, 1999; ATSDR, 1998; EPA, 2000; Fiedler et al., 2000; IARC, 1997; Scientific Committee on Food, 2000, 2001). The risks from exposure to DLCs outlined in these documents are based on population groups that received exposures exceeding the daily exposures estimated for the general population, so risks to the general population are not known. Efforts to regulate DLCs range from exposure limits and environmental emission regulations to guidelines and recommendations to limit DLC levels in feed and food. Efforts in the United States and other countries to monitor DLCs and gather more current data are described. Research on DLC levels in human foods indicates that the greatest contribution to exposure from the food supply is from animal fats in meat, dairy products, and fish.

REFERENCES

AEA Technology. 1999. *Compilation of EU Dioxin Exposure and Health Data.* Prepared for the European Commission DG Environment. Oxfordshire, England: AEA Technology.

ATSDR (Agency for Toxic Substances and Disease Registry). 1994. *Toxicological Profile for Chlorinated Dibenzofurans (CDFs).* Atlanta, GA: ATSDR.

ATSDR. 1998. *Toxicological Profile for Chlorinated Dibenzo-p-dioxins.* Atlanta, GA: ATSDR.

ATSDR. 2000. *Toxicological Profile for Polychlorinated Biphenyls (PCBs).* Atlanta, GA: ATSDR.

Axelson O, Sundell L. 1974. Herbicide exposure, mortality and tumor incidence. An epidemiological investigation on Swedish railroad workers. *Scand J Work Environ Health* 11:21–28.

Bertazzi PA, Zocchetti C, Guercilena S, Consonni D, Tironi A, Landi MT, Pesatori AC. 1997. Dioxin exposure and cancer risk: A 15-year mortality study after the "Seveso accident." *Epidemiology* 8:646–652.

Bertazzi PA, Bernucci I, Brambilla G, Consonni D, Pesatori AC. 1998. The Seveso studies on early and long-term effects of dioxin exposure: A review. *Environ Health Perspect* 106:625–633.

CDC (Centers for Disease Control and Prevention). 2001. *National Report on Human Exposure to Environmental Chemicals.* Atlanta, GA: CDC.

Chen PH, Gaw JM, Wong CK, Chen CJ. 1980. Levels and gas chromatographic patterns of polychlorinated biphenyls in the blood of patients after PCB poisoning in Taiwan. *Bull Environ Contam Toxicol* 25:325–329.

EPA (U.S. Environmental Protection Agency). 2000. *Exposure and Human Health Reassessment of 2,3,7,8-Tetrachlorodibenzo-p-Dioxin (TCDD) and Related Compounds.* Draft Final Report. Washington, DC: EPA.

FDA (U.S. Food and Drug Administration). 2000. Guidance for industry: Dioxin in anticaking agents used in animal feed and feed ingredients; Availability. *Fed Regist* 65:20996–20997.

Fiedler H, Hutzinger O, Welsch-Pausch K, Schmiedinger A. 2000. *Evaluation of the Occurrence of PCDD/PCDF and POPs in Wastes and Their Potential to Enter the Foodchain.* Prepared for the European Commission DG Environment. Bayreuth, Germany: University of Bayreuth.

Hass RJ, Stevens FM. 2001. Dealing with dioxin: The state of analytical methods. *Food Safety Mag* Dec 2000/Jan 2001.

Huisman M, Koopman-Esseboom C, Fidler V, Hadders-Algra M, van der Paauw CG, Tuinstra LGM, Weiglas-Kuperus N, Sauer PJJ, Touwen BCL, Boersma ER. 1995. Perinatal exposure to polychlorinated biphenyls and dioxins and its effect on neonatal neurological development. *Early Human Development* 41:111–127.

IARC (International Agency for Research on Cancer). 1997. *IARC Monographs on the Evaluation of Carcinogenic Risks to Humans. Volume 69: Polychlorinated Dibenzo-para-Dioxins and Polychlorinated Dibenzofurans.* Lyon, France: World Health Organization.

IOM (Institute of Medicine). 1991. *Nutrition During Lactation.* Washington, DC: National Academy Press.

Jacobson JL, Fein GG, Jacobson SW, Schwartz PM, Dowler JK. 1984. The transfer of polychlorinated biphenyls (PCBs) and polybrominated biphenyls (PBBs) across the human placenta and into maternal milk. *Am J Public Health* 74:378–379.

Koopman-Esseboom C, Huisman M, Weislas-Kuperus N, Van der Paauw CG, Tuinstra LGM, Boersma ER, Sauer PJJ. 1994a. PCB and dioxin levels in plasma and human milk of 418 Dutch women and their infants. Predictive value of PCB congener levels in maternal plasma for fetal and infant's exposure to PCBs and dioxins. *Chemosphere* 28:1721–1732.

Koopman-Esseboom C, Morse DC, Weisglas-Kuperus N, Lutkeschipholt IJ, Van der Paauw CG, Tuinstra LG, Brouwer A, Sauer PJ. 1994b. Effects of dioxins and polychlorinated biphenyls on thyroid hormone status of pregnant women and their infants. *Pediatr Res* 36:468–473.

Kuratsune M, Ikeda M, Nakamura Y, Hirohata T. 1988. A cohort study on mortality of Yusho patients: A preliminary report. In: Miller RW, Wantanabe S, Fraumeni JF, eds. *Unusual Occurrences as Clues to Cancer Etiology.* Tokyo: Japan Scientific Societies Press. Pp. 61–68.

Landi MT, Bertazzi PA, Consonni D. 1996. TCDD blood levels, population characteristics, and individual accident experience. *Organohalogen Compd* 30:290–293.

Landi MT, Consonni D, Patterson DG Jr, Needham LL, Lucier G, Brambilla P, Cazzaniga MA, Mocarelli P, Pesatori AC, Bertazzi PA, Caporaso NE. 1998. 2,3,7,8-Tetrachlorodibenzo-p-dioxin plasma levels in Seveso 20 years after the accident. *Environ Health Perspect* 106:273–277.

Lanting CI, Fidler V, Huisman M, Boersma ER. 1998. Determinants of polychlorinated biphenyl levels in plasma from 42-month-old children. *Arch Environ Contam Toxicol* 35:135–139.

Mocarelli P, Needham LL, Marocchi A, Patterson DG Jr, Brambilla P, Gerthoux PM, Meazza L, Carreri V. 1991. Serum concentrations of 2,3,7,8-tetrachlorodibenzo-p-dioxin and test results from selected residents of Seveso, Italy. *J Toxicol Environ Health* 32:357–366.

NCEH (National Center for Environmental Health). 2003. *Second National Report on Human Exposure to Environmental Chemicals.* NCEH Pub. No. 02-0716. Atlanta, GA: CDC.

Patandin S, Koopman-Esseboom C, de Ridder MA, Weisglas-Kuperus N, Sauer PJ. 1998. Effects of environmental exposure to polychlorinated biphenyls and dioxins on birth size and growth in Dutch children. *Pediatr Res* 44:538–545.

Patandin S, Dagnelie PC, Mulder PG, Op de Coul E, van der Veen JE, Weisglas-Kuperus N, Sauer PJ. 1999. Dietary exposure to polychlorinated biphenyls and dioxins from infancy until adulthood: A comparison between breast-feeding, toddler, and long-term exposure. *Environ Health Perspect* 107:45–51.

Riihimäki V, Asp S, Herberg S. 1982. Mortality of 2,4-dichlorophenoxyacetic acid and 2,4,5-trichlorophenoxyacetic acid herbicide applicators in Finland: First report of an ongoing prospective study. *Scand J Work Environ Health* 8:37–42.

Riihimäki V, Asp S, Pukkala E, Hernberg S. 1983. Mortality and cancer incidence among chlorinated phenoxyacid applicators in Finland. *Chemosphere* 12:779–784.

Scientific Committee on Food. 2000. *Opinion of the Scientific Committee on Food on the Risk Assessment of Dioxins and Dioxin-like PCBs in Food.* European Commission, Health and Consumer Protection Directorate-General. SCF/CS/CNTM/DIOXIN/8 Final. Brussels: European Commission.

Scientific Committee on Food. 2001. *Opinion of the Scientific Committee on Food on the Risk Assessment of Dioxins and Dioxin-like PCBs in Food. Update.* European Commission, Health and Consumer Protection Directorate-General. CS/CNTM/DIOXIN/20 Final. Brussels: European Commission.

Tran N, Wells C, Daniels C. 2002. *A White Paper on Existing Dioxin Regulations and Monitoring Program.* Prepared for the Committee on the Implications of Dioxin in the Food Supply. Washington DC: Novigen Sciences.

Vreugdenhil HJ, Slijper FM, Mulder PG, Weisglas-Kuperus N. 2002. Effects of perinatal exposure to PCBs and dioxins on play behavior in Dutch children at school age. *Environ Health Perspect* 110:A593–A598.

Weisglas-Kuperus N, Patandin S, Berbers GA, Sas TC, Mulder PG, Sauer PJ, Hooijkaas H. 2000. Immunologic effects of background exposure to polychlorinated biphenyls and dioxins in Dutch preschool children. *Environ Health Perspect* 108:1203–1207.

Winters DL, Anderson S, Lorber M, Ferrario J, Byrne C. 1998. Trends in dioxin and PCB concentrations in meat samples from several decades of the 20th century. *Organohalogen Compd* 38:75–78.

Wolfe WH, Michalek JE, Miner JC, Rahe A, Silva J, Thomas WF, Grubbs WD, Lustik MB, Karrison TG, Roegner RH, Williams DE. 1990. Health status of Air Force veterans occupationally exposed to herbicides in Vietnam. I. Physical health. *J Am Med Assoc* 264:1824–1831.

3

Sources of Dioxins and Dioxin-like Compounds in the Environment

Dioxin and dioxin-like compounds (referred to collectively as DLCs) are ubiquitous in the environment (ATSDR, 1998; Travis and Hattemer-Frey, 1989). People may be exposed to background levels (i.e., low concentrations) of DLCs by breathing air, by consuming food or beverages, or by their skin coming into contact with DLC-contaminated materials (ATSDR, 1998). However, the major route of human exposure to DLCs is through the food supply; the first step in that pathway is the introduction of DLCs into the environment.

This chapter provides a general overview of known sources of DLCs and how these compounds are transported from the environment into the pathways that lead to human foods. Five major sources of DLCs: combustion, metals smelting and refining, chemical manufacturing, biological and photochemical processes, and environmental reservoirs, are described. Quantitative information on DLCs released from each source is briefly presented and the environmental fate of DLCs, including how these compounds are transported within or to various geographic regions, is discussed. Finally, information is presented on how DLCs enter the food chain.

The information presented in this chapter is not an exhaustive summary of environmental sources of DLCs and their environmental fate and transport; such a summary is beyond the scope of this report. For more detailed information, the reader is referred to the Agency for Toxic Substances and Disease Registry's (ATSDR) toxicological profiles for chlorinated dibenzo-*p*-dioxins, chlorinated dibenzofurans, and polychlorinated biphenyls (PCBs) (ATSDR 1994, 1998, 2000); the U.S. Environmental Protection Agency's (EPA) draft *Exposure and Human Health Reassessment of 2,3,7,8-Tetrachlorodibenzo-p-Dioxin (TCDD) and Related Compounds* (EPA, 2000); and the original articles referenced in this chapter.

MAJOR SOURCES OF DLCs IN THE ENVIRONMENT

With the exception of the dioxin-like PCBs, DLCs are unintended byproducts of combustion. Combustion sources can be of anthropogenic (e.g., waste incineration) or natural origin (e.g., forest fires). Industrial (e.g., paper and chemical manufacturing) and biological processes also contribute to DLC production, although in smaller quantities. These DLCs are formed in trace quantities and released into the environment. Until 1977, when regulatory action prohibited further manufacture, PCBs were commercially produced in large quantities in the United States. The potential for human exposure to PCBs still exists, however, because they persist in air, soil, and water sediments for many years. Additionally, they are also found in older transformers, capacitors, fluorescent lighting fixtures, electrical devices, and appliances that are still in use.

Combustion

The primary environmental source of dioxins and furans is combustion (Zook and Rappe, 1994, as cited in ATSDR, 1998). Combustion processes include waste incineration (e.g., municipal solid waste, sewage sludge, medical waste, and hazardous waste), burning of various fuels (e.g., coal, wood, and petroleum products), other high-temperature sources (e.g., cement kilns), and poor or uncontrolled combustion sources (e.g., forest fires, volcanic eruptions, building fires, and residential wood burning) (Clement et al., 1985; EPA, 2000; Thoma, 1988; Zook and Rappe, 1994, as cited in ATSDR, 1998). At present, the major quantifiable source of DLC formation in the United States is from backyard burn barrels (Personal communication, D. Winters, EPA, April 2, 2002) (see Appendix Table A-28). Most of the direct releases of DLCs from combustion processes are into the air (Czuczwa and Hites, 1984, 1986a, 1986b, as cited in ATSDR, 1998). Estimates of polychlorinated dibenzo-*p*-dioxin and polychlorinated dibenzofuran sources made in EPA's draft inventory (EPA, 1998) indicate that domestic waste burning contributes more than 1,000 g toxicity equivalents (TEQ)/y, and likely contribute significantly to current DLC releases (Gullett et al., 2001; Wunderli et al., 2000).

DLCs are formed in chemical reactions that occur during the combustion of organic compounds in the presence of chlorinated materials (ATSDR, 1998). EPA (2000) outlines three mechanisms by which this occurs. First, stack emissions of DLCs result from the incomplete destruction of DLC contaminants present in materials delivered to the combustion chamber. Not all of the DLC components are destroyed by the combustion system, thus allowing trace amounts of DLCs to be emitted from the stack. Second, the formation of DLCs from aromatic precursor compounds occurs in the presence of a chlorine donor. The general reaction in this formation pathway is an interaction between an aromatic precursor compound and chlorine promoted by a transition metal catalyst on a reactive fly ash surface (Dickson and Karasek, 1987; Liberti and Brocco, 1982, as cited in EPA, 2000). Last, de novo synthesis promotes formation of DLCs in combustion processes from the oxidation of carbon particulate catalyzed by a transition metal in the presence of chlorine. Intermediate compounds, which are precursors to DLC formation, are produced during de novo synthesis. The formation of DLCs via either the precursor or de novo synthesis pathways requires the availability of gaseous chlorine (Addink et al., 1995, as cited in EPA, 2000; Luijk et al., 1994). The source of the chlorine is the materials (fuels or feed) in the combustion system.

Metals Smelting and Refining

There are several types of primary and secondary metal smelting and refining operations, including iron ore sintering, steel production, and scrap metal recovery. Such operations use both ferrous and nonferrous metals. Few data are available on the potential for the formation and environmental release of DLCs from primary nonferrous metal manufacturing operations. The contribution of DLCs produced from primary copper and aluminum smelters is thought to be minimal (Environmental Risk Sciences, 1995; Lexen et al., 1993, as cited in EPA, 2000). Titanium smelting and refining may be a source of DLCs (Bramley, 1998, as cited in EPA, 2000). Current information is insufficient to determine if primary magnesium smelting and refining releases DLCs into the environment. Secondary smelting and refining of nonferrous metals such as aluminum, copper, lead, and zinc may result in formation of DLCs, due to combustion of organic impurities (e.g., plastic, paints, and solvents) in the metals and chlorine-containing chemicals (e.g., sodium chloride and potassium chloride) used in the smelting process (Aittola et al., 1992; EPA, 1987, 1997, as cited in EPA, 2000).

Two types of operations used in primary ferrous metal smelting—iron sinter production and coke production—are potential sources of DLCs (Lahl, 1993, 1994; Lexen et al., 1993; Rappe, 1992a, as cited in EPA, 2000). Recycled dust and scraps from other processes in the sintering plant introduces traces of chlorine and organic compounds that generate DLCs (Lahl, 1993, 1994, as cited in EPA, 2000). Electric arc furnaces, used in secondary ferrous metal smelting and refining, have also been implicated as a source of DLCs.

Chemical Manufacturing

Three types of chemical manufacturing processes—bleaching of wood pulp in paper manufacturing, chlorine and chlorine-derivative manufacturing, and halogenated organic chemical manufacturing—lead to the production of DLCs.

DLCs, primarily the 2,3,7,8-tetrachlorodibenzo-*p*-dioxin (TCCD) and 2,3,7,8-tetrachlorodibenzofuran (TCDF) congeners, are present in effluent and sludge from pulp and paper mills that employ the bleached kraft process (Clement et al., 1989; EPA, 1991; Swanson et al., 1988, as cited in ATSDR, 1998). From 1988 to 1992, there was a 90 percent reduction in TEQs generated by pulp and paper mills for 2,3,7,8-TCCD and 2,3,7,8-TCDF (NCASI, 1993, as cited in EPA, 2000). To help reduce DLCs in effluents from pulp and paper mills, EPA promulgated effluent limitations guidelines and standards for certain segments of the pulp, paper, and paperboard industries (EPA, 1998, as cited in EPA, 2000). These industries are responsible for more than 90 percent of the bleached-chemical pulp production in the United States.

Chlorine gas is manufactured by electrolysis of brine electrolytic cells. The use of mercury cells that contain graphite electrodes has been shown to generate high levels of chlorodibenzofurans (Rappe, 1993; Rappe et al., 1990, 1991; Strandell et al., 1994, as cited in EPA, 2000). During the 1980s, graphite electrodes were replaced with titanium metal anodes at chlorine gas manufacturing facilities in the United States. Hutzinger and Fiedler (1991, as cited in EPA, 2000) reported low concentrations of DLCs in chlorine-derivative products.

DLCs can be formed during the manufacture and disposal of halogenated organic chemicals such as chlorophenols, chlorobenzenes, and PCBs (Ree et al., 1988; Versar, 1985, as cited in EPA, 2000). Chlorophenols, which have been used for various pesticide applications, may be released into the environment from industrial use of these compounds and from disposal of wastes from the manufacturing facilities that produce them. Production of chlorophenols has been limited to 2,4-dichlorophenol and pentachlorophenols (PCPs) since the late 1980s, and disposal of wastes generated from their manufacture is strictly regulated.

The production of chlorobenzenes (used as raw materials in the production of phenol, aniline, and various pesticides) by certain processes can inadvertently produce DLCs (Ree et al., 1988, as cited in EPA, 2000), but they are not believed to be generated in the production of mono-, di-, and trichlorobenzene, which are the only forms of chlorobenzene currently produced in the United States. The release of DLCs from the production of chlorobenzenes was estimated to be negligible in 1995 (EPA, 2000). Coplanar PCBs, which have dioxin-like activity, are entirely anthropogenic in origin and were produced in the United States from 1929 to 1977 (NRC, 2001). Most countries have banned the production of PCBs, but several potential sources for environmental release remain. These include continued use and disposal of PCB-containing products such as transformers, capacitors, and other electrical equipment that were manufactured before 1977; combustion of PCB-containing materials; recycling of PCB-contaminated products, such as carbonless copy paper; and releases of PCBs from waste storage and disposal (NRC, 2001). PCBs may still be manufactured in Russia and North Korea and therefore, PCBs may be entering the environment both in those countries and in other countries that buy their PCB-containing products (Carpenter,

1998; NRC, 2001). Other halogenated organic chemicals that, through their manufacture or disposal, have been associated with the release of DLCs into the environment include polyvinyl chloride, aliphatic chlorine compounds, and various dyes, pigments, and printing inks.

Biological and Photochemical Processes

Evidence suggests that DLCs can be formed under certain environmental conditions. DLCs have been found in various types of composts, possibly from atmospheric deposition onto plants that were subsequently composted or from uptake of DLCs from air by the compost (Krauss et al., 1994; Lahl et al., 1991, as cited in EPA, 2000). DLCs are also found in sewage sludge; specific sources include microbial biotransformation of chlorinated phenolic compounds, runoff to sewers from contaminated lands or urban surfaces, household or industrial wastewater, and chlorination operations within wastewater treatment facilities (Cramer et al., 1995; Horstmann and McLachlan, 1995; Horstmann et al., 1992; Rappe, 1992b; Rappe et al., 1994; Sewart et al., 1995, as cited in EPA, 2000). Evidence also suggests that DLCs can be generated by photolysis of PCPs, but this reaction has only been demonstrated under laboratory conditions (Lamparski et al., 1980; Vollmuth et al., 1994; Waddell et al., 1995, as cited in EPA, 2000), and it is not known if it occurs in the environment.

Reservoirs for DLCs

Reservoir sources also contribute to global DLC exposure. These reservoirs, representing a recirculation of past DLC generation, may be classified as long-, intermediate-, and short-term, or as static and dynamic sources. DLCs may be released from reservoirs by volatilization, particle and vapor deposition, suspension or resuspension into air and water sediments, and direct consumption by land and aquatic animals and humans. The impact of various reservoir sources on human DLC burdens is dependent on their direct or indirect contact with humans or with human food supplies.

DLC contamination of geological formations represents a long-term, static source. Ball clay deposits (Hayward et al., 1999) that are resurrected and put into circulation by mining and by feeding these deposits to land and aquatic animals is one example of reservoir activation. Other examples may be uncovered as various mineral or other geological deposits are mined and brought to the surface for further processing or use.

Intermediate reservoirs may be characterized as soils and sediments in waterways. Newly generated DLCs precipitate into these areas, dependent on the atmospheric conditions and point source generation capacities. These DLC deposits persist for many years without substantial degradation. Limited volatilization and sunlight-generated degradation may reduce DLC levels within these reservoirs, but the levels are relatively static or are cumulative in their concentrations. Soil and waterway exposures by grazing animals, direct soil contamination of fresh vegetables and fruits, and aquatic exposure to contaminated waterway sediments are examples of direct input of these reservoir sources into the human food supply.

Aerosol or environmental suspension represents a short-term reservoir as DLCs that are adsorbed onto particulate matter or that are in a volatilized state are transported until they precipitate from the atmosphere as rainfall or by particle aggregation and deposition. These means of transport may vary by geographical and temporal duration. This process is one method by which DLCs can be spread over large geographical areas, including intercontinental transport, from the point of generation to the point of entry into intermediate- and long-term reservoirs.

Fat deposits from animals and fish exposed to contaminated environmental and feed sources are an additional short-term reservoir and may be more dynamic than other sources. The generational periods of some species (e.g., bovines) may be long enough to characterize those species as intermediate reservoirs. DLC levels in adipose tissue are a function of the exposure experiences and species metabolic capacities for specific animal or aquatic populations. These fat deposits may be directly introduced to the human population through foods of animal or aquatic origins or by the use of their fats in processed-food preparation, or indirectly through the recycling of animal- and aquatic-origin fats in feed.

GENERAL INFORMATION ABOUT THE QUANTITATIVE INVENTORY OF RELEASES OF DLCs

The most recent analysis of emission estimates of DLCs for the United States have been developed for two reference years, 1987 and 1995, by EPA (2000). EPA chose to use data from 1987 because: (1) prior to 1987 there was little empirical data on source-specific emission estimates, (2) soon after 1987, there was widespread installation of DLC-specific emission controls at a number of facilities and, therefore, 1987 is the latest time representative of the emissions before those controls were installed, and (3) around 1987 there were significant advances in emissions measurement techniques and in techniques to measure low concentrations of DLCs in environmental samples. Data from 1995 were used because it is the latest time period that could be practically addressed by EPA in the time frame that its report was scheduled for completion. Since 1995, EPA has promulgated regulations limiting DLC emissions for several source categories that contribute to the DLC inventory (e.g., municipal waste combustors, waste incinerators, and pulp and paper facilities using chlorine bleach processes). Therefore, the 1995 data may not be indicative of current DLC releases. Because not every facility in the United States for each of the source categories has been tested for DLC releases and emissions, EPA developed an extrapolation method to estimate national DLC emissions from tested facilities. For a detailed description of that methodology, the reader is referred to EPA (2000). Some investigators have suggested that EPA's methodology may have led to the underestimation of releases (Bruzy and Hites, 1995; Harrad and Jones, 1992; Rappe, 1991, as cited in EPA, 2000). It is likely that unknown sources of DLCs still exist, also leading to the underestimation of releases.

EPA (2000) made several general conclusions about the quantitative inventory of sources for DLCs. These include:

- The best estimates of releases of DLCs to air, water, and land from reasonably quantifiable sources were approximately 3,300 g $TEQ_{DF\text{-}WHO98}$ (3,000 g I-TEQ) in 1995 and 14,000 g $TEQ_{DF\text{-}WHO98}$ (12,800 g I-TEQ) in 1987.

- A wide variety of sources are responsible for the environmental releases of DLCs in the United States; however, combustion sources are responsible for the majority of releases of TCDDs and TCDFs (approximately 85 percent of these compounds from quantifiable sources in 1995). Metals smelting and manufacturing and chemical manufacturing contributed approximately 10 percent and 5 percent, respectively, of environmental releases of DLCs from quantifiable sources in 1995.

- The primary environmental releases of dioxin-like PCBs are electrical equipment.

- There was an estimated decrease in releases of DLCs between 1987 and 1995 of approximately 76 percent. This decrease was primarily due to reductions in air emissions from municipal and medical waste incinerators.

- Insufficient data are available to comprehensively estimate point source releases of DLCs into water.

- The inventory includes only a limited set of activities that results in direct environmental releases to land. Direct releases to land include land application of sewage sludge, commercial sludge products, and pulp and paper mill wastewater sludge, use of 2,4-D pesticides, and manufacturing wastes from the production of ethylene dichloride and vinyl chloride.

- EPA regulates biosolids (sewage sludge) under 40 C.F.R. part 503, §405(d). A recent report by the National Research Council (NRC, 2002) concluded that "there is no documented scientific evidence that the Part 503 rule has failed to protect public health." The report does, however, recommend a need to update the scientific basis of the rule to assure the public and protect public health.

- Significant amounts of DLCs that are produced annually are not considered environmental releases (e.g., DLCs generated internal to a process but destroyed before release and waste streams that are disposed of in approved landfills) and, therefore, are not included in EPA's national inventory.

ENVIRONMENTAL FATE AND TRANSPORT OF DlCs

Transport in Air

Atmospheric transport is a major dispersal mechanism for DLCs in the environment. The presence of DLCs in remote, nonindustrial locations suggests that long-range atmospheric transport of the compounds occurs. The atmospheric distribution and transformation profile for a specific congener depends on its vapor pressure, the atmospheric temperature, and the particle concentration in the air. For a given congener, the less chlorinated it is (i.e., the lower its vapor pressure), the warmer the atmospheric temperature, and the fewer particles that are present, the greater the fraction that will be found in the vapor phase. The more chlorinated DLCs (i.e., pentachlorinated and greater) are associated almost exclusively with the particle-bound phase (ATSDR, 1998), whereas TCDD, which has an intermediate vapor pressure, is associated with both the vapor and particle-bound phases. A percentage of DLCs in the particle-bound phase appear to reach equilibrium between the particle-bound and vapor phases, while some portion remains irreversibly bound to the particles. Dioxin-like PCBs have greater volatility than other DLCs with similar chlorination and are associated more closely with the vapor phase. Like DLCs, they appear to travel great distances in the atmosphere.

DLCs are removed from the atmosphere by either dry or wet deposition. Dry deposition occurs for both the particle- and vapor-associated compounds. The more highly chlorinated DLCs adsorbed to atmospheric particles are removed primarily by gravity to the surface of soil, vegetation, or water. The less chlorinated congeners in the particle or vapor phase are removed by atmospheric turbulence and diffusion. Vapor-phase DLCs are generally deposited directly onto vegetation or soil covering (leaves and detritus). The greater volatility of PCBs means that they may volatilize from soil and water surfaces, thus acting as sources of atmospheric inputs, as demonstrated in the Great Lakes (EPA, 2000).

DLCs may be deposited onto soil, water, or vegetation via wet deposition, either suspended in the water or associated with precipitation particles. DLCs have been measured in precipitation in most locales (EPA, 2000). Wet deposition is the most efficient removal process for particle-bound DLCs from the atmosphere (ATSDR, 1998).

Transport in Soil

DLCs have low water solubility and are highly lipophilic and, as such, tend to partition from the atmosphere to soil and vegetation surfaces. As some DLCs are considered semivolatile, particularly the less chlorinated congeners, some small portion of deposited DLCs may reenter the atmosphere as a result of volatilization from these surfaces or bound to airborne soil particles (EPA, 2000).

Once below the soil surface, soil-bound DLCs do not appear to move up or down via volatilization without a carrier; this is particularly true for the tetra- and higher chlorinated DLCs. The presence of a solvent such as oil may facilitate the diffusive movement of DLCs through soil.

As most DLCs that attach to soil particles have little potential for leaching through or volatilizing from soil, they are removed from soil surfaces primarily by soil erosion (wind or water) and runoff to water bodies. For DLCs that are not eroded, burial is the major fate process. Without the presence of a carrier in the soil, as may occur at a hazardous waste site or spill, DLC contamination of the underlying groundwater from soil is unlikely; however, some leaching through soil may occur for the dioxin-like PCBs, particularly if the soil has a low organic content and is less likely to bind the compounds (ATSDR, 2000).

Suzuki and colleagues (1998) and Sinkkonen and Paasivirta (2000) modeled the transformation and fate of DLCs for various environmental media. These models estimate steady-state and nonsteady-state concentrations of DLCs for water, soil, and sediment, and they predict environmental fluxes. Although the models lack quantitative and qualitative precision, they do indicate that soil is an important reservoir source for DLCs, and that degradation rates in soil and sediment may be significant determinants of the environmental transformation and fate processes of these compounds.

Transport in Water

As DLCs enter the water column through soil runoff and erosion or atmospheric wet or dry deposition, they adhere to particles in the water column and are ultimately removed by sedimentation. A slight amount of volatilization from the water column back into the atmosphere is possible, especially for the lighter DLC congeners. However, volatilization from water is the predominant removal mechanism for the dioxin-like PCBs in the water column. The low water solubility of DLCs, including the dioxin-like PCBs, means that only a small portion of these compounds will dissolve in water.

Once adsorbed to particulate matter in the water column, the DLC-containing particles will settle out to the sediments and eventually be buried, although resuspension into the water column may occur as a result of sediment agitation due to floods, biological activity, or other phenomenon. Resuspended sediments may travel downstream for great distances before settling back out of the water column, accounting for the presence of DLCs in sediments at considerable distances from a source. The more volatile PCBs may also move from the sediments back into the water column and ultimately into the atmosphere, although this process is unlikely for other DLCs. This process is greater in the summer months and with shallow waters (ATSDR, 1998).

Transformation Processes and Persistence

DLCs are stable compounds and are highly resistant to most environmental degradation processes and to hydrolysis (EPA, 2000). Transformation processes for DLCs appear to be primarily through photooxidation in the atmosphere, producing hydroxyl radicals and some photolysis in air and on soil and water surfaces. Only a few studies are available on the transformation of DLCs under natural environmental conditions; they will not be discussed in this report.

In air, photooxidation via hydroxyl radical reaction appears to be a major degradation process for both particle-phase and vapor-phase DLCs. Tropospheric lifetimes are variable: 0.5 days for monochlorodibenzo-*p*-dioxins, approximately 2 days for 2,3,7,8-TCDD, 10 days for octachlorodibenzo-*p*-dioxins (OCDD), and 39 days for octachlorodibenzofurans (Atkinson, 1991, as cited in EPA, 2000). The reaction of vapor-phase PCBs with hydroxyl radicals results in a half-life that varies from 11 days for tetrachlorinated congeners to more than 94 days for heptachlorinated congeners (Atkinson, 1987, as cited in EPA, 2000). Other studies have indicated that the atmospheric lifetimes for vapor-phase reactions of DLCs with nitrate and ozone were approximately 5 days and greater than 330 days, respectively (Kwok et al., 1994). Atmospheric photooxidation rates increase in the summer months and decrease with increasing DLC chlorination.

Vapor-phase DLCs are subject to some photolysis, with rates decreasing as the DLC chlorination increases (EPA, 2000). The atmospheric half-life of 2,3,7,8-TCDD in summer sunlight was estimated to be 1 hour (Podoll et al., 1986). Photolysis of particle-bound congeners is much slower (8 percent of 2,3,7,8-TCDD adsorbed to fly ash was lost after 40 hours of irradiation) (Mill et al., 1987). In both cases, the transformation is photodechlorination from more to less chlorinated congeners. As the majority of these studies have been conducted under laboratory conditions, photolysis rates under actual environmental conditions may be different (EPA, 2000). For dioxin-like PCBs, as the chlorination increases, so does the photolysis rate. Unlike other DLCs, the less chlorinated PCBs are more resistant to photolysis and may be formed as products of photolysis of the more chlorinated congeners (EPA, 2000).

Photolysis is the major fate process for DLCs in water, but it is relatively slow. Dioxin congeners substituted in the 2,3,7, and 8 positions, rather than the 1,4,6, and 9 positions, are more readily degraded (ATSDR, 2000). In sunlight, 2,3,7,8-TCDD in natural waters is expected to have a half-life ranging from approximately 3 days in summer to 16 days in winter (EPA, 2000). The presence of organic molecules may increase or decrease the photolysis rates for the DLCs. Photolytic degradation of PCBs increases with chlorination of the compound; however, the OCDD congener is resistant to this process (ATSDR, 1998). Additionally, PCBs suspended in fresh surface water react with hydroxyl radicals with half-lives of 4 to 11 days (Sedlak and Andren, 1991, as cited in EPA, 2000).

In soils, DLCs bind strongly to organic matter with the result that degradation below the soil surface is virtually nonexistent. DLCs at the soil surface are subject to some photolysis, although at a slower rate than when in water (ATSDR, 1998). Degradation of the di- and trichlorinated DLCs in sandy loam soil has been reported at 15 months (Orazio et al., 1992, as reported in EPA, 2000), although no significant degradation was seen for the tetra- to octachlorinated DLCs.

Photolytic degradation of DLCs on soil surfaces has been determined for various soil types. DLCs appear to persist longer on soils with higher organic content, although the photolytic degradation rates for most soil types are equal after about 5 days (Kieatiwong et al., 1990, as cited in EPA, 2000). Paustenbach and colleagues (1992, as cited in EPA, 2000) estimated that 2,3,7,8-TCDD has a half-life of 25 to 100 years in subsurface soil and 9 to 15 years in the top 0.1 cm of soil, indicating that photolysis occurs only to the depth of penetration of ultraviolet light (EPA, 2000). DLCs applied to agricultural fields in sewage sludge showed no degradation after 43 days under natural sunlight in the late summer (Schwarz and McLachlan, 1993, as cited in EPA, 2000).

Although microbial degradation has been demonstrated for some DLC congeners, this does not appear to be an important transformation process. Hexachlorinated congeners were degraded by 70 to 75 percent by the white rot fungus *Phanerochaete sordida*, with 2,3,7,8-TCDD and the pentachlorinated congeners being more resistant (Takada et al., 1994, 1996, as cited in EPA, 2000). DLCs appear to be subject to the greatest reductive degradation under anaerobic conditions, although some aerobic degradation has been reported. Microorganisms from Passaic River sediments were able to degrade 2,3,7,8-TCDD by 30 percent within 7 months, with a resulting increase in the less chlorinated forms; similar results were seen with the incubation of OCDD, which was converted to less chlorinated forms (Barkovskii and Adriaens, 1995, 1996, as cited in EPA, 2000).

Unlike most DLCs, PCBs have been shown to be aerobically and anaerobically dechlorinated by several microbial species (ATSDR, 2000). Aerobic biodegradation decreases with increasing chlorination (EPA, 2000), but can still result in destruction of the PCB molecule. The half-life of tetra-PCB congeners in surface waters and soils exposed to ultraviolet light is estimated to be 7 to 60 days and 12 to 30 days, respectively; the half-lives of penta- and greater chlorinated PCBs in surface waters and soils exposed to ultraviolet light are estimated to be greater than 1 year (EPA, 2000). PCBs are associated strongly with anaerobic sediments and so anaerobic biodegradation of PCBs does occur (ATSDR, 2000). Although this process dechlorinates PCBs, it does not destroy the molecules (Liu et al., 1996; Sokol et al., 1994).

ENTRY OF DLCs INTO THE FOOD CHAIN

As discussed above, pathways of DLC entry into the food chain include atmospheric transport of emissions and their subsequent deposition on plants, soils, and water (Fries, 1995a) and industrial discharges directly into bodies of water or on land (Bobovnikova et al., 2000; Connolly et al., 2000). This section summarizes evidence suggesting that DLCs enter the food chain by these pathways.

Aerial transport of DLC-containing emissions is considered the primary pathway of entry of DLCs into the food chain (Fries, 1995a, 1995b) as a result of their deposition onto plants and soils. Another source of residues in the food chain is DLCs that are released from sediment and soil that were sinks for past introductions into the environment (Bushart et al., 1998). Minor introductions may also occur as a result of the application of sewage and paper-mill sludge to land (Rappe and Buser, 1989; Weerasinghe et al., 1986) and the use of wood treated with pentachlorophenol (Firestone et al., 1972; Ryan et al., 1985; Shull et al., 1981). More than 95 percent of contaminants deposited by air in terrestrial environments reach soil (Fries and Paustenbach, 1990).

The transfer of semivolatile lipophilic compounds such as DLCs is predicted to occur through volatilization and particulate deposition (Fries, 1995a, 1995b), although McCrady and colleagues (1990) and Bacci and colleagues (1992) found that no TCDD was detected in plants when suitable vapor barriers were provided. Uptake and translocation of TCDD from soils to plants is not believed to be a major route for DLCs. Isensee and Jones (1971) showed a lack of absorption and translocation after foliar application of TCDD, and several studies report that little or no TCDD was measured in seeds and aerial parts of crops grown in contaminated soil (Isensee and Jones, 1971; Jensen et al., 1983; Wipf et al., 1982). According to reviews by Esposito and colleagues (1980) and Norris (1981), approximately 0.1 percent of TCDD applied to soil was found in above-ground portions of oats and soybeans, although some absorption did occur from nutrient solutions. While volatilization is clearly a major factor in the worldwide distribution of DLCs, there was no evidence presented to the committee to confirm that transfer to forage crops was a major pathway of concern. While there may be some contamination of forage crops through particulate-bound DLCs, the lack of lipids in most forage crops makes it unlikely that vapor-phase DLCs will accumulate on plants.

In 1977, one year after the unintended release in Seveso, Italy, no traces of TCDD were found in the flesh of apples, pears, and peaches or in corncobs and kernels grown near the factory. However, approximately 100 pg/g of DLCs were detected in the peel of fruits (Wipf et al., 1982). Tree bark has been shown to be a useful environmental monitor of vapor-phase PCBs (Hermanson and Hites, 1990).

Little vapor transfer of the highly chlorinated congeners from soil is expected because the vapor pressure of a homologous series of compounds, such as DLCs, decreases with increasing chlorination (Fries, 1995a). Therefore, a predictable transport mechanism for the highly chlorinated congeners is by transfer of soil particles as dust or by splashing during precipitation. This mechanism does not appear, however, to be particularly important for animal forage crops

(Fries, 1995b). Concentrations of DLCs (primarily hepta and octa) in hay were determined to be unrelated to soil concentrations by Hulster and Marschner (1993), and residues of polybrominated biphenyls (which are similar in chemical composition to DLCs) were not detected in harvested forages grown in soil with concentrations as high as 300 ppb (Fries and Jacobs, 1986).

The predicted concentrations of TCDD on plants are highest in pastures due to average pasture yields that are lower than the average yields of other forages (Fries and Paustenbach, 1990). Also, the use of pasture as a roughage source for animals is an important factor because soil ingestion is added to direct plant contamination as a source of exposure (Fries, 1995b). In general, soil ingestion is related inversely to the availability of forage when pasture is the sole animal-feed source. For example, the amount of soil ingested by food animals is as little as 1 to 2 percent of dry-matter intake during periods of lush plant growth, but it rises to as much as 18 percent when forage is sparse (Mayland et al., 1975; Thorton and Abrahams, 1981).

This chapter provides a general overview of known sources of DLCs and how these compounds are transported from the environment into pathways leading to human foods. The data presented constitute the bulk of the general background information that the committee used to familiarize itself with issues related to environmental sources of DLCs and their entry into the food supply. Environmental releases of DLCs have been decreasing in recent decades. The sources contributing the majority of DLC releases are from combustion, in particular from nonregulated sources.

REFERENCES

Addink R, Bakker WCM, Olie K. 1995. Influence of HCl and Cl on the formation of 2 polychlorinated dibenzo-*p*-dioxins/dibenzofurans in a carbon/fly ash mixture. *Environ Sci Technol* 29:2055–2058.

Aittola J, Paasivirta J, Vattulainen A. 1992. Measurements of organochloro compounds at a metal reclamation plant. *Organohalogen Compd* 9:9–12.

Atkinson R. 1987. Estimation of OH radical reaction rate constants and atmospheric lifetimes for polychlorobiphenyls, dibenzo-*p*-dioxins, and dibenzofurans. *Environ Sci Technol* 21:305–307.

Atkinson R. 1991. Atmospheric lifetimes of dibenzo-*p*-dioxins and dibenzofurans. *Sci Total Environ* 104:17–33.

ATSDR (Agency for Toxic Substances and Disease Registry). 1994. *Toxicological Profile for Chlorinated Dibenzofurans*. Atlanta, GA: ATSDR.

ATSDR. 1998. *Toxicological Profile for Chlorinated Dibenzo-p-dioxins*. Atlanta, GA: ATSDR.

ATSDR. 2000. *Toxicological Profile for Polychlorinated Biphenyls (PCBs)*. Atlanta, GA: ATSDR.

Bacci E, Cerejeira MJ, Gaggi C, Chemello G, Calamari D, Vighi M. 1992. Chlorinated dioxins: Volatilization from soils and bioconcentration in plant leaves. *Bull Environ Contam Toxicol* 48:401–408.

Barkovskii AL, Adriaens P. 1995. Reductive dechlorination of tetrachloro-dibenzo-*p*-dioxin partitioned from Passaic River sediments in an autochthonous microbial community. *Organohalogen Compd* 24:17–21.

Barkovskii AL, Adriaens P. 1996. Microbial dechlorination of historically present and freshly spiked chlorinated dioxins and diversity of dioxin-dechlorinating populations. *Appl Environ Microbiol* 62:4556–4562.

Bobovnikova TI, Aleksseva LB, Dibtseva AV, Chernik GV, Orlinsky DB, Priputin LV, Pleskachevskaya GA. 2000. The influence of a capacitor plant in Serpukhov on vegetable contamination by polychlorinated biphenyls. *Sci Total Environ* 246:51–60.

Bramley MJ. 1998. *Dioxin and Hexachlorobenzene Releases from Magnesium Production in North America: Lessons from Noranda's Project in Asbestos, Quebec.* Toronto: Greenpeace Canada.

Bruzy LP, Hites RA. 1995. Estimating the atmospheric deposition of polychlorinated dibenzo-*p*-dioxins and dibenzofurans from soil. *Environ Sci Technol* 29:2090–2098

Bushart SP, Bush B, Barnard EL, Bott A. 1998. Volatilization of extensively dechlorinated polychlorinated biphenyls from historically contaminated sediments. *Environ Toxicol Chem* 17:1927–1933.

Carpenter DO. 1998. Polychlorinated biphenyls and human health. *Int J Occup Med Environ Health* 11:291–303.

Clement RE, Tosine HM, Osborne J, Ozvacic V, Wong G. 1985. Levels of chlorinated organics in a municipal incinerator. In: Keith LH, Rappe C, Choudhary G, eds. *Chlorinated Dioxins and Dibenzofurans in the Total Environment II.* Boston, MA: Butterworth Publishers. Pp. 489–514.

Clement RE, Suter SA, Reiner E, McCurrin D. 1989. Concentrations of chlorinated dibenzo-*p*-dioxins and dibenzofurans in effluents and centrifuged particles from Ontario pulp and paper mills. *Chemosphere* 19:649–954.

Connolly JP, Zahakos HA, Benaman J, Ziegler CK, Rhea JR, Russell K. 2000. A model of PCB fate in the Upper Hudson River. *Environ Sci Technol* 34:4076–4087.

Cramer PH, Heiman J, Horrigan M, Lester R, Armstrong S. 1995. Results of a national survey for polychlorinated dibenzo-*p*-dioxins, dibenzofurans, and coplanar polychlorinated biphenyls in municipal biosolids. *Organohalogen Compd* 24:305–308.

Czuczwa JM, Hites RA. 1984. Environmental fate of combustion-generated polychlorinated dioxins and furans. *Environ Sci Technol* 18:444–450.

Czuczwa JM, Hites RA. 1986a. Airborne dioxins and dibenzofurans: Sources and fates. *Environ Sci Technol* 20:195–200.

Czuczwa JM, Hites RA. 1986b. Sources and fates of PCDD and PCDF. *Chemosphere* 15:1417–1420.

Dickson LC, Karasek FW. 1987. Mechanism of formation of polychlorinated dibenzo-*p*-dioxins produced on municipal incinerator fly ash from reactions of chlorinated phenols. *J Chromatogr* 389:127–137.

Environmental Risk Sciences, Inc. 1995. *An Analysis of the Potential for Dioxin Emissions in the Primary Copper Smelting Industry.* Prepared for the National Mining Association. New York: Environmental Risk Sciences, Inc.

EPA (U.S. Environmental Protection Agency). 1987. *EPA Contract Laboratory Program: Statement of Work for Organic Analysis.* Washington, DC: EPA.

EPA. 1991. Proposed regulation of land application of sludge from pulp and paper mills using chlorine and chlorine derivative bleaching processes. *Fed Regist* 56:21802.

EPA. 1997. Standards of performance for new stationary sources and emission guidelines for existing sources: Hospital/medical/infectious waste incinerators; Final rule. *Fed Regist* 62:48348.

EPA. 1998. National emission standards for hazardous air pollutants for source category: Pulp and paper production; Effluent limitations guidelines, pretreatment standards, and new source performance standards; Pulp, paper, and paperboard category; Final rule. *Fed Regist* 63:18504–18751.

EPA. 2000. *Exposure and Human Health Reassessment of 2,3,7,8-Tetrachlorodibenzo-p-Dioxin (TCDD) and Related Compounds.* Draft Final Report. Washington, DC: EPA.

Esposito MP, Tiernan TO, Drydent FE. 1980. *Dioxins.* Prepared for the U.S. Environmental Protection Agency, EPA-600/2-80-197. Springfield, VA: National Technical Information Service.

Firestone D, Ress J, Brown NL, Barron RP, Damico JN. 1972. Determination of polychlorodibenzo-*p*-dioxins and related compounds in commercial chlorophenols. *J Assoc Off Anal Chem* 55:85–92.

Fries GF. 1995a. A review of the significance of animal food products as potential pathways of human exposures to dioxins. *J Anim Sci* 73:1639–1650.

Fries GF. 1995b. Transport of organic environmental contaminants to animal products. *Rev Environ Contam Toxicol* 141:71–109.

Fries GF, Jacobs LW. 1986. *Evaluation of Residual Polybrominated Biphenyl Contamination Present on Michigan Farms in 1978.* Research Report 477. East Lansing, MI: Michigan State University Agricultural Experiment Station.

Fries GF, Paustenbach DJ. 1990. Evaluation of potential transmission of 2,3,7,8-tetrachlorodibenzo-*p*-dioxin-contaminated incinerator emissions to humans via food. *J Toxicol Environ Health* 29:1–43.

Gullett BK, Lemieux PM, Lutess CC, Winterrowd CK, Winters DL. 2001. Emissions of PCDD/F from uncontrolled, domestic waste burning. *Chemosphere* 43:721–725

Harrad SJ, Jones KC. 1992. A source inventory and budget for chlorinated dioxins and furans in the United Kingdom environment. *Sci Total Environ* 126:89–107.

Hayward DG, Nortrup D, Gardner A, Clower M Jr. 1999. Elevated TCDD in chicken eggs and farm-raised catfish fed a diet with ball clay from a southern United States mine. *Environ Res* 81:248–256.

Hermanson MH, Hites RA. 1990. Polychlorinated biphenyls in tree bark. *Environ Sci Technol* 24:666–671.

Horstmann M, McLachlan MS. 1995. Concentrations of polychlorinated dibenzo-*p*-dioxins (PCDD) and dibenzofurans (PCDF) in urban runoff and household wastewaters. *Chemosphere* 31:2887–2896.

Horstmann M, McLachlan M, Reissinger M. 1992. Investigation of the origin of PCDD/CDF in municipal sewage. *Organohalogen Compd* 9:97–100.

Hulster A, Marschner H. 1993. Soil-plant transfer of PCDD/PCDF to vegetables of the cucumber family (*Cucurbitaceae*). *Environ Sci Technol* 28:1110–1115.

Hutzinger O, Fiedler H. 1991. Formation of dioxins and related compounds in industrial processes. In: Bretthauer EW, Kraus HW, di Domenico A, eds. *Dioxin Perspectives. A Pilot Study on International Information Exchange on Dioxins and Related Compounds.* New York: Plenum Press. Pp. 435–516.

Isensee AR, Jones GE. 1971. Absorption and translocation of root and foliage applied 2,4-dichloro-phenol, 2,7-dichlorodibenzo-*p*-dioxin, and 2,3,7,8-tetrachlorodibenzo-*p*-dioxin. *J Agric Food Chem* 19:1210–1214.

Jensen DJ, Getzendaner ME, Hummel RA, Turley J. 1983. Residue studies for (2,4,5-trichloro-phenoxy) acetic acid and 2,3,7,8-tetrachlorodibenzo-*p*-dioxin in grass and rice. *J Agric Food Chem* 31:118–122.

Kieatiwong S, Nguyen LV, Hebert VR, Hackett M, Miller GC, Miille MJ, Mitzel R. 1990. Photolysis of chlorinated dioxins in organic solvents and on soils. *Environ Sci Technol* 24:1575–1580.

Krauss T, Krauss P, Hagenmaier H. 1994. Formation of PCDD/PCDF during composting? *Chemosphere* 28:155–158.

Kwok ESC, Arey J, Atkinson R. 1994. Gas-phase atmospheric chemistry of dibenzo-*p*-dioxin and dibenzofuran. *Environ Sci Technol* 28:528–533.

Lahl U. 1993. Sintering plants of steel industry—The most important thermical PCDD/CDF source in industrialized regions. *Organohalogen Compd* 11:311–314.

Lahl U. 1994. Sintering plants of steel industry—PCDD/F emission status and perspectives. *Chemosphere* 29:1939–1945.

Lahl U, Wilken M, Zeschmar-Lahl B, Jager J. 1991. PCDD/PCDF balance of different municipal waste management methods. *Chemosphere* 23:1481–1489.

Lamparski LL, Stehl RH, Johnson RL. 1980. Photolysis of pentachlorophenol-treated wood. Chlorinated dibenzo-*p*-dioxin formation. *Environ Sci Technol* 14:196–200.

Lexen K, De Wit C, Jansson B, Kjeller LO, Kulp SE, Ljung K, Söderstrom G, Rappe C. 1993. Polychlorinated dibenzo-*p*-dioxin and dibenzofuran levels and patterns in samples from different Swedish industries analyzed within the Swedish dioxin survey. *Chemosphere* 27:163–170.

Liberti A, Brocco D. 1982. Formation of polychlorodibenzodioxins and polychlorodibenzofurans in urban incinerator emissions. In: Hutzinger O, Frei RW, Merian E, Pocchiari F, eds. *Chlorinated Dioxins and Related Compounds*. New York: Pergamon Press. Pp. 245–251.

Liu X, Sokol RC, Kwon O-S, Bethoney CM, Rhee G-Y. 1996. An investigation of factors limiting the reductive dechlorination of polychlorinated biphenyls. *Environ Toxicol Chem* 15:1738–1744.

Luijk R, Akkerman DM, Slot P, Olie K, Kapteljn F. 1994. Mechanism of formation of polychlorinated dibenzo-*p*-dioxins and dibenzofurans in the catalyzed combustion of carbon. *Environ Sci Technol* 28:312–321.

Mayland HF, Florence AR, Rosenau RC, Lazar VA, Turner HA. 1975. Soil ingestion by cattle on semiarid range as reflected by titanium analysis of feces. *J Range Manage* 28:448–452.

McCrady JK, McFarlane C, Gander LK. 1990. The transport and fate of 2,3,7,8-TCDD in soybeans and corn. *Chemosphere* 21:359–376.

Mill T, Rossi M, McMillen D, Coville M, Leung D, Spang J. 1987. *Photolysis of Tetrachlorodioxin and PCBs Under Atmospheric Conditions*. Internal report prepared by SRI International for the Office of Health and Environmental Assessment, U.S. Environmental Protection Agency. Washington, DC: SRI International.

NCASI (National Council of the Paper Industry for Air and Stream Improvement). 1993. *Summary of Data Reflective of Pulp and Paper Industry Progress in Reducing the TCDD/TCDF Content of Effluents, Pulps and Wastewater Treatment Sludges*. New York: NCASI.

Norris LA. 1981. The movement, persistence, and fate of the pheonxy herbicides and TCDD in forests. *Residue Rev* 80:65–135.

NRC (National Research Council). 2001. *A Risk-Management Strategy for PCB-Contaminated Sediments*. Washington, DC: National Academy Press.

NRC. 2002. *Biosolids Applied to Land: Advancing Standards and Practices*. Washington, DC: National Academy Press.

Orazio CE, Kapila S, Puri RK, Yanders AF. 1992. Persistence of chlorinated dioxins and furans in the soil environment. *Chemosphere* 25:1469–1474.

Paustenbach D, Wenning RJ, Lau V, Harrington NW, Rennix DK, Parsons AH. 1992. Recent developments on the hazards posed by 2,3,7,8-TCDD in soil: Implications for setting risk-based cleanup levels at residential and industrial sites. *J Toxicol Environ Health* 36:103–149.

Podoll RT, Jaber HM, Mill T. 1986. Tetrachlorodibenzodioxin: Rates of volatilization and photolysis in the environment. *Environ Sci Technol* 20:490–492.

Rappe C. 1991. Sources of human exposure to CDDs and PCDFs. In: Gallo M, Scheuplein R, van der Heiden K, eds. *Biological Basis for Risk Assessment of Dioxin and Related Compounds*. Banbury Report No. 35. Plainview, NY: Cold Spring Harbor Laboratory Press. Pp. 121–131.

Rappe C. 1992a. Sources of exposure, environmental levels and exposure assessment of PCDDs and PCDFs. *Organohalogen Compd* 9:5–8.

Rappe C. 1992b. Sources of PCDDs and PCDFs. Introduction. Reactions, levels, patterns, profiles and trends. *Chemosphere* 25:41–44.

Rappe C. 1993. Sources of exposure, environmental concentrations and exposure assessment of PCDDs and PCDFs. *Chemosphere* 27:211–225.

Rappe C, Buser HR. 1989. Chemical and physical properties, analytical methods, sources and environmental levels of halogenated dibenzodioxins and dibenzofurans. In: Kimbrough RD, Jensen AA, eds. *Halogenated Biphenyls, Terphenyls, Napthalenes, Dibenzodioxins and Related Products*, 2nd ed. Amsterdam: Elsevier Science. Pp. 71–102.

Rappe C, Glas B, Kjeller LO, Kulp SE. 1990. Levels of PCDDs and PCDFs in products and effluent from the Swedish pulp and paper industry and chloralkali process. *Chemosphere* 20:1701–1706.

Rappe C, Kjeller L, Kulp S, deWit C, Hasselsten I, Palm O. 1991. Levels, profile and pattern of PCDDs and PCDFs in samples related to the production and use of chlorine. *Chemosphere* 23:1629–1636.

Rappe C, Andersson R, Karlaganis G, Bonjour R. 1994. PCDDs and PCDFs in samples of sewage sludge from various areas in Switzerland. *Organohalogen Compd* 20:79–84.

Ree KC, Evers EHG, Van Der Berg M. 1988. Mechanism of formation of polychlorinated dibenzo-(*p*)dioxins (PCDDs) and polychlorinated dibenzofurans (PCDFs) from potential industrial sources. *Toxicol Environ Chem* 17:171–195.

Ryan JJ, Lizotte R, Sakuma T, Mori B. 1985. Chlorinated dibenzo-*p*-dioxins, chlorinated dibenzo-furans, and pentachlorophenol in Canadian chicken and pork samples. *J Agric Food Chem* 33:1021–1026.

Schwarz K, McLachlan MS. 1993. The fate of PCDD/F in sewage sludge applied to an agricultural soil. *Organohalogen Compd* 12:155–158.

Sedlak DL, Andren AW. 1991. Aqueous-phase oxidation of polychlorinated biphenyls by hydroxyl radicals. *Environ Sci Technol* 25:1419–1427.

Sewart A, Harrad SJ, McLachlan MS, McGrath SP, Jones KC. 1995. PCDD/Fs and non-o-PCBs in digested U.K. sewage sludges. *Chemosphere* 30:51–67.

Shull LR, Foss M, Anderson CR, Feighner K. 1981. Usage patterns of chemically treated wood on Michigan dairy farms. *Bull Environ Contam Toxicol* 26:561–566.

Sinkkonen S, Paasivirta J. 2000. Degradation half-times of PCDDs, PCDFs and PCBs for environmental fate modeling. *Chemosphere* 40:943–949.

Sokol RC, Bethoney CM, Rhee GY. 1994. Effect of hydrogen on the pathway and products of PCB dechlorination. *Chemosphere* 29:1735–1742.

Strandell ME, Lexen KM, deWit CA, Järnberg UG, Jansson B, Kjeller LO, Kulp E, Ljung K, Söderström G, Rappe C. 1994. The Swedish Dioxin Survey: Summary of results from PCDD/F and coplanar PCB analyses in source-related samples. *Organohalogen Compd* 20:363–366.

Suzuki N, Yasuda M, Sakurai T, Nakanishi J. 1998. Model simulation of environmental profile transformation and fate of polychlorinated dibenzo-*p*-dioxins and polychlorinated dibenzofurans by the multimedia environmental fate model. *Chemosphere* 37:2239–2250.

Swanson SE, Rappe C, Malmstrom J, Krinsgstad KP. 1988. Emissions of PCDDs and PCDFs from the pulp industry. *Chemosphere* 17:681–691.

Takada S, Nakamura M, Matsueda T, Kurokawa Y, Fukamati K, Kondo R, Sakai K. 1994. Degradation of PCDDs/PCDFs by ligninolytic fungus *Phanerochaete sordida* YK-624. *Organohalogen Compd* 20:195–198.

Takada S, Nakamura M, Matsueda T, Kondo R, Sakai K. 1996. Degradation of polychlorinated dibenzo-*p*-dioxins and polychlorinated dibenzofurans by the white rot fungus *Phanerochaete sordida* YK-624. *Appl Environ Microbiol* 62:4323–4328.

Thoma H. 1988. PCDD-F concentrations in chimney soot from house heating systems. *Chemosphere* 17:1369–1379.

Thorton I, Abrahams P. 1981. Soil ingestion as a pathway of metal intake into grazing livestock. In: *Proceedings of the International Conference of Heavy Metals in the Environment*. Edinburgh, Scotland: CEP Consultants. Pp. 267–272.

Travis CC, Hattemer-Frey HA. 1989. Human exposure to dioxin from municipal solid waste incineration. *Waste Manag* 9:151–156.

Versar, Inc. 1985. *List of Chemicals Contaminated or Precursors to Contamination with Incidentally Generated Polychlorinated and Polybrominated Dibenzodioxins and Dibenzofurans*. EPA Contract No. 68-02-3968, Task No. 48. Springfield, VA: Versar.

Vollmuth S, Zajc A, Niessner R. 1994. Formation of polychlorinated dibenzo-*p*-dioxins and polychlorinated dibenzofurans during the photolysis of pentachlorophenol-containing water. *Environ Sci Technol* 28:1145–1149.

Waddell DS, Boneck-Ociesa H, Gobran T, Ho TF, Botton JR. 1995. PCDD/PCDF formation by UV photolysis of pentachlorophenol with and without the addition of hydrogen peroxide. *Organohalogen Compd* 23:407–412.

Weerasinghe NCA, Schecter AJ, Pan JC, Lapp RL, Giblin DE, Meehan JL, Hardell L, Gross ML. 1986. Levels of 2,3,7,8-tetrachlorodibenzo-*p*-dioxin (2,3,7,8-TCDD) in adipose tissue of Vietnam veterans seeking medical assistance. *Chemosphere* 15:1787–1794.

Wipf HK, Homberger E, Neuner N, Ranalder UB, Vetter W, Vuilleumier JP. 1982. TCDD levels in soil and plant samples from the Seveso area. In: Hutzinger O, Frei RW, Merian E, Pocchiari P, eds. *Chlorinated Dioxins and Related Compounds: Impact on the Environment.* New York: Pergamon Press. Pp. 115–126.

Wunderli S, Zennegg M, Dolezal IS, Gujer E, Moser U, Wolfenberger M, Hasler P, Noger D, Studer C, Karlaganis G. 2000. Determination of polychlorinated dibenzo-p-dioxins and dibenzo-furans in solid residues from wood combustion by HRGC/HRMS. *Chemosphere* 40:641–649.

Zook DR, Rappe C. 1994. Environmental sources, distribution, and fate of polychlorinated dibenzodioxins, dibenzofurans, and related organochlorines. In: Schecter A, ed. *Dioxins and Health.* New York: Plenum Press. Pp. 80–113.

4

Animal Production Systems

DIRECT AND INDIRECT DLC PATHWAYS INTO FOOD PRODUCTS

As discussed in Chapter 3, dioxin and dioxin-like compounds (referred to collectively as DLCs) may enter the animal feed to human food chain through both direct and indirect pathways. The direct environmental pathways include: air-to plant/soil, air-to plant/soil-to animal, and water/sediment-to fish (EPA, 2000). Whether newly produced or from reservoirs, DLCs deposit on vegetation, soils, and in water sediments from the atmosphere or through agricultural pesticides, fertilizers, and irrigation, and are retained on plant surfaces and in the surrounding soil and sediment in waterways. It is estimated that 5 percent of aerial deposits of DLCs in terrestrial environments are retained on plants and the remaining 95 percent ultimately reaches the soil (Fries and Paustenbach, 1990). The soil-borne DLCs then become a reservoir source that could reach plants used for animal feeds by volatilization and redeposition or as dust. Modeling studies by Trapp and Matthies (1997) indicated that volatilization of polychlorinated dibenzo-p-dioxins (PCDDs) from soil into vegetation is only significant in the case of highly contaminated soils. DLCs from contaminated plant products that are consumed by animals bioaccumulate in the animals' lipid tissues.

DLCs can enter aquatic systems via direct discharge into water, by deposition onto soil, and by runoff from watersheds. Aquatic animals accumulate these compounds through direct contact with the water, suspended particles, and bottom sediments, and through their consumption of other aquatic organisms. Limited mass balance studies in dairy animals indicate that air and water are negligible sources of DLCs (McLachan et al., 1990). Thus, both terrestrial and aquatic food animals may be exposed to DLCs primarily through soil-based ecological pathways.

In addition to environmental pathways, animal agriculture practices in the United States may incorporate indirect pathways of DLC exposure that lead to contamination of plant and animal by-products used to formulate animal diets and manufacture animal feeds. These indirect pathways have the potential to produce elevated DLC levels in animals. Exposure to a contaminated commercial agricultural environment, such as through animal contact with pentachlorophenol-(PCP) treated wood used in animal housing (a practice now banned), through animal-feed contamination episodes (e.g., DLCs in poultry in the United States [EPA, 2000; FDA, 1997; FSIS, 1997, 2002]), and through contaminated citrus-pulp products in Belgium (Allsopp et al., 2000), have resulted in isolated groups of animals with high exposure levels (van Larebeke et al., 2001). When point-source contamination episodes, such as those mentioned above, were identified, they were removed once causation was determined.

Figure 4-1 shows the pathways through which DLCs enter into animal feed and human food systems. The figure shows sources and major routes (dark arrows) and minor routes (light arrows) by which DLCs can cycle between compartments and ultimately reach humans.

In Figure 4-1, the environmental reservoir represents the major production source and recyclable reservoir for DLCs. As stated in the EPA draft reassessment (see Chapter 2), there are a number of sources from which DLCs are derived (chemical processing, incineration processes, and other human activities), and most DLCs are stored in environmental reservoirs such as soils and sediments. DLCs are transported through atmospheric routes to animal forage, feed, and grasses used for feed; to vegetables, fruits and cereals consumed by humans; and to terrestrial and aquatic animals consumed by humans. The route of exposure through vegetables, fruits, and cereals consumed by humans is generally considered a minor pathway, but, surface contamination by soils may increase exposure. Atmospheric contamination may also occur in plant products intended for animal rather than for human consumption, and may be eaten directly by a terrestrial food animal or used in animal feed, and thus may become a major source of DLC exposure to animals.

The pathway to aquatic animals also is a major route by which DLCs can enter the human food supply. Aquatic organisms can accumulate elevated DLC levels from recent atmospheric deposition or historical reservoirs of DLCs in sediments or terrestrial drainage areas. These DLCs can enter the aquatic food web and concentrate in commercially important aquatic species, although levels of DLCs vary within this environment. Direct human exposure occurs through eating fish or shellfish that contain elevated levels of DLCs. This pathway represents the exposure scenario that may predominate in subsistence fishers or specific ethnic groups (see Chapter 5).

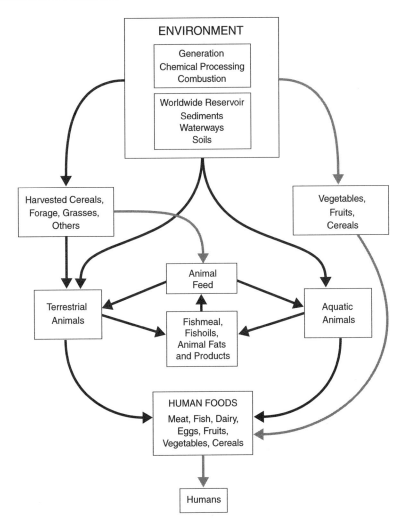

FIGURE 4-1 Pathways leading to exposure to dioxin and dioxin-like compounds through the food supply. Boxes depict point sources in the pathways. Dark arrows refer to pathways with a greater relative DLC contribution than the pathways with light arrows.

Aquatic animal by-products may be used in animal feeds. The feeds may include cereals, forages, and terrestrial animal by-products. These feeds may then be fed to other terrestrial and aquatic animals, potentially contributing to their DLC load. This loop may provide an important intervention step in interrupting the DLC cycle. Terrestrial and aquatic animals represent the principal pathways for the production of meat, dairy products, eggs, and fish for human consumption and are therefore the primary route for introduction of DLCs to humans.

The focus of this chapter is the identification of potential steps in agricultural production where interventions can be put in place to reduce DLC exposure to humans through the food supply.

LIVESTOCK PRODUCTION SYSTEMS

Livestock, for the purposes of this section, are defined as mammalian and avian species raised for human food production. Livestock production encompasses a wide range of management systems, from predominantly extensive range and pasture systems for cow-calf and sheep production, to intensive production of poultry, pork, and dairy products. Extensive production will be used here to indicate systems in which animals have direct access to soils, including animals held in unpaved feedlot settings. Conversely, intensive production will be limited to animal production systems in which direct access to soil is eliminated.

Livestock are recognized as DLC accumulators based on the amount of exposure they receive through their environment and diet. DLC exposure levels may be influenced by the production system employed and by other local environmental factors. The primary sources for DLC contamination of livestock can be categorized as environmental exposures, water sources, and feed rations. Current data suggest that animals raised on pasture grasses and roughage will be more likely to have higher DLC levels than concentrate-fed animals (Fries, 1995a). However, the relative weights of these factors may be influenced by the selected production system.

Identification of Points of Exposure

The DLC exposure risk to animal production systems can be predicted, to some extent, because environmental sources and chemical characteristics that allow these compounds to persist are known. However, quantitative, and sometimes qualitative, data about DLC levels in specific production systems and local environments may be limited. Therefore, estimations of the relative risks for various production scenarios have been used where data are not available.

Grain-Based Diets

Grains that are traditionally fed to livestock (e.g., corn, wheat, oats, and barley) are not likely to acquire DLC contamination during production. Grains that are produced in pods, in inedible sheaths, or in shells are considered to have minimal opportunities for deposition or aerosol contamination pathways (Travis and Hattemer-Frey, 1991). These characteristics are common in all the indigenous grain feedstuffs used in North America. However, grain by-products, bran or middling, and recycled grains and vegetable by-products may contain residual DLC levels as a result of concentrated surface contamination contributed by local incinerators or by persistent soil contamination from past herbicide application (Roeder et al., 1998). These products may comprise a substantial portion of the rations fed to individual groups of animals and thus increase exposure risk, whereas they represent minimal risks for larger animal populations.

The significance of DLC contamination of grains and forage may be better understood by comparing relative risks. Travis and Hattemer-Frey (1991) predicted a total concentration for 2,3,7,8-tetrachlorodibenzo-p-dioxin (TCDD) (wet weight) of 59 pg/kg in forages compared with 4.1 pg/kg in grains and protected produce. They also demonstrated the relatively higher exposure risk for forages compared with grains as a DLC source, although DLC contamination levels for concentrate-fed animals may vary in response to differing feed ingredients.

Grasses and Forage Diets

Intake of pasture grasses or roughage is considered to be the most important DLC exposure factor in extensive animal production systems. Grasses and forage represent a recognized pathway for organic contamination by air to leaf vapor transfer, deposition, and root uptake. Plant-root uptake of DLCs from soils has been shown to be minimal for most plants, except for members of the cucumber family (Fries, 1995a). Volatilization is not believed to be a major pathway for DLC contamination because of the relative vapor pressure for these highly chlorinated compounds (Shui et al., 1988, as reported in Roeder et al., 1998). These relationships have been demonstrated by the observation that the levels of PCDD and polychlorinated dibenzofuran (PCDF) congeners in forage were not related to the concentration of DLCs in the soils in which they were grown (Hulster and Marschner, 1993).

Plant DLC concentrations are a reflection of the environmental contamination levels in the areas where, and at times when, grown. Deposition of TCDD on the outer surface of plants is the primary route of contamination (Travis and Hattemer-Frey, 1991). Aerial deposition efficacy depends on particle size, leaf area and roughness, and plant biomass and density (Fries, 1995a). Rain may move DLC particles to the soil or to the lower portions of the plants, where animals would be exposed to them during grazing, but they would be excluded from forage that was harvested.

Commoner and colleagues (1998) observed that concentrations of DLCs in air and concurrently grown vegetation were linearly proportional. However, on eight farms located in two states they observed that for grazing dairy cows there was a greater than tenfold difference in DLC levels (0.027–0.346 pg toxicity equivalents [TEQ]/g) in the plant-based diets consumed by the cows, thus demonstrating diversity in the retention of DLC contaminants on the plants (Commoner et al., 1998). The density of grasses (spring pastures versus summer and fall) available for grazing or the provision of supplemental feeds may reduce DLC uptake by reducing soil ingestion (Fries, 1995b). Roughages that are harvested and stored reflect the environmental contaminant levels during their growing periods, although some forage processing techniques, such as hay drying, may reduce DLCs through volatilization (Archer and Crosby, 1969, as reported in Fries, 1995b).

Some livestock production systems utilize nearly equivalent amounts of grain and roughage in animal diets, particularly for lactating dairy cattle. In these mixed diets, DLC levels would be intermediate between those of grain- and roughage-based diets.

Soil

Soil represents a significant reservoir for DLC contamination under grazing conditions and is a source of run-off and sediments that contaminate waterways. The bioavailability of DLCs in this reservoir varies from 20 to 40 percent, depending on the source from which they were generated (Fries, 1995a). Ninety-five percent of aerial contamination will eventually reach the soil (Fries and Paustenbach, 1990); therefore, soil will reflect the environmental load from all sources for the area, both current and historic.

It has been estimated that grazing dairy and beef animals may receive at least 20 and 29 percent, respectively, of DLCs per day through soil ingestion (Travis and Hattemer-Frey, 1991), and pasture conditions through the grazing season may significantly influence this uptake (Fries, 1995b). Animals in unpaved feedlots also consume small amounts of soil that may lead to detectable residues (Fries et al., 1982, as reported in Fries, 1995a). In addition, soil erosion and sediment production will contaminate surface water sources, which may further enhance total daily DLC exposure levels under range conditions.

Water

The strongly lipophilic nature of DLCs reduces the potential for contamination of water except through soil contamination. Filtration water systems (municipal or private) or wells with no surface contamination likely contain minimal DLC levels, whereas sediment particles in other water systems may contain adsorbed DLCs. This is important in the case of aquatic environmental contamination because surface water, used by grazing livestock, may represent another DLC exposure route as animals stir up sediments when they enter the waterways to drink. The contribution, however, of surface water to DLC accumulation in livestock is unknown.

Aerosols

Aerosol contamination contributes to environmental DLC sources through particle deposition onto plants and soils. Although limited data are available on the effects of inhalation of DLCs in livestock production, balance studies in lactating cows have shown inhalation exposure and water contamination to be negligible sources for DLCs (McLachlan et al., 1990).

Manure

Manure contamination is a reflection of the DLC intake of the animal that may add to soil burdens if used in compost. Coprophagous activities, particularly in swine and poultry, may also contribute to DLC recycling. However, there is not enough available data to adequately characterize DLC exposure risks from manure, particularly as related to animal recycling effects.

Point-Source Contaminants

Animal housing and handling facilities may be a source for DLC contamination because of the materials used in their construction. Prior to the 1980s, woods treated with pentachlorophenol (PCP) were used in feed bunks, fencing, and other structural components for livestock buildings, particularly for ruminants. Dioxin and furan contamination of the PCP-treated woods resulted in a point-source reservoir. Animals that licked the wood structure or came in contact with exposed feeds developed detectable DLC residue levels in their fat stores. Once identified, the use of these treated woods in animal contact areas stopped, and some remediation of existing facilities was completed. As older facilities have been remodeled, additional sources have been removed.

Other products, such as greases, oils, or other organic chemicals that come into contact with animals or animal feeds, may present additional opportunities for point-source contamination. Inadvertent or purposeful contamination of high-fat animal feeds has been reported. For example, inadvertent DLC production was recently discovered in the chelation process of a mineral supplement (Personal communication, H. Carpenter, Minnesota Department of Health, April 2, 2002). Such occurrences illustrate the spectrum of potential point-source contamination events that must be considered in DLC reduction efforts.

The contamination levels found in deposits of ball clay used in animal feeds from one region demonstrate that DLCs can appear in a wide range of natural environments (Hayward et al., 1999), which represent another inadvertent point source of contamination. These deposits, the result of natural combustion processes that occurred centuries ago, remain as a reservoir until uncovered.

Management Practices

The relative importance of the various pathways of exposure for animals is influenced by the production system under which the animals are raised. As described earlier, animals have direct access to soils during their residence in an extensive production system (which includes unpaved feedlots), and animals are limited to facilities or are under conditions where direct access to soil is eliminated in an intensive production system. The percentage of products in each category (beef, lamb, pork, poultry, eggs, and dairy) produced under the two management systems varies, based on economic conditions, resources available, market opportunities, geography, and animal health and management concerns. Ruminants (e.g., cattle and sheep) are the species most likely to be produced in extensive systems. Extensive systems are expected to generate greater DLC exposure to food animals than intensive systems, due to direct contact with soil and greater consumption of forage products by the animals, however, there are limited data regarding geographic variations in soil contamination levels with which to quantify the differences.

Extensive Production

In extensive production systems, environmental media are potential sources of DLC contamination. The levels of DLC contamination in these sources, and in grazing livestock, will reflect the local history of environmental releases. Soils and vegetation may accumulate DLCs on their surfaces, but little migration or absorption into the plants is expected to occur.

Sediments in ponds and streams may be DLC sources to the extent that livestock have direct access to these waters and are able to stir and ingest sediment while drinking. Forage and grasses harvested and supplied as supplemental animal feeds may also contribute to DLC contamination levels; however, since soil ingestion is reduced, so is the total daily DLC exposure. Thus, poultry, swine, and other animals that ingest soil during food foraging activities will have higher DLC exposure potentials than those not exposed to soil.

Intensive Production

Intensive operations remove an animal's access to soils and ground water sources and thus limit their opportunities for DLC exposure. Thus, both aerosol exposure and surface water contact are minimized by the facilities in which the animals are raised. In most intensive animal production operations, feeds are provided in processed forms, and, for monogastric animals, are primarily grain-based in composition. Because of these factors, environmental DLC contamination is less likely to occur in intensive than in extensive operations. As analytical methods improve and costs are reduced, air and water quality sampling will permit more precise comparisons of exposure between intensive and extensive operations. If the rations fed to animals do not include forages or grasses, exposure is further reduced. However, a shift in animal production practices from extensive to intensive to reduce exposure may have economic, sustainability, and animal welfare consequences that are beyond the scope of this report.

ANIMAL HUSBANDRY PRACTICES

As discussed in the previous section, foods of animal origin in commercial channels may be derived under either extensive or intensive production systems. In many cases, the production system will not be readily identifiable at the meat counter, although specialized or niche products may offer this information to differentiate themselves from other suppliers. When this is not the case, a general knowledge of the dominant production types for these products may provide guidance as to the likely production system, which will in turn enable a better evaluation of the relative risks of DLC contamination from various animal food sources. The paucity of analytical data to characterize DLC contamination within food animal species, however, makes the assessment of risks difficult, but based on general production systems risks, it can be expected that a range of DLC exposures exists within each class of meats, milk, fish, and eggs.

Ruminants

Beef and Lamb Production

Grazing livestock and those fed contaminated forages can be expected to reflect the environmental burdens for these localized areas of access. This observation has a direct impact on beef production at the cow-calf and stocker-calf stages where the primary source of nutrients is forage. Similar concerns can be raised with range-lamb production. Since DLCs are not readily taken up by plants and deposited onto grains and other edible plant parts, it is possible for beef and lamb feeder animals placed on predominately grain diets in an extensive feedlot system to reduce DLC intake levels after pasture exposures (Lorber et al., 1994). Removal of known point-source contaminants, reduced atmospheric DLC production, and the subsequent reduced soil contamination levels will, in turn, reduce expected DLC levels in extensively managed animals. However, because animal-based fats and protein ingredients may be added to the diets of animals in intensive production systems, DLCs can accumulate in the animals' tissues, although DLC levels are likely to be much lower than those found in animals on pasture-based diets.

The second major source for contamination for beef and lamb is feeds and feedstuffs. As discussed previously, forages are the major source of DLC exposures for ruminant species. Nearly 35 percent of U.S. land area (788 million acres) is devoted to grazed forest land, grassland pasture and range, or cropland pasture (Vesterby and Krupa, 2001). Because of the great expanses of land in the western United States, grazing is the predominant agricultural land use in this area (Figure 4-2). Intensive forage production can be found throughout the midwestern and eastern regions and primarily on irrigated lands in some western areas of the United States. More concentrated grazing and forage production practices are found in southern, midwestern, and eastern regions.

Cow-calf, sheep, and lambs are predominantly raised under extensive conditions in all regions (Figures 4-3, 4-4, and 4-5). Large feedlots for finishing cattle and sheep are found predominately in the plains states of Texas, Oklahoma, Nebraska, and Colorado (Figure 4-6). These feedlots are extensive but concentrated operations. Forage supplies are generally of local origin, but grains may be imported into the area. Smaller feedlots are found predominately in the states east of the Missouri River and in California. A few intensive (total confinement) finishing lots can be found in the midwestern and eastern regions, but they represent a minimal percentage of total beef and lamb production. These smaller feedlot operations, irrespective of production system, use locally produced feedstuffs.

1 Dot = 100,000 Acres

United States Total
491,046,004

FIGURE 4-2 Geographic distribution of pastureland in the United States, 1997. SOURCE: NASS (1999).

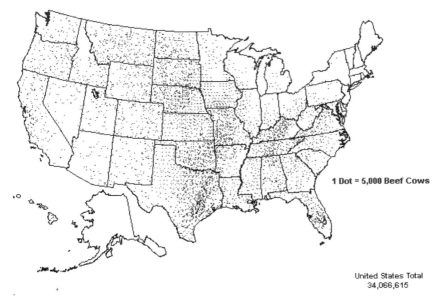

FIGURE 4-3 Geographic distribution of beef cows in the United States, 1997. SOURCE: NASS (1999).

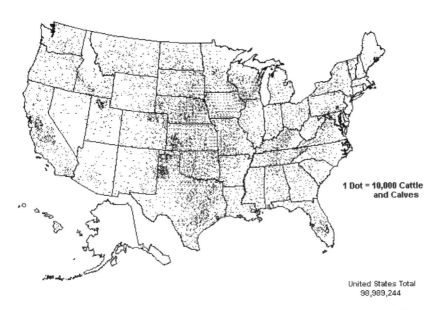

FIGURE 4-4 Geographic distribution of cattle and calves in the United States, 1997. SOURCE: NASS (1999).

FIGURE 4-5 Geographic distribution of sheep and lambs in the United States, 1997. SOURCE: NASS (1999).

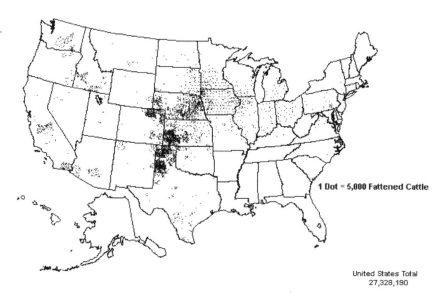

FIGURE 4-6 Geographic distribution of fattened cattle sold in the United States, 1997. SOURCE: NASS (1999).

Dairy Cattle

Dairy operations have two routes through which they may contribute DLCs to the human food supply: milk and meat. Milk levels reflect the DLC body burden of the herd and the fat content of the milk. Mobilization of body fat during lactation likely causes variances in milk DLC levels, but these differences may be leveled through the practice of staggered lactations of individual cows, which is designed to maintain a steady daily herd production level. One exception to this staggered production is selected herds, where the entire herd is brought into simultaneous lactation to take maximum advantage of pasture grass growth to sustain production levels. As in beef operations, a higher proportion of grains in the ration may reduce the total DLC exposure potentials, irrespective of the type of operation. Dairy operations in the western, southwestern, and plains states are primarily concentrated operations (Figure 4-7). Many of these operations import feedstuffs, both forage and grains, as well as replacement animals from outside the immediate area. Therefore, DLC exposure from meat and milk contamination is predominately a function of DLC levels in these source areas.

Large intensive dairy operations increasingly are found in the Midwest and parts of the Southeast, where forages are grown locally and grains may or may not be imported from outside the immediate area. Water supplies for these inten-

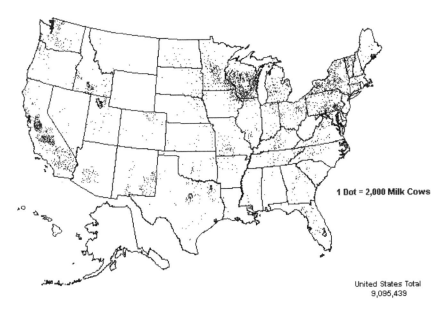

1 Dot = 2,000 Milk Cows

United States Total
9,095,439

FIGURE 4-7 Geographic distribution of lactating dairy cattle in the United States, 1997. SOURCE: NASS (1999).

sive units are generally from wells or filtered sources, rather than from surface water.

In these same areas, as well as in the Northeast, smaller extensive dairies have traditionally predominated. In these extensive operations, cows graze pastures during the summer months, but are confined during the winter. In all cases, rations maintain a forage-grain configuration, based on production needs.

Intensive operations are more inclined to feed increased proportions of grains, grain and animal by-products, and other supplements in an attempt to achieve higher production targets. The feedstuffs and replacement animals for intensive and extensive groups are generally locally raised. Therefore, DLC levels in these operations are more reflective of local environmental conditions.

Veal

Veal production is a specialized program that places approximately 750,000 calves annually into intensive production systems. Veal production is concentrated in the eastern Midwest and the Northeast, and is largely a method for utilizing the male dairy calves for a specialty market. These calves receive colostrum prior to placement and are generally maintained in individual crates for the feeding period. Veal calves are fed a commercial milk-based liquid diet for the 16 to 21 weeks of production. These animals receive no forage during the feeding period and have limited access to soils to maintain the pale meat color desired by consumers. Thus, the level of DLC exposure to these calves is dependent upon the DLC levels in the milk used to formulate their diet.

Swine and Poultry

Swine and poultry (monogastric) food animals are fed predominately grain-based diets. Additional animal fats and meat products, fish meal, grain by-products, and other supplements may be added to the diets to meet nutritional or best-performance production goals, and DLCs may accumulate in animal tissues that are utilized as feed. Minimal access to grasses or forages occurs except in some extensive production situations where soil access is a major source of DLC contamination.

Swine consume soil as a part of their normal rooting activities when given access to extensive systems. Fries and coworkers (1982) estimated that swine consume from 3.3 to 8 percent of their diet in pastures. Similarly, poultry with soil access ingest DLCs from the environment during their normal feeding behaviors. Geese were found to consume 8 percent and wild turkey 9 percent of their diets as soil (Beyer et al., 1994). Free-range poultry may exhibit similar behavior. Laying hens exposed to soils produce eggs with DLC contamination levels that are reflective of the soil contamination levels (Schuler et al., 1997).

Swine

Swine operations are concentrated in the midwestern areas east of or immediately west of the Missouri River and in North Carolina (Figure 4-8). Swine operations are predominantly intensive throughout these regions. Intensive operations raise swine in enclosed facilities on concrete, wire, or other slatted materials to separate them from their manure. In some midwestern areas, the use of intensive but bedded systems for finishing swine is increasingly popular. These bedding materials range from wood chips to mixtures of corn fodder and recycled paper products. Bedding materials are produced locally and generally are replaced after each group of swine is marketed.

Feed supplies may be produced locally, or in the case of the southern and mid-Atlantic states, shipped from the Midwest. Feeds are predominately vegetable and grain-based diets. Water sources are generally from wells or other controlled sources, rather than surface waters. Therefore, exposures are primarily limited to dietary introductions.

A small but growing population of swine is raised in extensive production systems. Many of these operations are small producers with local or specialty markets that desire to have free-range products. These animals have access to

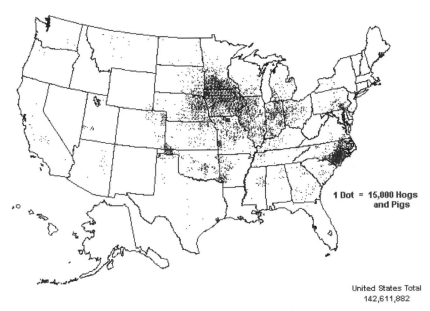

1 Dot = 15,000 Hogs and Pigs

United States Total
142,611,882

FIGURE 4-8 Geographic distribution of hogs and pigs sold in the United States, 1997. SOURCE: NASS (1999).

soils and pasture. They may have access to surface water sources, but are more likely to receive water from wells or other controlled sources. Their diets are grain-based, similar to the intensive operations, but DLC levels may vary as other sources of exposure (soil and forage) are introduced into their environment.

Poultry

Egg production is concentrated in scattered geographic regions (Figure 4-9). Poultry operations (broiler, turkey, and egg production) are predominantly intensive production models. Broiler production is concentrated in the southern and mid-Atlantic states (Figure 4-10). Turkey production is found in these areas and in the upper Midwest states (Figure 4-11).

Broiler and turkey operations, while intensive in production, are predominantly litter-based systems. Litter materials include wood chips, ground corncobs, and other suitable bedding materials and may be used for several groups of animals before being recycled. Egg production predominantly occurs on wire structures or in cages that limit the bird's exposure to soil and fecal material. Feed supplies for all production types may be locally produced, or in the case of the southern and mid-Atlantic states, shipped from the Midwest. Feeds are predominately vegetable and grain-based diets, supplemented with animal and grain by-products in best-cost formulations.

1 Dot = 60,000 Layers and
Pullets 13 Weeks
Old and Older

United States Total
366,989,851

FIGURE 4-9 Geographic distribution of layers and pullets in the United States, 1997. SOURCE: NASS (1999).

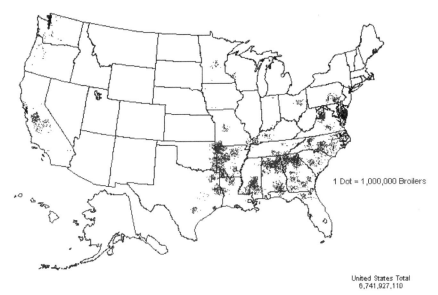

FIGURE 4-10 Geographic distribution of broilers and other meat-type chickens sold in the United States, 1997.
SOURCE: NASS (1999).

FIGURE 4-11 Geographic distribution of turkeys sold in the United States, 1997.
SOURCE: NASS (1999).

As with swine, a small but growing population of broilers and layers are being housed in extensive production systems for the free-range markets. These animals have access to soil and, in some cases, to pasture. Feeds are predominately grain-based diets, and water sources are generally from wells or other controlled sources. Rarely, however, do they have access to standing surface water.

FISH AND SEAFOOD

There are a number of ways in which aquatic organisms can accumulate persistent organic pollutants (POPs) such as DLCs. Two common scenarios that lead to the accumulation of POPs in aquatic organisms are presented below.

Wild-Caught Seafood

Persistent compounds accumulate in the wild populations of fish and shellfish that are caught in freshwater, estuaries, and the open ocean. This is because of their lipophilicity, reliance on passive transport, and resistance to metabolism. In contaminated aquatic environments, POPs are generally associated with particles that settle in sediment or are carried downstream to the delta or depositional areas, where they then are deposited. Fish or aquatic organisms may come in direct contact with POPs if they live near the bottom or have contact through their gills.

The aquatic food chain, just as with the human food chain, is the predominant route of DLC exposure for fish, with the DLCs accumulating in the larger fish that eat the small crustaceans or smaller fish. The most critical step in exposure is desorption of the DLCs from particulate matter and subsequent entry into the food chain. Once in the food chain, the DLCs associate with lipids, where the rate of accumulation is directly related to lipid uptake.

Once DLCs are ingested, they distribute to different tissues in fish, based primarily on tissue lipid content, and, to some extent, nonspecific binding proteins. The liver, gonads, and fat are the three highest fat tissues and the most highly contaminated. The muscle is generally less contaminated, but that is partially determined by the lipid content in the muscle tissue of a specific type of fish. In general, older, larger, and oily fish have higher DLC levels than other fish.

In addition, different fish harvested from the same location may have different DLC levels. Wild fish are migratory and therefore may on one day eat food that is highly contaminated and then the next day, eat the same type of food from another location that is less contaminated. Different species also have preferred habitats or behaviors that may expose them to higher DLC levels. The major sources for aquatic DLC contamination are air deposition over large watersheds and regional point sources. Since their deep-water food sources have lower DLC levels, ocean fish have lower contamination levels than near-shore species.

In the case of other seafood, such as crabs, crawfish, oysters, clams, and scallops, the environment in which they grow determines their DLC body burden. Lobsters that inhabit near-shore areas may accumulate DLCs in high concentrations. Unfortunately, the most productive lobster trapping areas are along the coasts, where watersheds drain and point sources are located.

Regional variations in methods for preparing and eating fish can have a direct impact on the levels of DLCs to which the consumer is exposed. This is the reason that states such as New Jersey have developed recommendations for preparing fish that reduce the exposure to high-fat-containing tissues (i.e., fat underlying the skin of the fish). If the fish is used whole, such as in some types of soups, the average exposure is higher than if just the filets are used. Soups made from whole organisms (e.g., lobster) contain both low-contamination muscle meat and high-contamination organ meats. Food preparation methods to decrease DLC exposure are discussed further in Chapter 5.

Aquaculture

Aquaculture in the United States and other countries supplies an increasing proportion of fresh fish and certain shellfish that is consumed in the United States. Some of the major aquaculture species sold include salmon, catfish, tilapia, striped bass, crawfish, and prawns.

Management practices in aquaculture are designed to maximize the animal's growth and the quality of its meat. This is accomplished by modifying the composition of the feeds used during different stages in an animal's life. Extensive research has been carried out on the nutrient requirements and feeding of catfish and salmonids, and consequently, their nutrient requirements and feeding characteristics are well understood. The diets for intensively cultured species provide, in a water-stable, readily digestible formulation, all known nutrients at required levels and energy necessary for their utilization. Because of the intensive culture conditions, the majority of the food is supplied by the producers. The feed produced for fish includes protein, vitamins, minerals, lipids, and energy for normal growth and other physiological functions.

As an example, the percent composition of feed for channel catfish at different times in the growth cycle is presented in Table 4-1. The composition of the feed is varied to meet the physiological requirements of the fish at each life stage. The higher the percentage of protein, the more rapid the growth target. At the fry stage, the highest amount of menhaden meal is used in the feed, allowing for rapid growth in the earliest stages of culturing.

Because of this intensive management approach, even low levels of DLCs can accumulate into fish tissues. A reduction in the use of animal-based fats as an

TABLE 4-1 Examples of Ingredient Composition of Typical Channel Catfish Fry, Fingerling, and Food Fish Feeds

Ingredients	Fry Feed (50%)[a]	Fingerling Feed (35%)	Food Fish (32%)	Food Fish (28%)
Soybean meal (48%)[a]	—	38.8	35.0	24.4
Cottonseed meal (41%)	—	10.0	10.0	10.0
Menhaden meal (61%)	60.2	6.0	4.0	4.0
Meat/bone/blood (65%)	15.3	6.0	4.0	4.0
Corn grain	—	16.1	29.9	35.5
Wheat middling	19.0	20.0	15.0	20.0
Dicalcium phosphate	—	1.0	0.5	0.5
Catfish vitamin mix[b]	Included	Included	Included	Included
Catfish mineral mix[b]	Included	Included	Included	Included
Fat/oil[c]	5.0	2.0	1.5	1.5

[a]Values in parentheses represent the percentage of protein.
[b]Commercial mix that meets or exceeds all requirements for channel catfish.
[c]Fat or oil is sprayed on the finished feed pellets to reduce feed dust (fines).

SOURCE: Robinson et al. (2001).

ingredient in fish feeds would generally reduce the uptake of minor contaminants present in the fat. There is a need to examine all of the feed ingredients to ensure that DLCs are not being inadvertently introduced. In the case of pond-reared crawfish, catfish feed is often used. Similar concerns exist for salmonids and other finfish species raised under intensive culture conditions.

Feeds that are dependent on animal fats are potential sources of DLC exposure. However, reformulation of fish diets requires the development of a defined diet that meets the nutritional needs of the fish and reduces the potential introduction of DLCs into the food system. The use of defatted (rendered) fishmeal offers a potential means to reduce entry of DLCs into animal feeds. Diets for fatty fish also require formulations that include omega-3 fatty acids.

WILD ANIMALS

For some population subgroups such as Northern Dwellers (i.e., Native populations living in Canada and Alaska) and Native American fishers, the consumption of wild-caught animals is an important source of nutrition, and hunting and fishing activities are important cultural activities. However, the consumption of food products from wild-caught animals may result in increased DLC exposure.

As with commercial extensively raised livestock, wild animals are exposed to DLCs through the plants and animals they eat. Exposure may be due to either the generalized presence of DLCs in the environment, as is the case throughout much of the Arctic region (Arctic Monitoring and Assessment Programme, 2001), or to the presence of environmental hot spots. DLC levels in animal tissues vary, depending on the animal's fat content and its length of exposure. Arctic animals may be of special concern because of their high fatty makeup (e.g., seals). Animals in other areas of the United States (e.g., deer in Michigan) may be of lesser concern because products from them are relatively low in fat.

DLC exposure may also vary due to the migratory habits of animals and to the contamination levels in areas through which they migrate. For example, in the Arctic region, birds that migrate south along the Eastern seaboard of the United States have higher levels of several POPs than do birds from farther west (Arctic Monitoring and Assessment Programme, 2001). Exposure of wild animals, and of humans who consume them, to DLCs is also affected by point sources of pollution, as is the case among the Akwesasne people living along the St. Lawrence River. As a result, DLC levels found in particular animals (e.g., deer, elk, caribou, seals, and salmon) may vary considerably, resulting in localized hot spots.

In contrast to commercially produced animals, the "production" of wild animals is relatively unmanaged and unregulated and has few intervention points that can significantly affect DLC loads. Reduction of DLC levels in wild animals is almost entirely dependent on reductions in environmental DLC loads.

ANIMAL FEEDS

As discussed in previous sections, for livestock raised intensively and seafood produced by aquaculture, animal feeds used to supplement or replace natural dietary components may be an important source of DLC exposure. Feeds are carefully formulated to provide complete and balanced nutrition to animals with the greatest efficiency. Not all feeds contain DLCs, but when they do, it is likely due to isolated ingredients and is often, but not exclusively, limited to feeds that contain animal products.

Feed Ingredients

Collective terms are classes of feed ingredients grouped by ingredient origin, as jointly defined by the Association of American Feed Control Officials and the Food and Drug Administration (FDA) (AAFCO, 2002). Collective terms are permitted on all animal feed labels (rather than the listing of each individual ingredient within that feed term) in all states except for California and Florida. In California, each individual feed ingredient must be listed in the ingredient statement in descending order of relative presence. In Florida, the collective terms can only be used on the feed labels of certain species. Table 4-2 presents the collective terms and the ingredients most commonly used in them (AAFCO, 2002).

Many other feed ingredients that are used in animal feeds are not classified into the collective terms. These ingredients are listed individually in the ingredi-

TABLE 4-2 Animal Feed Collective Terms and Illustrative Ingredients

Collective Term	Illustrative Ingredient
Animal protein products	Fish meal
	Meat and bone meal
	Milk, dried, whole
	Poultry by-product meal
	Whey, dried
	Hydrolyzed poultry feathers
	Animal blood, dried
	Casein
Forage products	Alfalfa meal, dehydrated
	Corn plant, dehydrated
	Soybean hay, ground
Grain products (i.e., whole,	Barley
ground, cracked, flaked)	Corn
	Grain, sorghum
	Oats
	Wheat
	Rice
Plant protein products	Canola meal
	Cottonseed meal
	Peanut meal
	Soybean meal
	Sunflower meal
	Yeast, dried
Processed grain by-products	Aspirated grain fractions
	Brewers dried grains
	Corn gluten meal
	Corn distillers dried grains
	Wheat bran
	Wheat middling
Roughage products	Apple pomace, dried
	Beet pulp, dried
	Citrus pulp, dried
	Corn cob fractions
	Cottonseed hulls
	Oat hulls
	Peanut hulls
	Rice hulls
	Soybean mill run
	Straw, ground
Molasses products	Beet molasses
	Cane molasses

ent statement on the animal-feed product label, but can be grouped into feed ingredient categories, as shown in Table 4-3. The potential for DLC exposure varies for each of the categories shown, although animal fat is considered to have the greatest potential for containing DLCs. Forage and roughage products vary geographically depending upon air exposure to DLCs and the season of the cut, although more data is needed to confirm which regions are likely to have high DLC levels. Processed grain by-products may contain DLCs depending upon the type of processing. Some animal medications used in the past may have had DLC contaminants (nitrofurans), but none of the presently approved medications are known to have DLC contaminants.

The annual production of animal fats (white and yellow tallow, greases, and poultry fat) is estimated to be 3.6 billion pounds of inedible tallow, 3 billion pounds of grease, and 1.4 billion pounds of recycled fat (PROMAR International, 1999). Animal fats have been identified as the greatest potential source of DLC contamination; however, removal of this by-product from feed formulations may create secondary problems with disposal of the unused fat. Implementation of appropriate and efficient disposal routes for contaminated fat would serve as an acceptable alternative to eliminating the recycling of fat, permanently or temporarily. The transformation of unused fats and oils into alternative uses, such as biofuels (Pearl; 2002, Tyson, 1998; Wiltsee, 1998), is an intriguing alternative that may become more important as the technology develops.

In 2000, prompted by the 1997 findings of the entry of DLCs into animal feed products through contaminated ball clay (described in Ferrario et al., 2002; see also Box 4-1), FDA's Center for Veterinary Medicine initiated the Prelimi-

TABLE 4-3 Animal Feed Categories and Illustrative Ingredients

Category	Illustrative Ingredient
Fats and oils	Animal fat
	Vegetable fat or oil
	Fat product, feed grade
Microingredients	Vitamin A supplement
	Vitamin E supplement
	Calcium oxide
	Cobalt carbonate
Medication	Amprolium
	Carbadox
Fermentation products	Cereals/grain fermentation solubles, condensed
	Direct-fed microorganisms, such as *Aspergillus niger*
Enzymes and herbals	Lactase
	Phytase

BOX 4-1 Unintended Releases of Dioxins and Dioxin-like Compounds (DLCs)

Portions of the human food supply have been affected by unintended and unexpected contamination with DLCs. Three major episodes are described below.

During a nationwide survey performed by the U.S. Department of Agriculture's Food Safety and Inspection Service and the U.S. Environmental Protection Agency in 1997, two samples of poultry products were found to be highly contaminated with DLCs. It was discovered that contaminated ball clay was added to soybean meal as an anticaking agent, which was then added to animal feed provided to poultry and catfish. The ball clay used in the contaminated feed was traced back to a Mississippi bentonite mine, which seemed to be naturally, and not accidentally, contaminated. The Food and Drug Administration has since banned the use of ball clay in animal feed. This incident provoked the development of extensive testing programs for DLC contamination in catfish, eggs, and poultry intended for human consumption (EPA, 2000; FDA, 1997; FSIS, 1997, 2002).

In 1997–1998, high levels of DLCs discovered in European cow's milk samples were traced back to contaminated lime (quicklime and hydrated lime) used in the production of citrus pulp pellets (CPP), a by-product of orange juice production. Lime is used in the preparation of CPP to raise the pH and facilitate the drying process. The contaminated CPP, used as an additive in dairy cow feed, was traced back to a Belgian-based chlorine producer operating in Brazil. A European Union scientific panel has recommended a fixed maximum level of DLCs in citrus pulp imports (Allsopp et al., 2000; EC, 1999; Greenpeace, 1999a, 1999b; Reuters, 1998).

In another incident during the period of January 1999–June 1999, European officials discovered that DLC-contaminated animal fats had been used in animal feeds for chickens, cows, and pigs. A Belgian processing company sold the animal fats to various livestock farms in Belgium, the Netherlands, France, and Germany. Upon investigation, the animal fat was found to have been stored in tanks previously used to store transformer oil. The contamination of the animal feed, and consequently the animals and animal products, led to a European Union ban on all Belgian beef, chicken, eggs, and pork exports. The European Commission's Veterinary Committee ordered the farms involved to destroy all poultry, poultry products, and livestock that may have been given the contaminated feeds. Belgian food retailers cleared their shelves of all suspect food items, including chicken and egg products, as well as those containing eggs as an ingredient (e.g., baked goods and mayonnaise) (Allsopp et al., 2000; Tyler, 1999; van Larebeke et al., 2001).

---

nary National Survey of DLCs in Animal Fats, Animal Meals, Oilseed Deodorizer Distillates, and Molasses (CVM, 2000), with a follow-up survey in 2001 (CVM, 2001). The purpose of the surveys was to determine background levels of DLCs in fatty and other ingredients commonly used in animal feeds. The results, expressed as TEQ parts per trillion (ppt) dry weight, showed the following congener concentrations for groupings of samples:

1. Fish meal (used as an ingredient in animal diets) samples varied by species. Meal derived from Pacific whiting had the lowest DLC values (< 0.5 ppt), whereas menhaden had the highest values (average 0.9 ppt). The DLC levels in fish from the Atlantic and Gulf areas are considered to be uncontrollable in that there is presently no known intervention that can lower the DLC levels in these fish.

2. Deodorized distillates showed DLC values of 1.4 to 7.1 ppt (average 4.4 ppt). Oil from fish-oil distillates is used as a feed ingredient (and is also used in the manufacture of dietary supplements intended for consumption by humans).

3. Other animal feed ingredients, including carcass-animal fat (including pork, beef, and mixed species) showed DLC values of 0.2 to 3.95 ppt (average 0.5 ppt), while meat and bone meal from mixed species, by-product meal from poultry, and eggs showed < 0.3 ppt. For animal diets, the DLC level found in rendered fat may represent a small percent of the total dietary fat because the final diet likely contains a mixture of both animal and vegetable fats. However, animal fats are a significant source of recycled DLCs, and can be removed from animal diets. For example, meat and bone meal (a protein source derived from the same rendering process), can be defatted to reduce the risk for DLC contamination (Ferrario et al., 2002).

4. Results are pending for mineral additives such as copper sulfate. DLCs have been found where kelp and heat are used in manufacturing and processing. The DLCs produced in this process are considered to be controllable through the drying process during manufacture.

5. Anti-caking agents (from a 1998 survey) showed DLC values of 0.4 to 22.5 ppt. The broad range of DLC levels may indicate hot pockets of environmental contamination. There also may be other clays contaminated with DLCs in addition to the ball clay that has already been banned from animal feeds. Some of these clays are used for bleaching in both animal and human food processing.

Although there have sporadic events in which very high levels of DLCs have been detected as shown in Boxes 4-1 and 4-2, more representative analytical results for specific DLC congeners are shown in Table 4-4.

Feed Formulation

The feeding objective in food animals is weight gain in the shortest period of time at the lowest cost. There are breeding issues that are also addressed with feed formulations, but formula cost is the driver of the feed formulation consistent with the feeding objective.

BOX 4-2 Dioxin Formation During Feed Processing

Low levels of dioxins and dioxin-like compounds (DLCs) were found in animal feed supplement products manufactured by the Minnesota-based QualiTech Inc. and exported to Ireland. The products implicated were protected minerals and mineral premixes, sold under the names Sea-Questra-Min and Carbosan, that were intended to provide essential micronutrients for use in animal feeds. Qual-iTech issued a voluntary recall of the mineral products, as well as of feeds that may have been produced using the mineral supplements. Although the exact cause of the DLC contamination has not been determined, preliminary test results suggest that the DLCs were formed during a heating process used to manufacture these products (CVM, 2002).

Least-Cost Formulations

In the least-cost formulation process, the goal is to select ingredients for a formulation that will provide the desired nutritional characteristics at the lowest cost. The complexity of the various systems available for developing formulations range from a simple percentage system to extremely complicated systems with many ingredients and ingredient constraints.

Using computer software, information about each available ingredient in a formulation is entered into the system, including the nutritional value of the ingredient (e.g., protein, fat, vitamins, minerals, fiber), the ingredient cost, and amount to be used (depending on species and age). Other factors may be included in the ingredient selection process, depending upon the sophistication of the system being used.

Most systems include:

- ingredient profiles
- commodity prices
- nutrient specifications for the products
- manufacturing capabilities
- interface with tagging/labeling systems
- links with feed-manufacturing plants updates on the nutrient content of the ingredients (some systems automatically update changes)
- drug validation
- allocation of limited or excess available ingredients
- history storage systems
- links with pricing systems
- links with accounting systems
- links with purchasing systems

TABLE 4-4 Summary Data of Dioxins and Dioxin-like Compounds Congener Concentrations (ppt) for Feed Ingredients

Description[a]	Animal-Based Ingredients[b]							Plant-Based Ingredients[c]	
	AF-SL	AF-MX	MBM-MX	PBM & E	FM-WHI	FM-CAT	FM-MEN	DD	MOL
Number of samples	6	3	3	5	3	1	4	11	8
2,3,7,8-TCDD	0.04	0.08	0.01	0.01	0.01	0.09	0.37	0.23	0.02
1,2,3,7,8-PeCDD	0.12	0.36	0.03	0.04	0.04	0.46	0.62	0.92	0.04
1,2,3,4,7,8-HxCDD	0.13	0.53	0.03	0.04	0.03	0.61	0.17	1.31	0.04
1,2,3,6,7,8-HxCDD	0.61	2.99	0.10	0.04	0.08	0.91	0.66	3.27	0.04
1,2,3,7,8,9-HxCDD	0.19	0.66	0.06	0.04	0.03	0.64	0.46	3.45	0.04
1,2,3,4,6,7,8-HpCDD	3.17	16.26	1.66	0.61	0.29	6.62	3.11	62.55	1.12
OCDD	12.03	76.74	15.06	7.29	2.80	47.84	54.07	511.15	11.61
2,3,7,8-TCDF	0.04	0.73	0.04	0.01	0.20	0.05	2.32	0.68	0.02
1,2,3,7,8-PeCDF	0.06	0.11	0.02	0.04	0.03	0.02	0.31	0.40	0.04
2,3,4,7,8-PeCDF	0.11	0.32	0.02	0.04	0.08	0.04	0.58	0.77	0.04
1,2,3,4,7,8-HxCDF	0.13	0.63	0.04	0.04	0.03	0.02	0.12	0.70	0.04
1,2,3,6,7,8-HxCDF	0.11	0.42	0.02	0.04	0.03	0.02	0.06	0.51	0.04
2,3,4,6,7,8-HxCDF	0.06	0.22	0.04	0.04	0.04	0.05	0.12	0.65	0.04
1,2,3,7,8,9-HxCDF	0.06	0.06	0.02	0.04	0.03	0.02	0.02	0.24	0.05
1,2,3,4,6,7,8-HpCDF	0.17	2.87	0.24	0.19	0.09	0.06	0.28	6.54	0.50
1,2,3,4,7,8,9-HpCDF	0.17	0.25	0.07	0.11	0.08	0.05	0.06	0.57	0.12
OCDF	0.50	2.25	0.42	0.27	0.25	0.02	0.33	10.55	1.63
PCB 77	2.06	22.15	1.68	119.06	2.01	2.37	82.10	133.70	0.71
PCB 126	1.62	6.60	0.32	0.50	1.42	0.37	11.63	9.73	0.02
PCB 169	0.54	1.29	0.04	0.13	0.40	0.08	1.97	0.44	0.04
TEQ–D	0.28	1.02	0.04	0.07	0.07	0.84	1.15	2.63	0.08
TEQ–F	0.10	0.40	0.03	0.04	0.07	0.04	0.57	0.76	0.05
TEQ–P	0.17	0.68	0.03	0.05	0.15	0.04	1.19	0.99	<0.01
TEQ–DF	0.38	1.43	0.11	0.11	0.15	0.87	1.72	3.38	0.13
TEQ–DF; ND = 0	0.26	1.41	0.07	0.01	0.11	0.87	1.72	3.35	0.02
TEQ–DFP	0.54	2.10	0.14	0.16	0.29	0.91	2.91	4.39	0.13
TEQ–DFP; ND = 0	0.43	2.09	0.10	0.06	0.26	0.91	2.91	4.34	0.02
Min TEQ–DFP	0.08	1.07	0.13	0.09	0.20	0.91	2.14	1.43	0.02
Max TEQ–DFP	1.49	3.95	0.16	0.30	0.44	0.91	3.33	7.08	0.18

[a]TCDD = tetrachlorodibenzo-p-dioxin, PeCDD = pentachlorodibenzo-p-dioxin, HxCDD = hexachlorodibenzo-p-dioxin, HpCDD = heptachlorodibenzo-p-dioxin, OCDD = octachlorodibenzo-p-dioxin, TCDF = tetrachlorodibenzofuran, PeCDF = pentachlorodibenzo-p-furan, HxCDF = hexachlorodibenzo-p-furan, HpCDF = heptachlorodibenzo-p-furan, OCDF = octachlorodibenzo-p-furan, PCB = polychlorinated biphenyls.
[b]AF-SL = animal fat from slaughtered animals, AF-MX = animal fat from mixed animal species, MBM-MX = meat and bone meal from mixed animal species, PBM&E = poultry by-product meal and eggs, FM-WHI = Pacific whiting meal, FM-CAT = catfish meal, FM-MEN = menhaden meal.
[c]DD = deodorizer distillates, MOL = molasses, corn oil, and canola meal.

NOTE: Nondetects (ND) = ½ limit of detection (LOD); levels of 0.01–0.04 ppt indicate ND for all or most samples; average toxicity equivalents (TEQ) concentrations in ppt when ND = ½ detection limit and at ND = 0.
SOURCE: Ferrario et al. (2002).

Through the use of such least-cost formulation systems, limits on DLC levels for individual ingredients can be set. The major obstacle in the use of a computer system for DLC-level control is obtaining the level of DLCs in an ingredient because of sampling, assay, and timing issues. In addition, as more limitations are placed on ingredients during the selection process, the cost of the animal feed increases. If the use of a cost-effective feed ingredient is limited or eliminated because of a concern for DLC contamination, then the computer must select alternate ingredients. Such alternate selections may come at a high price and may trigger other undesirable effects, such as adverse impacts on food quality, quantity, or production duration. For example, different sources of fat may affect the meat quality of swine or may decrease milk production in dairy animals.

Typical Feed Formulations

Animal diet formulations are based on the species and stage of life of the animal. Tables 4-5 and 4-6 provide two least-cost feed formulations. These formulations include the typical feed ingredients used as the primary animal diet fed in the United States at each stage of life and the effect on costs if animal products were removed.

Replacement of animal products can cause either an increase or decrease in feed cost. This can also cause increases or decreases in secondary costs, such as the relative costs of ingredients due to changes in demand and the time it takes to raise an animal to term. Replacement of animal products in feeds may also cause an unacceptable decrease in the nutrient content or quality of the final food animal product (Cameron and Enser, 1991).

TABLE 4-5 Example of a Least-Cost Feed Formulation for Dairy Cattle

Ingredient	Typical Amount (%)	Amount Without Animal Products (%)
Corn or grain	22.0	32.5
Midds or ground soybean hull	53.5	42.0
Cottonseed meal	4.0	7.0
Soybean meal (could be some sun meal)	10.0	10.0
Calcium carbonate	2.0	2.0
Salt	1.0	1.0
Animal fat	2.0	0.0
Molasses	5.0	5.0
Trace minerals and vitamins pack	0.5	0.5
Total	100.0	100.0
Total ingredient cost	$106.00	$107.50

TABLE 4-6 Example of a Least-Cost, Starter-Phase Feed Formulation for Hogs

Ingredient	Typical Amount (%)	Amount Without Animal Products (%)
Corn	51.5	63.0
Wheat byproducts	11.0	0.0
Soybean meal	30.0	33.0
Fish meal	1.0	0.0
Calcium carbonate	1.0	1.0
Dical or mono dical phosphate	1.5	2.0
Salt	0.5	0.5
Animal fat	3.0	0.0
Trace minerals and vitamins pack	0.5	0.5
Corn	51.5	63.0
Total percent	100	100
Total ingredient cost[a]	$127.00	$126.00

[a]A hog-starter product without animal fat or other animal products is much lower in protein so the time to complete grow-out will be increased.

Certification Programs

The U.S. Department of Agriculture (USDA) has some certification programs for various feed ingredients, although none are specific to DLCs.

Distribution

The distribution mode for feed ingredients and products should be considered as a potential means for distributing DLCs. The distribution conveyances used for animal-feed ingredients require that all containers be cleansed and sequenced to protect feed ingredients from harmful contaminants, including DLCs.

FOOD PROCESSING AND PACKAGING

Most food sources have the potential to contain some level of DLC contamination, depending on the area from which they originated and the agricultural practices under which they were grown or raised. Once these foods sources are destined for the food supply (e.g., harvested, collected, slaughtered, caught), they are prepared for market and then for consumption. For some foods, this requires minimal processing and packaging; for others, significant opportunities for additional DLC contamination exist.

Composite and Processed Foods

Processed foods, including foods containing significant levels of animal fat, such as sausage, bacon, fondues, and products fried in animal fat (e.g., fried snack foods), contribute to DLC intakes. Processed foods may contain varying levels of DLCs depending on the DLC content of each ingredient in the composite food. Therefore, all mixtures of processed foods containing animal, dairy, or fish fats should be considered as potential sources of DLCs.

Water, used in processing and contained in the products themselves, probably does not contribute to the overall DLC load. In a study of the persistence of TCDD metabolites in lake water and sediment (under laboratory conditions), Ward and Matsumura (1978) determined that most of the TCDD spiked into a mixed water-sediment sample is partitioned with the sediment, leaving less than 4 percent of the metabolites in the water itself.

Food Processing and Packaging

Information on the entry or generation of DLCs in the processing and packaging of foods is limited. However, analysis of current practices and procedures may be useful in predicting potential sources of entry of DLCs into the food supply by these routes (Table 4-7).

Processing

There are numerous ways that food is processed, some of which may alter the DLC content in foods, and some of which may not. The processes that are not likely to alter the DLC content in foods are:

- Processing, mixing, and blending foods (e.g., a blend of cereal grains), and forming and molding foods (e.g., bread, pie crusts, biscuits, confections) at ambient temperatures (although much higher temperatures have been found to decompose DLCs) (Zabik and Zabik, 1999)
- Other ambient processing techniques such as sorting, cutting, and separating debris
- The process of flame peeling used in onion processing may generate DLCs due to the high processing temperature, but they may also be washed away in the process
- Blanching, pasteurization, heat sterilization, baking, and roasting
- Freeze processing products sold and maintained as frozen foods (e.g., frozen fruits, vegetables, and meat products) (Larsen and Facchetti, 1989) and chilled and refrigerated storage of fresh and processed foods (freeze-drying or freeze-concentration of certain substances may increase the ratio of DLCs relative to the mass of the resulting processed food)

TABLE 4-7 Effects of Food Processing Methods on Levels of Dioxin and Dioxin-like Compounds (DLCs)

Processing Methods	Effect on DLC Levels
Raw material preparation	
Cleaning	
Wet cleaning	May reduce or remove DLCs
Dry cleaning	No effect on DLCs
Removing contaminants/foreign bodies	No effect on DLCs
Sorting	
Shape and size sorting	No effect on DLCs
Color sorting	No effect on DLCs
Weight sorting	No effect on DLCs
Peeling	
Flash steam peeling	May reduce or remove DLCs (adsorbed to surface)
Knife peeling	May reduce or remove DLCs (adsorbed to surface)
Abrasion peeling	May reduce or remove DLCs (adsorbed to surface)
Flame peeling	Process used for onions, may generate DLCs, but may also be washed away during process, no fat content in onions to hold DLCs
Mixing and forming	
Mixing	
Solids mixing	No effect on DLCs
Liquids mixing	No effect on DLCs
Forming	
Bread molders	No effect on DLCs
Pie and biscuit formers	No effect on DLCs
Confectionery molders	No effect on DLCs
Separation and concentration of food components	Could concentrate and reduce existing DLCs in lipid and aqueous phases, respectively (e.g., milk separation, more concentrated in cream, less concentrated in skim milk)
Centrifugation	May concentrate or reduce existing DLCs
Filtration	May concentrate or reduce existing DLCs
Expression	May concentrate existing DLCs as in fish-oil production
Extraction using solvents	May concentrate existing DLCs or introduce DLCs from solvent residues
Membrane concentration	May concentrate existing DLCs

continued

TABLE 4-7 Continued

Processing Methods	Effect on DLC Levels
Fermentation and enzyme technology	
Fermentation	No effect on DLCs
Enzyme technology	No effect on DLCs
Blanching	
Theory	No effect on DLCs
Effect on foods	No known effect on DLCs
Pasteurization	
Theory	No effect on DLCs
Heat Sterilization	
In-container sterilization	No effect on DLCs
Ultra high-temperature (UHT)/aseptic processes	No effect on DLCs
Evaporation and distillation	
Evaporation	May concentrate existing DLCs
Distillation	No effect on DLCs
Baking and roasting	
Theory	No effect on DLCs
Frying	
Shallow (or contact) frying	No effect on existing DLCs, could be introduced from contaminated oils or fats
Deep-fat frying	No effect on existing DLCs, could be introduced from contaminated oils or fats
Chilling	
Fresh foods	No effect on DLCs
Processed foods	No effect on DLCs
Cook-chill systems	No effect on DLCs
Chill storage	No effect on DLCs
Freezing	
Theory	No effect on DLCs
Freeze drying and freeze concentration	
Freeze drying (lyophilization)	May concentrate existing DLCs
Freeze concentration	May concentrate existing DLCs
Packaging	
Interactions between packaging and foods	May introduce DLCs from packaging
Printing	May introduce DLCs from inks and pigments
Filling and sealing of containers	
Filling	No effect on DLCs
Sealing	No effect on DLCs
Labeling	No effect on DLCs

- Frying meats, fruits, or vegetables, unless DLCs are introduced into the food by contaminated fats and oils during frying.

Processes that may alter DLC content of foods are:

- Heat processing of meats, which has been shown to reduce DLC levels through the loss of fats (Petroske et al., 1998; Stachiw et al., 1988). (However, because high temperatures and other favorable conditions can produce DLCs, research should be undertaken to determine if high-temperature processing in baking, extrusion, puffing, and short-time, high-temperature pasteurization has an impact on DLC levels in the finished food product.)
- Extraction and drying during food processing, specifically, the extraction of fat or moisture. For example, a food that had its moisture content reduced would exhibit an increased DLC content as a percent by weight, even though it would contain the same total DLC content as the original food (e.g., dehydrated peas); expression of oils from food products may concentrate DLCs from the original intact food into the oil intended for consumption (e.g., fish oil production); and certain forms of extraction using solvents may concentrate existing DLCs or may introduce them from solvent residues.
- Separation of raw milk, during which existing DLCs will be reduced in the skim milk (aqueous phase) and increased in the cream (lipid phase). (Products made from skim milk and standardized low-fat milk will have lower DLC concentrations than those produced with full-fat milk or cream. Products with a higher fat content, such as cheese, in which much of the aqueous content has been removed in the form of whey, would have a relatively higher DLC concentration than the same volume of food with more water and less fat content.)
- Trimming and cutting fat from meat products, similar to extraction, reduces the DLC content of the prepared food.
- Filter processing may involve the use of filtering agents that contain DLCs and therefore could become a source of food contamination. Examples include the frequently cited ball-clay incident in chicken feeds (Hayward et al., 1999) and the identification of 1,2,7,8-tetrachlorodibenzofuran (TCDF) and 2,3,7,8-TCDD as low-level contaminates introduced into coffee filter papers in Japan (Hashimoto et al., 1992). (Currently, maximum levels of PCDDs and PCDFs in filter paper have been established at 0.00038 to 3.6 pg TEQ/g of paper, with about one-third of the total PCDD and PCDF contamination being eluted from the filter paper during coffee brewing. Hot water elutes low levels of DLCs; therefore, the existing low level of contamination could be avoided by rinsing the filter prior to use.)

Food safety food-processing procedures such as irradiation, ozonation, ultraviolet light, sunlight, and chlorination have not been examined for their impacts on DLC levels.

Packaging

With the exception of fresh vegetables, virtually all foods sold to the public are packaged. Generally, the packaging material is glass, metal, paperboard, or films. Paperboard may also have a film layer. Films may be made from various density polymers or flexible metals of several layers, which support moisture control, migration of specific molecules, gas barriers, physical support, and package labeling and graphics.

There are no reported incidences of glass or metal packaging that alter DLC levels in the products they contain. Furthermore, there is no reason to suspect they would alter DLC levels given their composition and stability.

There have been incidences of chemicals migrating from paperboard packaging to foods (Cramer et al., 1991; Garattini et al., 1993; LaFleur et al., 1991; Ryan et al., 1992). Other researchers have attempted to predict this migration (Chung et al., 2002; Franz, 2002; Furst et al., 1989). The Dow Chemical Company has undertaken the development of analytical methodology capable of detecting the presence of 2,3,7,8-TCDD and 2,4,7,8-TCDF in low-density polyethylene matrices at concentrations between the range of 200 and 400 parts per quadrillion (Nestrick et al., 1991). Other analytical methods, currently in use, are discussed in Chapter 2.

While attention regarding human exposure to DLCs has concentrated on the ingestion of animal products as the most likely source of human exposure, food processing and packaging might play a minor role DLC contribution.

IMPORTED FOODS

Food imports to the United States have increased steadily in the last two decades (see Table 4-8). The average share of imports in U.S. food consumption has risen from 6.8 percent to 8.8 percent. Consumption of imported cereal, fruits, and vegetables has risen from 10 to 12 percent since the early 1980s, and imported animals product (including fish and seafood) consumption has risen from 3 to 4 percent (Jerardo, 2002). More reliable sources, reversed seasonality, improved shipping and storage technology, wider ethnic food preferences, and various economic factors have contributed to these trends.

As import shares increase, ensuring the safety of the U.S. food supply becomes more challenging. The U.S. government regulates and monitors, from farm to table, the production, processing, and transportation of foods produced in the United States. The targeted monitoring and regulation of the overseas production of foods destined for the United States, while theoretically possible, would

TABLE 4-8 Summary of Import Shares of U.S. Food Consumption (%)[a]

Food Groups	Years 1981–1985	1986–1990	1991–1995	1996	1997	1998	1999	2000
Total food consumption	6.8	7.3	7.4	8.1	8.5	8.8	8.8	8.8
Animal products[b]	3.2	3.4	3.2	3.2	3.2	4.0	4.2	4.2
Red meat	6.7	8.1	7.3	6.4	7.1	7.7	8.2	8.9
Dairy products	1.9	1.8	1.9	2.0	1.9	2.9	2.9	2.7
Fish and shellfish	50.9	56.0	56.0	58.5	62.1	64.7	68.1	68.3
Animal fat	0.5	0.7	1.4	1.4	2.3	2.3	2.5	2.8
Crops and products[c]	9.9	10.6	10.6	11.9	12.5	12.4	12.1	12.3
Fruits, juices and nuts	12.0	16.5	15.5	14.9	16.7	16.9	18.2	18.7
Vegetables	4.8	6.1	5.9	7.8	8.0	9.0	8.9	8.8
Vegetable oils	15.7	19.7	19.3	19.2	20.9	21.0	17.9	20.2
Grain cereals	1.6	3.1	6.7	7.2	7.0	7.4	6.5	6.3
Sweeteners and candy	19.8	9.8	9.1	14.8	14.8	10.4	8.5	8.0

[a]Calculated from units of weight, weight equivalents, or content.
[b]Import shares of poultry and eggs are included, but negligible. Red meats are estimated from carcass weights.
[c]Includes coffee, cocoa, and tea for which import shares are 100 percent. Also includes crop content of beer and wine.

DATA SOURCE: Jerardo (2002).

tax the system beyond its capacity. There is significant variation in DLC environmental contamination levels and agricultural, processing, and packaging practices among countries.

Some monitoring programs, such as FDA's Total Diet Study, may include some imported foods; however, the country of origin is not currently documented. As a result, current food safety laws require that imported foods meet the same safety standards as domestically produced foods, although it is difficult to speculate on the relative levels of contaminants in imported foods.

Mirroring the differences in their domestic control programs, FDA and USDA's Food Safety and Inspection Service (FSIS) rely on different systems to reach food safety goals. FDA relies primarily on physical inspection and chemical analysis of port-of-entry samples of a small proportion of imported foods, particularly from those countries where food safety systems do not meet U.S. standards. FSIS enforces the Federal Meat Inspection Act, the Poultry Products Inspection Act, and the Egg Products Inspection Act. These laws are applicable to domestic and imported products, which must meet the same standards for safety, wholesomeness, and labeling.

Meat, poultry, and egg products may be imported into the United States only from countries that FSIS has evaluated and found to have equivalent science-based systems of food inspection that include mandatory HACCP processing systems. The equivalence evaluation process has two parts: analysis of an application from a prospective exporting country followed by an on-site audit in the applicant country. Once a country is deemed to be equivalent, it may certify establishments for export to the United States. FSIS verifies the continuing equivalence of exporting countries through annual on-site audits of foreign inspection systems and daily port-of-entry inspections in the United States.

All shipments of imported meat, poultry and egg products are reinspected by FSIS at ports-of-entry. Each shipment is inspected for proper documentation and condition of containers. Additional types of inspection may be made on randomly selected individual lots of product, including product examinations for physical defects or laboratory analyses for chemical residues or microbiological contamination. Products that fail FSIS inspection are refused entry to the United States and must be reexported, reconditioned (if approved by FSIS), converted to non-human food use (if approved by FDA), or destroyed under FSIS supervision.

Absent a system for monitoring DLCs in imported foods, reduction of exposure from this food source will be difficult to achieve. Testing for DLCs, especially in animal products, must rely on the standard domestic surveillance programs. Since there are no current standards for allowable DLC levels in foods, rejection of those sampled foods that are determined to be high in DLCs can only be based on pesticide limits set in or on raw agricultural commodities. Other potential interventions are not applicable to imported foods.

SUMMARY

As stated in the introduction, DLCs are undesirable contaminants in the environment that serve no beneficial purpose and have a number of adverse biological effects in a wide variety of organisms, including humans. The focus of this chapter has been on identifying and describing DLC entry into agricultural pathways and subsequent exposure of the general U.S. population, as illustrated in Figure 4-1. Individuals or specially identified groups may be exposed to higher or lower DLC levels through alternative pathways, and it is possible that an unintended release or production of DLCs could result in high levels of contamination in any one of these pathways.

Terrestrial and aquatic animal management practices, animal feed formulations, and food processing and packaging present the primary potential intervention opportunities. Because of the reuse of animal products through feed manufacturing and the potential for bioconcentration of DLCs, animal feed practices may be especially important in reducing exposure to DLCs through foods.

REFERENCES

AAFCO (Association of American Feed Control Officials). 2002. *AAFCO Official Publication, 2002.* Oxford, IN: AAFCO.

Allsopp M, Erry B, Stringer R, Johnston P, Santillo D. 2000. *Recipe for Disaster: A Review of Persistent Organic Pollutants in Food.* Amsterdam: Greenpeace International.

Archer TE, Crosby DG. 1969. Removal of DDT and related chlorinated hydrocarbon residues from alfalfa hay. *J Agric Food Chem* 16:623–626.

Arctic Monitoring and Assessment Programme. 2001. *Persistent Organic Pollutants.* Online. Available at http://www.amap.no/assess/soaer6.htm. Accessed December 9, 2002.

Beyer WN, Connor EE, Gerould S. 1994. Estimates of soil ingestion by wildlife. *J Wildl Manage* 58:375–382.

Cameron ND, Enser MB. 1991. Fatty acid composition of lipid in longissimus dorsi muscle of Duroc and British Landrace pigs and its relationship with eating quality. *Meat Sci* 29:295–307.

Chung D, Papadakis SE, Yam KL. 2002. Simple models for assessing migration from food-packaging films. *Food Addit Contam* 19:611–617.

Commoner B, Richardson J, Cohen M, Flack S, Bartlett PW, Cooney P, Couchot K, Eisl H, Hill C. 1998. Dioxin sources, air transport and contamination in dairy feed crops and milk. Flushing, NY: Center for Biology of Natural Systems, Queens College.

Cramer G, Bolger M, Henry S, Lorentzen R. 1991. USFDA assessment of exposure to 2,3,7,8-TDCC and 2,3,7,8-TCDF from foods contacting bleached paper products. *Chemosphere* 23:1537–1550.

CVM (Center for Veterinary Medicine). 2000. *Preliminary National Survey of Dioxin-like Compounds in Animal Fats, Animal Meats, Oilseed Deodorizer Distillates, and Molasses.* Online. U.S. Food and Drug Administration (FDA). Available at http://www.fda.gov/cvm/index/dioxin/dioxin_survey.html. Accessed January 15, 2003.

CVM. 2001. *Preliminary National Survey of Dioxin-like Compounds in Oilseed Meals, Fat-soluble Vitamins, Complete Feeds, Milk Products, Minerals, and Wood Products.* Online. FDA. Available at http://www.fda.gov/cvm/index/dioxin/dioxin_survey1.html. Accessed January 15, 2003.

CVM. 2002. *Contaminated Animal Feed Supplements Recalled.* Online. FDA. Available at http://www.fda.gov/cvm/index/updates/dioxinup.htm. Accessed May 5, 2003.

EC (European Commission). 1999. *Mission Report on a Mission Carried Out in Brazil from 11 to 15 January 1999 Concerning the Organisation of Official Inspections in the Field of Animal Nutrition: Dioxin Contamination of Citrus Pulp Pellets.* Online. Available at http://europa.eu.int/comm/food/fs/inspections/fnaoi/reports/contaminants/brazil/fnaoi_rep_braz_1005-1999_en.pdf. Accessed September 25, 2002.

EPA (U.S. Environmental Protection Agency). 2000. *Exposure and Human Health Reassessment of 2,3,7,8-Tetrachlorodibenzo-p-Dioxin (TCDD) and Related Compounds.* Draft Final Report. Washington, DC: EPA.

FDA (U.S. Food and Drug Administration). 1997. *FDA Stops Distribution of Some Eggs and Catfish Because of Dioxin-Contaminated Animal Feed.* Online. HHS News. Available at http://www.fda.gov/bbs/topics/NEWS/NEW00574.html. Accessed January 9, 2003.

Ferrario J, Lovell R, Gardner P, Lorber M, Winters D, Byrne C. 2002. *Analysis of Animal- and Plant-Derived Feed Ingredients for Dioxin-like Compounds.* Abstract presented at Dioxin 2002 meeting, Barcelona, Spain, August 11–16.

Franz R. 2002. Programme on the recyclability of food-packaging materials with respect to food safety considerations: Polyethylene terephthalate (PET), paper and board, and plastics covered by functional barriers. *Food Addit Contam* 19:93–110.

Fries GF. 1995a. A review of the significance of animal food products as potential pathways of human exposures to dioxins. *J Anim Sci* 73:1639–1650.

Fries GF. 1995b. Transport of organic environmental contaminants to animal products. *Rev Environ Contam Toxicol* 141:71–109.

Fries GF, Paustenbach DJ. 1990. Evaluation of potential transmission of 2,3,7,8-tetrachlorodibenzo-*p*-dioxin-contaminated incinerator emissions to humans via foods. *J Toxicol Environ Health* 29:1–43.

Fries GF, Marrow GS, Snow PA. 1982. Soil ingestion by dairy cattle. *J Dairy Sci* 65:611–618.

FSIS (Food Safety and Inspection Service). 1997. *Advisory Letter to: Owners and Custodians of Poultry, Livestock and Eggs.* Online. U.S. Department of Agriculture (USDA). Available at http://www.fsis.usda.gov/OA/topics/dioxinlt.htm. Accessed January 8, 2003.

FSIS. 2002. *FSIS Dioxin Survey for Meat and Poultry.* Online. USDA. Available at http://www.fsis.usda.gov/OA/background/dioxins02.htm. Accessed January 8, 2003.

Furst P, Kruger C, Meemken H-A, Groebel W. 1989. Interaction between sample and packaging material—A potential source of contamination with PCDDs and PCDFs. *Chemosphere* 18:891–896.

Garattini S, Mariani G, Benfenati E, Fanelli R. 1993. Preliminary survey on 2,3,7,8-TCDD in cellulose-containing consumer products on the Italian market. *Chemosphere* 27:1561–1564.

Greenpeace. 1999a. *Solvay Chemical Plant in Brazil Identified as Source of European Food Contamination.* Online. Available at http://archive.greenpeace.org/pressreleases/toxics/1999mar25.html. Accessed January 8, 2003.

Greenpeace. 1999b. *Solvay Investigated for Dioxin Contamination.* Online. Available at http://archive.greenpeace.org/pressreleases/toxics/1999jul13.html. Accessed January 8, 2003.

Hashimoto S, Ito H, Morita M. 1992. Elution of polychlorinated dibenzo-para-dioxins and dibenzofurans from coffee filter papers. *Chemosphere* 25:297–305.

Hayward DG, Nortrup D, Gardner A, Clower M Jr. 1999. Elevated TCDD in chicken eggs and farm-raised catfish fed a diet with ball clay from a southern United States mine. *Environ Res* 81:248–256.

Hulster A, Marschner H. 1993. Transfer of PCDD/PCDF from contaminated soils to food and fodder crop plants. *Chemosphere* 27:439–446.

Jerardo A. 2002. *The Import Share of U.S.-Consumed Food Continues to Rise.* Electronic Outlook Report from the Economic Research Service. FAU-66-01. Washington, DC: USDA.

LaFleur T, Bosquet T, Ramage K, Davis T, Mark M, Lorusso D, Woodrow D, Saldana T. 1991. Migration of 2,3,7,8-TCDD and 2,3,7,8-TCDF from paper based food-packaging and food contact products. *Chemosphere* 23:1575–1579.

Larsen BR, Facchetti S. 1989. Effects of long-term freezing on the extraction of fat and 2,3,7,8-TCDD from milk. *Chemosphere* 19:1017–1018.

Lorber M, Cleverly D, Schaum J, Phillips L, Schweer G, Leighton T. 1994. Development and validation of an air-to-beef food chain model for dioxin-like compounds. *Sci Total Environ* 156:39–65.

McLachlan MS, Thoma H, Ressinger M, Hutzinger O. 1990. PCDD/Fs in an agricultural food chain. Part 1: PCDD/F mass balance of a lactating cow. *Chemosphere* 20:1013–1020.

NASS (National Agricultural Statistics Service). 1999. *1997 Census of Agriculture.* Online. USDA. Available at http://www.nass.usda.gov/census/census97/atlas97/index.htm. Accessed December 19, 2002.

Nestrick TJ, Lamparski LL, Cramm RH. 1991. Procedure and initial results for the determination of low part per quadrillion levels of 2,3,7,8-tetrachlorodibenzo-*p*-dioxin and 2,3,7,8-tetrachlorodibenzofuran in polyethylene. *Chemosphere* 22:215–228.

Pearl GG. 2002. Tech topics: Bioenergy 2002. *Render Mag* Sept:48–51.

Petroske E, Zaylskie RG, Feil VJ. 1998. Reduction in polychlorinated dibenzodioxin and dibenzofuran residues in hamburger meat during cooking. *J Agric Food Chem* 46:3280–3284.

PROMAR International. 1999. *Profile of the Recycled Grease Market.* Report for the USB Domestic Marketing Committee. Alexandria, VA: PROMAR International.

Reuters. 1998, July 17. EU recommends dioxin ceilings on citrus pulp. Brussels Newsroom.

Robinson EH, Li MH, Manning BB. 2001. *A Practical Guide to Nutrition, Feeds, and Feeding of Catfish*. Bulletin 113. Mississippi State, MS: Office of Agricultural Communications, Mississippi State University.

Roeder RA, Garber MJ, Schelling GT. 1998. Assessment of dioxins in foods from animal origins. *J Anim Sci* 76:142–151.

Ryan JJ, Shewchuck C, Lau BPY, Sun WF. 1992. Polychlorinated dibenzo-para-dioxins and polychlorinated dibenzofurans in Canadian bleached paperboard milk containers (1988–1989) and their transfer to fluid milk. *J Agric Food Chem* 40:919–923.

Schuler F, Schmid P, Schlatter C. 1997. The transfer of polychlorinated dibenzo-*p*-dioxins and dibenzofurans from soil into eggs of foraging chicken. *Chemosphere* 34:711–718.

Shui WY, Doucette W, Gobas FAPC, Andren A, Mackay D. 1988. Physical-chemical properties of chlorinated dibenzo-*p*-dioxins. *Environ Sci Technol* 22:651–658.

Stachiw NC, Zabik ME, Boorem AM, Zabik MJ. 1988. Tetrachlorodibenzo-*p*-dioxin residue reduction through cooking/processing of restructured carp fillets. *J Agric Food Chem* 36:848–852.

Trapp S, Matthies M. 1997. Modeling volatilization of PCDD from soil and uptake into vegetation. *Environ Sci Technology* 31:71–74.

Travis CC, Hattemer-Frey HA. 1991. Human exposure to dioxin. *Sci Total Environ* 104:97–127.

Tyler R. 1999. *Dioxin Contamination Scandal Hits Belgium: Effects Spread through European Union and Beyond*. Online. Available at http://www.wsws.org/articles/1999/jun1999/belg-j08_prn.shtml. Accessed January 8, 2003.

Tyson KS, ed. 1998. *Biodiesel Research Progress 1992–1997*. NREL/TP-580-24433. Golden, CO: National Renewable Energy Laboratory, U.S. Department of Energy.

van Larebeke N, Hens L, Schepens P, Covaci A, Baeyens J, Everaert K, Bernheim JL, Vlietinck R, De Poorter G. 2001. The Belgian PCB and dioxin incident of January–June 1999: Exposure data and potential impact on health. *Environ Health Perspect* 109:265–273.

Vesterby M, Krupa KS. 2001. *Major Uses of Land in the United States, 1997*. Statistical Bulletin No. 973. Washington, DC: Resources Economics Division, Economic Research Service, USDA.

Ward CT, Matsumura F. 1978. Fate of 2,3,7,8-tetrachlorodibenzo-*p*-dioxin (TCDD) in a model aquatic environment. *Arch Environ Contam Toxicol* 7:349–357.

Wiltsee G. 1998. *Urban Waste Grease Resource Assessment*. NREL/SR-570-26141. Report prepared by Appel Consultants, Inc. Golden, CO: National Renewable Energy Laboratory, U.S. Department of Energy.

Zabik ME, Zabik MJ. 1999. Polychlorinated biphenyls, polybrominated biphenyls, and dioxin reduction during processing/cooking food. *Adv Exp Med Biol* 459:213–231.

5

Human Foods and
Food-Consumption Patterns

Food is the major route for human exposure to dioxins and dioxin-like compounds (referred to collectively as DLCs). Most DLCs in foods are contained in the lipid component of foods of animal origin. While DLC exposure through fruits, vegetables, and grains also occurs, it is thought to result primarily from the adhesion of soil to plant material. DLCs have relatively long half-lives in the human body, and the rate of elimination of these compounds from the body is inversely related to age and increased adiposity (Kreuzer et al., 1997), presumably on the basis of the age-associated increase in adipose tissue depots. Thus, the overall body burden tends to increase over time, even if exposure levels do not change.

The ability to estimate dietary exposure and the within-population variability in exposure to DLCs is limited by the availability of appropriate data. The most recent National Health and Nutrition Examination Survey (NHANES) included testing of serum dioxin levels on a subsample of individuals and dietary measures on all participants, which should provide a clearer picture of exposures. (Dioxin measurements were added to this survey in 1999–2000, and the first data were released in January 2003, too late for inclusion in this report.) At this writing, the best available approach to estimating DLC exposure through diet is to combine information from two sources: the U.S. Department of Health and Human Services, Food and Drug Administration's (FDA) Total Diet Study (TDS), which has been accumulating information on the DLC content of selected commonly consumed food items since 1999, and the national food-consumption monitoring surveys that are conducted repeatedly by the U.S. Department of Agriculture (USDA). The consumption data selected for this report were based on actual intakes rather than on disappearance data in order to obtain a more accurate perspective when the data was applied to exposure estimates. The committee commissioned an analysis based on these data; the analysis is discussed later in this chapter and in Appendix B. Aspects of current dietary patterns in the United States with regard to fat intake and to consumption of products that contribute the largest share of exposure to DLCs (i.e., meats, poultry, dairy products, and fish) are reviewed briefly below.

DIETARY-INTAKE PATTERNS

Overall, American adults derive about 34 percent of their dietary energy from fat and one-fourth to one-third of that is from saturated fat, which is largely animal fat. USDA survey data (1977–1978, 1989–1991, and 1994–1996) show that the average adult intake of fat in total grams increased slightly (from an average of 78.5 g/d to 83.1 g/d) between the 1989–1991 and 1994–1996 surveys (although the percent of dietary energy derived from fat decreased from 34.9 percent to 32.9 percent, a function of an overall increase in dietary energy intake). Saturated fat intake was more stable over the same time period at an average of 27.6 to 28.3 g/d (although again, the percent of dietary energy from saturated fat decreased). Fat intakes were substantially higher for men than for women (Kennedy et al., 1999).

The stability of the average absolute intakes, however, masks real changes in the food sources of fat over the same time period. A steady substitution of low-fat and skim milk for whole milk and yogurt occurred between 1970 and 1990, but consumption of cheese and of cream products (including sour cream, half and half, cream-based dips, and light cream) increased (Enns et al., 1997).

A trend toward more poultry and fish consumption and less red-meat consumption occurred over the same time period. Based on national food supply data (Putnam and Allshouse, 1993), the U.S. population per capita consumed an average of 18 lb less red meat, 26 lb more poultry, and 3 lb more fish and shellfish in 1992 than in 1970.

Adults (both men and women) consume more meat and fish than children, but children are major consumers of dairy products. On average, children consume almost twice as much milk and dairy products as do adults. Fish consumption is of particular interest for several reasons. One is that fish consumption has been increasing in the population, and this trend is generally a good one from a health standpoint. However, the level of DLCs in fish is variable; fish from contaminated waterways and those that receive feed that contains DLCs have higher DLC levels. Furthermore, there are relatively small subgroups of the population whose reliance on fish may be more important than that for the general population. These subgroups have the potential to have substantially higher DLC exposures than the general population.

American Indian groups in traditional fishing cultures (primarily in the Northeast and northern Midwest) and Alaska Native and northern Canadian groups (Northern Dwelling populations) that traditionally consume large amounts of fish and marine mammals, and rural subsistence and sport fishers in various parts of the country, are potentially at risk. A number of rivers and lakes that are important for fishing, including the Great Lakes, are near or downstream from industrial sources of contamination. While data on dietary exposures to DLCs in these groups are limited, a few studies are relevant.

Outside of American Indian groups, there is a paucity of data on the intakes of subsistence or sport fishers. Frequent consumption of game fish is of concern because many of these fish are particularly high sources of DLCs (Anderson et al., 1998). Subsistence fishing is a cultural tradition in many African-American families and in Southeast-Asian immigrants.

DLC Analyses of Foods Consumed

Due primarily to the high cost per sample to analyze DLCs, there is relatively little information on the DLC content of the U.S. food supply. Recently however, FDA conducted DLC analyses for selected foods collected as market basket samples in the 1999–2000 and 2001 TDS. The TDS survey samples foods purchased by FDA personnel from supermarkets or grocery stores four times per year; one sample is collected from each of three cities from each of four geographic regions. Each market basket contains similar foods purchased in the three cities in a given region. The market basket food samples are then prepared as for consumption, and the three samples of a food from a region are composited into a single sample, prepared in duplicate, and analyzed for DLCs. Representative ranges of DLC content in foods from a portion of the 2001 analysis are presented in Table 5-1 (the complete analysis is presented in Appendix B). The ranges on the table reflect the estimates of the absolute DLC levels in foods—not exposure levels, which are determined by the amount of food consumed. Furthermore, the analysis was based on whole food or wet weight rather than on a lipid basis in order to maintain consistency with consumption data and to preserve data quality, although inclusion of the lipid content might have been useful in comparisons among reports.

Table 5-1 shows that DLCs may be found in all foods, with a range of content within each food group. However, only in animal products, specifically red meat, fish, and dairy products, do concentrations reach more than 0.10 ppt. In general, higher DLC concentrations are associated with higher fat contents, particularly with animal fats. Foods representing the upper levels of these ranges include high-fat cuts of beef, bacon, frankfurters, full-fat cheeses, fatty fish (e.g., salmon), and butter. On the other hand, within each of these categories, foods containing relatively low levels of DLCs can be selected, including lean steak (0.03 ppt), lean ham (0.04 ppt), cottage cheese (0.03 ppt), shrimp (0.06 ppt), and

TABLE 5-1 Dioxin and Dioxin-like Compound (DLC) Content Per Gram in Selected Food Groups from the 2001 Total Diet Study

Food Group	DLC Content (ppt toxicity equivalents)
Meat, fish, poultry, eggs	
Beef, pork, lamb	0.005–0.46
Processed meats	0.01–0.21
Fish, shellfish	0.01–0.33
Poultry	0.004–0.06
Eggs	0.01–0.05
Dairy foods	
Cheese	0.002–0.24
Cream/ice cream	0.0001–0.06
Milk	0.0006–0.01
Fats, oils, nuts	
Butter	0.22
Vegetable oils	0.002–0.06
Nuts	0.003–0.006
Breads, cereals	
Breads	0.001–0.05
Crackers	0.001–0.02
Pasta	0.0001–0.02
Breakfast cereal	0.0007–0.01
Fruits, vegetables	
Fruit	0.0007–0.01
Vegetables	0.0001–0.05

NOTE: Mean toxicity equivalents at nondetects = 0.
SOURCE: Douglass and Murphy (2002).

margarine (0.03 ppt). Fruit, vegetable, and grain products have considerably lower concentrations of DLCs, but because large volumes are consumed, these foods also contribute to total DLC exposure (Douglass and Murphy, 2002).

USDA's Agricultural Research Service conducts a nationally representative survey, formerly known as the Continuing Survey of Food Intakes by Individuals (CSFII), which has now been merged with NHANES. This survey provides important information about the types of foods Americans are consuming and the quantities consumed (NCHS, 2003).

The 1994–1996 CSFII was conducted between January 1994 and January 1997. In each of the three survey years, data were collected from a nationally representative sample of noninstitutionalized individuals of all ages in the United States. The 1998 CSFII was a survey of children ages 0 through 9 years. It was conducted using the same sampling and dietary intake methodologies as the 1994–1996 CSFII so that the survey results could be merged to increase the total sample size for children. In both surveys, dietary intakes were collected through in-person interviews using 24-hour recalls on two nonconsecutive days, approximately one week apart. A total of 21,662 individuals provided data for the first day; of those individuals, 20,607 provided data for the second day. Each dietary recall included a record of all foods and beverages consumed in the previous 24 hours, including the gram weight of each food or beverage consumed.

Data obtained from the sequential CSFIIs over the past 30 years indicate a trend in food consumption toward a lower total saturated fat intake and a lower total fat intake as a percent of energy consumed, although with a higher total energy intake. A summary of 1994–1996 CSFII intake data for foods that are most likely to contain high amounts of DLCs is presented in Figures 5-1 through 5-3. The figures demonstrate that differing age and sex groups tend to be exposed to DLCs through intakes of foods from different food groups. For example, male adolescents and men consume considerably more beef than women, with little consumed by children under 6 years of age. More fish is consumed by adults than by children, and more by men than by women. On the other hand, the average milk and dairy consumption is highest for young children; adults consume considerably less.

The committee commissioned an analysis (see Appendix B) to estimate exposures to DLCs from the U.S. food supply for various age and sex groups. The analysis utilized data from the TDS for samples collected in 2001 (which were

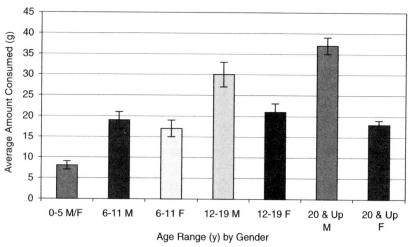

FIGURE 5-1 Beef consumption by age and sex, 1-day average intake, 1994–1996. Values represent population intake averages of absolute amounts of beef consumed per individual for 1 day. Error bars are standard errors of the mean. M = male, F = female. SOURCE: Food Surveys Research Group (1997).

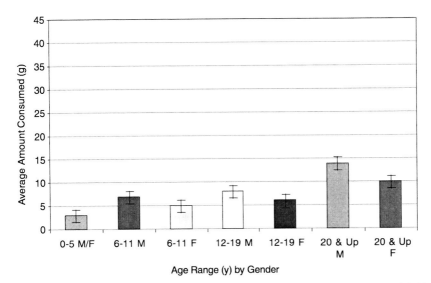

FIGURE 5-2 Fish consumption by age and sex, 1-day average intake, 1994–1996. Values represent population intake averages of absolute amounts of fish consumed per individual for 1 day. Error bars are standard errors of the mean. M = male, F = female.
SOURCE: Food Surveys Research Group (1997).

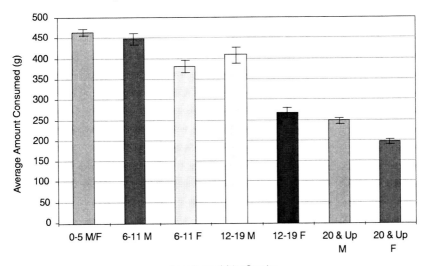

FIGURE 5-3 Dairy consumption by age and sex, 1-day average intake, 1994–1996. Values represent population intake averages of absolute amounts of milk and dairy foods consumed per individual for 1 day. Error bars are standard errors of the mean. M = male, F = female.
SOURCE: Food Surveys Research Group (1997).

Errata

Replace page 115 with the new page on the other side of this sheet.

Figures 5-2 and 5-3 are reversed in the book and have been correctly positioned on the errata page.

Dioxins and Dioxin-like Compounds in the Food Supply: Strategies to Decrease Exposure
The National Academies Press, Washington, D.C. 2003
ISBN 0-309-08961-1

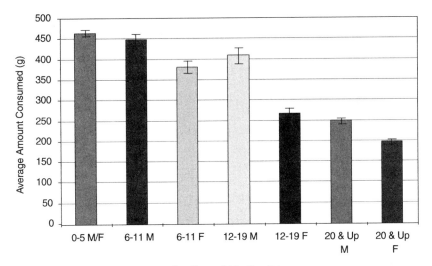

FIGURE 5-2 Fish consumption by age and sex, 1-day average intake, 1994–1996. Values represent population intake averages of absolute amounts of fish consumed per individual for 1 day. Error bars are standard errors of the mean. M = male, F = female.
SOURCE: Food Surveys Research Group (1997).

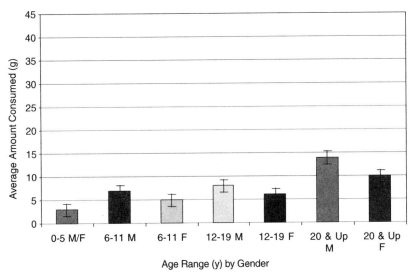

FIGURE 5-3 Dairy consumption by age and sex, 1-day average intake, 1994–1996. Values represent population intake averages of absolute amounts of milk and dairy foods consumed per individual for 1 day. Error bars are standard errors of the mean. M = male, F = female.
SOURCE: Food Surveys Research Group (1997).

analyzed for 17 dioxin congeners). The TDS data were linked to intake data from the 1994–1996 and 1998 CSFII to estimate dietary exposure to DLCs, through foods actually consumed. The TDS weighting factors developed for use in estimating exposure based on analyte concentrations in TDS foods are based on results of the 1987–1988 USDA Nationwide Food Consumption Survey, and therefore are not likely to reflect current patterns of intake. Therefore, the analysis commissioned by the committee used results from the 1994–1996, 1998 CSFII.

Because relatively few foods were sampled for the TDS relative to the thousands of foods in the CSFII database, CSFII foods were grouped to estimate DLC content from the most similar analyzed food. This method generates substantial errors for individual intake estimates, but because FDA sampled the foods it considered most representative of the U.S. diet, the results should provide reasonable average DLC intake estimates. These limitations must, however, be taken into consideration when interpreting the results.

The results of this merged analysis (presented in Figures 5-4 through 5-11) allow a view of the relative contribution, within the limitations of the data, to total DLC intake from various food groups based on actual U.S. dietary intake. The analysis did not utilize "usual intakes" (commonly used for nutrient intake levels from foods) because of the extensive variability of levels of contaminants (e.g., DLCs) in foods.

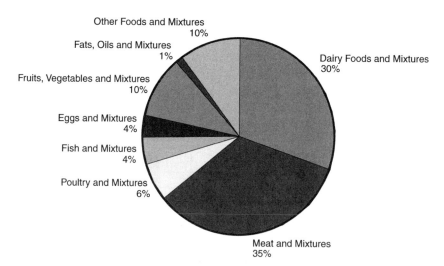

FIGURE 5-4 Estimated percent contribution of foods to intakes of dioxins and dioxin-like compounds (DLCs), children ages 1 through 5 years. Nondetects = 0. Percentages were calculated based on mean 2-day average DLC intake/kg of body weight.

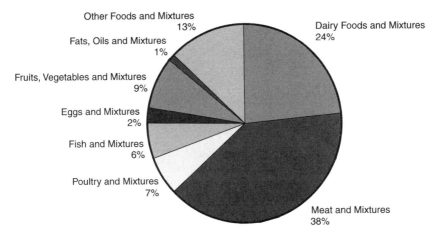

FIGURE 5-5 Estimated percent contribution of foods to intakes of dioxins and dioxin-like compounds (DLCs), children ages 6 through 11 years. Nondetects = 0. Percentages were calculated based on mean 2-day average DLC intake/kg of body weight.

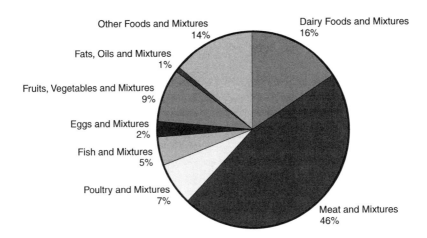

FIGURE 5-6 Estimated percent contribution of foods to intakes of dioxins and dioxin-like compounds (DLCs), boys ages 12 through 19 years. Nondetects = 0. Percentages were calculated based on mean 2-day average DLC intake/kg of body weight.

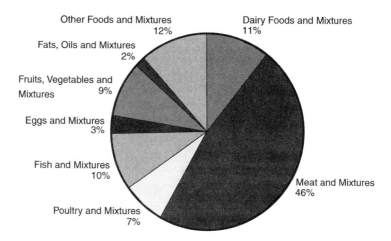

FIGURE 5-7 Estimated percent contribution of foods to intakes of dioxins and dioxin-like compounds (DLCs), men ages 20 years and older. Nondetects = 0. Percentages were calculated based on mean 2-day average DLC intake/kg of body weight.

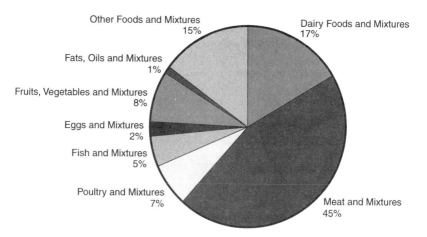

FIGURE 5-8 Estimated percent contribution of foods to intakes of dioxins and dioxin-like compounds (DLCs), girls ages 12 through 19 years. Nondetects = 0. Percentages were calculated based on mean 2-day average DLC intake/kg of body weight.

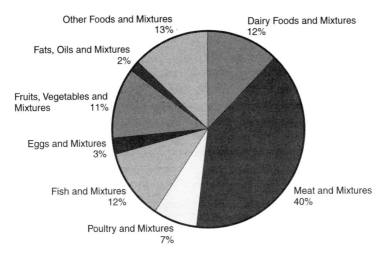

FIGURE 5-9 Estimated percent contribution of foods to intakes of dioxins and dioxin-like compounds (DLCs), women ages 20 years and older. Nondetects = 0. Percentages were calculated based on mean 2-day average DLC intake/kg of body weight.

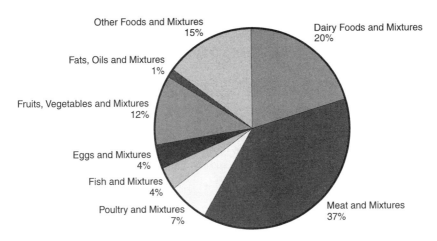

FIGURE 5-10 Estimated percent contribution of foods to intakes of dioxins and dioxin-like compounds (DLCs), pregnant and/or lactating females ages 12 years and older. Nondetects = 0. Percentages were calculated based on mean 2-day average DLC intake/kg of body weight.

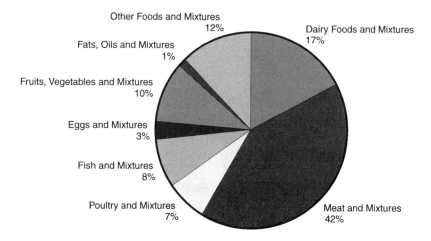

FIGURE 5-11 Estimated percent contribution of foods to intakes of dioxins and dioxin-like compounds (DLCs), males and females ages 1 year and older. Nondetects = 0. Percentages were calculated based on mean 2-day average DLC intake/kg of body weight.

Because it was not possible to more closely define the food categories, the "fruits, vegetables and mixtures" and "other foods and mixtures" categories contain an array of items. Foods that fell within the general category of fruits and vegetables included raw and cooked vegetables and fruits, along with vegetable items prepared as mixtures that contained some milk, eggs, cheese, and meat. The others foods and mixtures category included food combinations and mixtures, some of which were made with animal food products. The imputation of foods in these two groups to data from the CSFII also required making generalizations about specific intakes.

The data in the figures are relative exposures, with DLC nondetect levels assumed to be zero. The committee chose this reference value because it felt that, under these circumstances, this value gave the most accurate representation of the DLC intake. The common alternative methods (assuming nondetects = ½ of the limit of detection or nondetects = the limit of detection) are more conservative, but they tend to artifactually indicate a greater contribution of substances known to be low in DLCs (grains, fruits, and vegetables) because of their relative roles in the total diet.

The results of the analysis, discounting the other foods and mixtures category, suggest that DLC exposure levels to the general adult population from animal-food products and based on estimated amounts consumed by population groups, are greatest from meat, followed by dairy foods and fish. The estimates for dairy foods do not take into account the consumption of low-fat and skim milk because only whole milk was analyzed. The dairy foods estimates do, however, include cheese and other high-fat dairy products, except for butter, which is included under fats, oils and mixtures. A discussion of DLC-intake estimation scenarios for skim versus whole-milk consumption appears later in this chapter (supporting data is provided in Appendix B).

Because DLC content is strongly associated with the fat content of animal products, individuals who select lean cuts of beef, fish, and low-fat dairy products (including low fat cheeses) and who utilize lean animal products in combination foods will have considerably less exposure to DLCs than those who choose higher-fat versions of these foods.

Table 5-2 shows the estimated exposure to DLCs through foods for all age and sex groups combined, that is, males and females ages 1 year and older. When age groups are combined, fish and mixtures move down in ranking because young children consume very little fish. This combined ranking, however, supports the finding that most DLC intake comes from animal-based foods, particularly meat (other than lean cuts) and full-fat dairy foods. As previously stated, DLCs are found primarily in the fat component of these foods. Lower-fat, animal-based foods may be consumed in place of high-fat selections to reduce DLC exposure and not alter nutrient intake, except for energy from fat.

Dietary Intake Scenario for Meat, Poultry, and Fish

The commissioned analysis also included an estimation scenario of DLC exposure through food that compared low (less than 1 oz) meat, poultry, and fish consumption with high (greater than 1 oz) consumption for each age and sex

TABLE 5-2 Ranked Dioxin and Dioxin-like Compound (DLC) Sources from Foods Consumed Among All Age Groups in the U.S. Diet

Food Category	Percent Estimated Per Capita Contribution of Foods to DLC Intake[a]
Meat and mixtures	42
Dairy foods and mixtures	17
Other foods and mixtures	12
Fruits, vegetables and mixtures	10
Fish and mixtures	8
Poultry and mixtures	7
Eggs and mixtures	3
Fats, oils and mixtures	1

[a]Percentages represent 50th percentile intakes at nondetects = 0.

SOURCE: Douglass and Murphy (2002).

TABLE 5-3 Estimated Intakes of Dioxin and Dioxin-like Compounds (DLCs) by Consumers of High versus Low Amounts of Meat, Poultry, and Fish

Population	Consumption of Meat, Poultry, and Fish	n	DLC Intake (pg TEQ[a]/kg body weight/d)			Dietary Comparison Measures		
			Nondetects[b] = 0	Nondetects = 0.5 (LOD)[c]	Nondetects = LOD	Energy (kcal/d)	Protein (g/d)	Total Fat (g/d)
Males and females, 1–5 y, not breastfeeding	High	5,306	1.17	1.76	2.35	1,554	56.6	57.1
	Low	1,035	0.76	1.26	1.77	1,241	39.7	42.3
Males and females, 6–11 y	High	1,735	0.72	1.14	1.55	1,906	67.3	69.6
	Low	166	0.39	0.77	1.15	1,576	45.9	51.2
Males, 12–19 y	High	675	0.53	0.89	1.25	2,737	99.4	101.9
	Low	18	0.18	0.47	0.77	2,119	63.2	69.3
Males, 20+ y	High	4,608	0.42	0.69	0.96	2,422	95.5	92.1
	Low	110	0.19	0.40	0.61	1,772	51.4	56.0
Females, 12–19 y, not pregnant or lactating	High	615	0.39	0.69	1.00	1,884	66.7	68.9
	Low	71	0.17	0.41	0.66	1,435	40.2	45.6
Females, 20+ y, not pregnant or lactating	High	4,128	0.34	0.59	0.83	1,620	63.8	60.0
	Low	281	0.16	0.38	0.60	1,303	37.7	40.2
Females, pregnant or lactating	High	108	0.38	0.65	0.91	2,092	78.7	76.7
	Low	3	0.29	0.54	0.79	2,014	62.5	69.7
Males and females, 1+ y, includes pregnant and/or lactating women	High	17,175	0.47	0.78	1.08	2,015	77.5	75.4
	Low	1,684	0.33	0.64	0.95	1,420	42.0	45.5

[a]TEQ = toxicity equivalents.
[b]Reflects treatment of samples for which no dioxin congener was detected.
[c]LOD = limit of detection.

NOTE: Data represent 2-day averages generated using U.S. Department of Agriculture (USDA) sample weights. Low combined meat, poultry, and fish consumption defined as 2-day average intake less than 1 oz (28 g).

DATA SOURCE: Dioxin concentrations: U.S. Food and Drug Administration Total Diet Study (1999–2001); energy, protein, and fat intakes: 1994–1996, 1998 group. The results are presented in Table 5-3 and supporting data is provided in Appendix B. The intake difference for each group can be summarized as a percentage comparison of low with high toxicity equivalents (TEQ) intake. Using the assumption that nondetects = 0, for adolescent (12–19 years old) and adult (20+ years old) males, DLC-intake exposure for low-meat consumers is 34 and 45 percent, respectively, of that for high-meat consumers. For nonpregnant adolescent and adult females, DLC-intake exposure for low-meat consumers is 44 and 47 percent, respectively, of that for high-meat consumers.

These results suggest that reducing the consumption of foods that contain animal fats can reduce DLC intake exposure. Although the analysis did not compare trimmed with untrimmed meats, poultry, and fish, it is fair to assume that trimming and discarding their excess fats will reduce DLC intake. There is an array of nutrients that can be obtained from these foods, including iron, niacin, vitamin B_{12}, and protein; lean cuts of meat and poultry will provide levels of these nutrients that are comparable to untrimmed versions while reducing intake exposure to DLCs.

Dietary Intake Scenario for Skim versus Whole Milk

Another scenario produced in the commissioned analysis was an estimation of exposure to DLCs through the consumption of skim milk compared with whole milk (3.5 percent fat). This was an area of interest to the committee because the subgroups within the population that are high-milk consumers are young children and girls who have not entered their childbearing years, and whole milk and full-fat dairy products are a source of DLCs. The intake estimation scenario compares the expected levels of DLC intake exposure for (1) all foods if whole milk versus skim milk is consumed, and (2) dairy foods and mixtures if whole milk versus skim milk is consumed.

The results are presented in Tables 5-4 through 5-6. Using the assumption that nondetects = 0, a mean reduction in DLC intake exposure of at least 60 percent from dairy foods and mixtures can be predicted if skim rather than whole milk is consumed. However, when dairy intake is compared with intake from all foods, the mean reduction is approximately 10 to 20 percent. The results of this estimation scenario suggest that DLC intake exposure can be reduced by consuming skim rather that whole milk, but the reduction would be more significant if combined with a reduction in intake of other sources of DLC exposure, such as animal fats from meat, poultry, and fish.

TABLE 5-4 Estimated Intake of Dioxins and Dioxin-like Compounds (DLCs) from Food by Boys and Girls, 1–5 Years Old (*n* = 6,409)

Food Category	Percent of Population Group Consuming	Food Intake (g/kg body weight/d)	Consumers' Dioxin Toxicity Equivalents Intake (pg/kg body weight/d)											
			Nondetects[a] = 0				Nondetects = 0.5 (LOD)[b]				Nondetects = LOD			
				Percentile				Percentile				Percentile		
			Mean	10	50	90	Mean	10	50	90	Mean	10	50	90
All foods	100.0	90.5	1.09	0.43	0.93	1.89	1.67	0.85	1.50	2.63	2.24	1.18	2.04	3.51
If skim milk consumed[c]	100.0	90.5	0.85	0.25	0.69	1.61	1.42	0.64	1.24	2.34	1.98	0.93	1.78	3.21
Dairy foods and mixtures	98.3	29.5	0.34	0.08	0.27	0.67	0.42	0.12	0.34	0.81	0.50	0.14	0.41	0.94
If skim milk consumed[c]	98.3	29.5	0.10	0.01	0.04	0.24	0.16	0.01	0.09	0.40	0.23	0.01	0.13	0.56
Meat and mixtures	80.3	4.7	0.46	0.03	0.30	1.02	0.51	0.06	0.35	1.09	0.56	0.08	0.39	1.19
Poultry and mixtures	61.1	3.6	0.11	0.01	0.05	0.31	0.13	0.02	0.08	0.33	0.16	0.03	0.10	0.36
Fish and mixtures	17.0	2.8	0.28	—[d]	0.05	0.89	0.33	0.03	0.12	0.90	0.38	0.04	0.18	0.93
Eggs and mixtures	33.2	2.7	0.12	0.02	0.10	0.24	0.13	0.04	0.10	0.24	0.13	0.05	0.11	0.25
Fruits, vegetables, and mixtures	98.9	21.9	0.11	—	0.05	0.31	0.33	0.05	0.25	0.70	0.54	0.09	0.40	1.16
Fats, oils, and mixtures	59.8	0.4	0.02	—	—	0.06	0.02	—	0.01	0.06	0.03	—	0.01	0.07
Other foods and mixtures[e]	99.9	32.3	0.11	0.02	0.08	0.25	0.33	0.11	0.27	0.62	0.55	0.18	0.45	1.05

[a]Reflects treatment of samples for which no dioxin congener was detected.

[b]LOD = limit of detection.

[c]Dioxin TEQ intake from milk assuming that all plain milk consumed is skim milk rather than whole milk, as was otherwise assumed due to lack of analytical data on milks other than whole milk. Dioxin TEQ for skim milk was estimated assuming that all dioxin congeners concentrate in milk fat, that whole milk contains 3.34% fat, and that skim milk contains 0.18% fat (Nutrient Data Laboratory, 2002).

[d]< 0.0005 pg/kg body weight/d.

[e]Grains and mixtures, legumes and mixtures, beverages (other than milk and juice), candy.

NOTE: Data represent 2-day averages generated using U.S. Department of Agriculture (USDA) sample weights. Breastfeeding children are excluded.

DATA SOURCE: Dioxin concentrations: U.S. Food and Drug Administration Total Diet Study (1999–2001), food consumption: 1994–1996, 1998 USDA Continuing Survey of Food Intakes by Individuals.

TABLE 5-5 Estimated Intake of Dioxins and Dioxin-like Compounds (DLCs) from Food by Boys and Girls, 6–11 Years Old (n = 1,913)

Food Category	Percent of Population Group Consuming	Food Intake (g/kg body weight/d)	Consumers' Dioxin Toxicity Equivalents Intake (pg/kg body weight/d)											
			Nondetects[a] = 0				Nondetects = 0.5 (LOD)[b]				Nondetects = LOD			
				Percentile				Percentile				Percentile		
			Mean	10	50	90	Mean	10	50	90	Mean	10	50	90
All foods	100.0	53.4	0.69	0.24	0.57	1.24	1.10	0.53	0.99	1.76	1.51	0.76	1.38	2.43
If skim milk consumed[c]	100.0	53.4	0.58	0.17	0.46	1.11	0.99	0.44	0.88	1.64	1.40	0.66	1.27	2.31
Dairy foods and mixtures	98.1	14.4	0.17	0.04	0.13	0.33	0.23	0.07	0.20	0.43	0.29	0.08	0.25	0.58
If skim milk consumed[c]	98.1	14.4	0.06	—[d]	0.03	0.15	0.12	0.01	0.08	0.28	0.18	0.01	0.13	0.41
Meat and mixtures	84.8	3.4	0.32	0.02	0.21	0.65	0.36	0.05	0.25	0.72	0.40	0.07	0.29	0.82
Poultry and mixtures	57.1	2.4	0.08	—	0.03	0.20	0.09	0.01	0.05	0.21	0.11	0.02	0.07	0.25
Fish and mixtures	15.6	2.2	0.25	—	0.04	0.83	0.29	0.02	0.08	0.83	0.33	0.03	0.13	0.87
Eggs and mixtures	23.9	1.6	0.07	0.01	0.05	0.14	0.07	0.02	0.06	0.14	0.08	0.03	0.06	0.15
Fruits, vegetables and mixtures	98.4	10.1	0.06	—	0.02	0.17	0.18	0.02	0.12	0.41	0.30	0.04	0.20	0.69
Fats, oils and mixtures	66.8	0.4	0.01	—	—	0.04	0.02	—	0.01	0.04	0.02	—	0.01	0.04
Other foods and mixtures[e]	100.0	24.0	0.09	0.02	0.06	0.18	0.27	0.10	0.23	0.48	0.45	0.16	0.38	0.80

[a]Reflects treatment of samples for which no dioxin congener was detected.
[b]LOD = limit of detection.
[c]Dioxin TEQ intake from milk assuming that all plain milk consumed is skim milk rather than whole milk, as was otherwise assumed due to lack of analytical data on milks other than whole milk. Dioxin TEQ for skim milk was estimated assuming that all dioxin congeners concentrate in milk fat, that whole milk contains 3.34% fat, and that skim milk contains 0.18% fat (Nutrient Data Laboratory, 2002).
[d]< 0.0005 pg/kg body weight/d.
[e]Grains and mixtures, legumes and mixtures, beverages (other than milk and juice), candy.

NOTE: Data represent 2-day averages generated using U.S. Department of Agriculture (USDA) sample weights.
DATA SOURCE: Dioxin concentrations: U.S. Food and Drug Administration Total Diet Study (1999–2001), food consumption: 1994–1996, 1998 USDA Continuing Survey of Food Intakes by Individuals.

TABLE 5-6 Estimated Intake of Dioxins and Dioxin-like Compounds (DLCs) from Food by Females, 12–19 Years Old, Not Pregnant or Lactating (n = 692)

Food Category	Percent of Population Group Consuming	Food Intake (g/kg body weight/d)	Consumers' Dioxin Toxicity Equivalents Intake (pg/kg body weight/d)											
			Nondetects^a = 0				Nondetects = 0.5 (LOD)^b				Nondetects = LOD			
				Percentile				Percentile				Percentile		
			Mean	10	50	90	Mean	10	50	90	Mean	10	50	90
All foods	100.0	30.6	0.37	0.11	0.29	0.66	0.66	0.28	0.58	1.08	0.96	0.40	0.85	1.67
If skim milk consumed^c	100.0	30.6	0.33	0.09	0.26	0.66	0.63	0.25	0.54	1.07	0.92	0.92	0.81	1.63
Dairy foods and mixtures	88.7	5.3	0.07	0.01	0.05	0.15	0.10	0.02	0.07	0.20	0.13	0.02	0.09	0.27
If skim milk consumed^c	88.7	5.3	0.03	0.00	0.01	0.08	0.06	0.00	0.03	0.14	0.09	0.09	0.05	0.21
Meat and mixtures	78.6	2.2	0.21	0.01	0.13	0.46	0.24	0.03	0.15	0.52	0.26	0.04	0.17	0.61
Poultry and mixtures	53.1	1.7	0.05	—^d	0.03	0.12	0.06	0.01	0.04	0.13	0.07	0.01	0.05	0.15
Fish and mixtures	17.4	1.4	0.10	—	0.01	0.31	0.14	0.01	0.06	0.31	0.17	0.02	0.10	0.34
Eggs and mixtures	23.4	0.9	0.04	—	0.03	0.07	0.04	0.01	0.04	0.07	0.04	0.02	0.04	0.07
Fruits, vegetables and mixtures	96.8	5.4	0.03	—	0.01	0.10	0.11	0.01	0.07	0.25	0.18	0.02	0.11	0.45
Fats, oils and mixtures	60.1	0.3	0.01	—	—	0.02	0.01	—	—	0.02	0.01	—	0.01	0.02
Other foods and mixtures^e	100.0	17.4	0.06	0.01	0.04	0.13	0.22	0.07	0.18	0.43	0.38	0.31	0.31	0.74

^aReflects treatment of samples for which no dioxin congener was detected.

^bLOD = limit of detection.

^cDioxin TEQ intake from milk assuming that all plain milk consumed is skim milk rather than whole milk, as was otherwise assumed due to lack of analytical data on milks other than whole milk. Dioxin TEQ for skim milk was estimated assuming that all dioxin congeners concentrate in milk fat, that whole milk contains 3.34% fat, and that skim milk contains 0.18% fat (Nutrient Data Laboratory, 2002).

^d< 0.0005 pg/kg body weight/d.

^eGrains and mixtures, legumes and mixtures, beverages (other than milk and juice), candy.

NOTE: Data represent 2-day averages generated using U.S. Department of Agriculture (USDA) sample weights.

DATA SOURCE: Dioxin concentrations: U.S. Food and Drug Administration Total Diet Study (1999–2001), food consumption: 1994–1996, 1998 USDA Continuing Survey of Food Intakes by Individuals.

CURRENT DIETARY PATTERNS

Dietary Exposure to DLCs

General Population

Trends from 1977–1978 through 1994–1995 show that there has been a decline in the consumption of whole milk, but there have been increases in the consumption of grains and meat-based mixtures (Enns et al., 1997). Results of the most recently reported CSFII (1994–1996) show that, on average, Americans consumed 4.9 servings of meat or meat substitutes daily. Within the meat group, average intakes included 1.9 servings of red meat, 1.2 of poultry, 0.8 of processed meat, 0.4 of fish, 0.4 of eggs, and 0.1 of nuts or seeds. Average intakes of other foods included 3.1 servings of vegetables, 1.5 of fruit, 1.5 of dairy foods, and 6.8 of grain-based products (Smiciklas-Wright et al., 2002). These data supports the observation from the commissioned analysis that DLC exposure through animal-based foods comes primarily from meat and meat mixtures.

Sensitive and Highly Exposed Subgroups of the General Population

Some life stage groups within the general population, particularly developing fetuses and infants, may be more sensitive to DLC exposure because their developmental immaturity. These life stage groups also may be more highly exposed as a result of smaller body size and, in the case of infants, through breastfeeding.

Certain cultural or behavioral subgroups of the population may have higher exposures to DLCs than the general population, based upon differences in food-consumption patterns. Such populations include some American Indian groups, Alaska Native and northern Canadian groups, and subsistence or sport fishers. These populations may consume greater amounts of fish known to have relatively high concentrations of DLCs due to the location where the fish are caught. The committee took these groups into consideration as it explored potential ways to reduce human dietary exposure to DLCs.

In Utero Exposure. DLCs, although lipophilic, are known to cross the placenta during pregnancy (Holladay, 1999). The total DLC body burden of the developing fetus has been estimated using cord blood samples obtained at birth, based on the assumption that this value represents in utero exposure. Koopman-Esseboom and colleagues (1994a) determined that concentrations of polychlorinated biphenyls (PCB) in cord blood, collected on a cohort in the Netherlands in 1990, were approximately 20 percent of maternal plasma values, expressed on a concentration (volume) basis. In contrast to these findings, in the Michigan Maternal/Infant Cohort Study, Jacobson and colleagues (1984) found that PCB levels expressed on a lipid basis were equivalent between maternal plasma and cord serum samples. This contrast is because fetal plasma contains only approximately 20 percent of the lipid level of maternal plasma. Both studies, however, demonstrate that fetal exposure to DLCs is a function of the mother's body burden. Total fetal content of PCBs is comparable on a lipid basis with the mother. However, on the basis of total body weight, the relative DLC content is less because the average fraction of body weight represented by lipids in the human newborn is lower than the average lipid content of adults.

Various developmental health outcomes have been described that are attributed to different levels of exposure to DLCs and PCBs. Jacobson and Jacobson (2002) described impaired cognitive functioning in infants born to mothers who consumed PCB-contaminated fish from Lake Michigan. Other health outcomes to infants attributed to in utero exposure to DLCs includes poor psychomotor skills (Koopman-Esseboom et al., 1996), altered thyroid hormone levels (Koopman-Esseboom et al., 1994b), and reduced neurological optimality (Huisman et al., 1995). The health effects of DLC exposure to infants and young children are further described in Chapter 2.

A woman's DLC body burden is an accumulation of her intake exposure to DLCs from birth, beginning with her consumption of breast milk during the first months of life and continuing up to the time that she becomes pregnant. As a result of her lifetime-accumulated exposure and the long half-life of DLCs, a woman's dietary modifications that are intended to reduce DLC intake during pregnancy will not have an impact on exposure to the developing fetus.

Breastfeeding Infants. The primary postpartum DLC exposure route to infants is through breast milk. The American Academy of Pediatrics (AAP) promotes breastfeeding of infants as the foundation of good feeding practices and recognizes the critical role of breastfeeding as primary in achieving optimal infant and child health, growth, and development. AAP recommends exclusive breastfeeding for the first six months of life and continuation of breastfeeding through 1 year of age (Work Group on Breastfeeding, 1997). There are well-documented advantages of breastfeeding, not just to the infant, but to the mother, the family, and to society, including health, nutritional, immunological, developmental, psychological, social, economical, and environmental benefits (Work Group on Breastfeeding, 1997). Besides being the ideal nutritional source for human infants, human milk contains growth factors that aid in the development of the neonate's immature gastrointestinal tract and in repair following insults such as common diarrheal illness (Bernt and Walker, 1999; Playford et al., 2000); hormones and enzymes with multiple functions; and numerous immune factors that protect against many infections (Wold and Adlerberth, 2000). In addition, breastfed infants are leaner (Dewey et al., 1995) and have a somewhat lower risk of later childhood obesity (Dewey, 2003). Psychological and cognitive development also appears to be favorably affected by breastfeeding (Jacobson et al., 1999; Lucas et al., 1992).

A report from the U.S. Department of Health and Human Services (HHS) Office on Women's Health, *HHS Blueprint for Action on Breastfeeding* (HHS, 2000b), echoes the recommendations of AAP. In addition to its support, however, the Blueprint identifies certain conditions that warrant caution toward breast-feeding, including exposure to contaminants such as DLCs. These cautions are aimed specifically at high-risk circumstances, such as high-level exposures to contaminants that result in poisonings and the consumption of contaminated noncommercial fish and wildlife (HHS, 2000b). Aside from these specific cir-cumstances, the Blueprint supports the goals of *Healthy People 2010: Objectives for Improving Health:* to increase the proportion of mothers who breastfeed to 75 percent in the early postpartum period, 50 percent at 6 months, and 25 percent at 1 year (HHS, 2000a).

Two ongoing, nationwide, cross-sectional surveys, NHANES and the Ross Laboratories Mothers Survey, have contributed to the knowledge of rates and trends in breastfeeding patterns in the United States. Results from NHANES III, covering the years 1988–1994, indicate that 47 percent of newborns were exclu-sively breastfed at 7 days after birth, and that 10 percent were still exclusively breastfed at age 6 months. Nonexclusive breastfeeding was seen in 52 and 22 percent of infants at 7 days and 6 months, respectively. Similarly, the Ross survey, from data collected in 2001, indicated that 46.3 and 17.2 percent of newborns were exclusively breastfed in-hospital (birth) and 6 months respec-tively, whereas nonexclusive breastfeeding was seen in 69.5 and 32.5 percent of infants in-hospital and at 6 months, respectively. An additional and important finding from the Ross survey was that the greatest increase in both initiation and continuation of breastfeeding between 1989 and 1995 occurred among those most at risk for poor nutritional status: women of color receiving benefits from the Special Supplemental Nutrition Program for Women, Infants, and Children (WIC), women with low education, first-time mothers, and women employed full time and residing in regions where breastfeeding is not routinely practiced (Ryan, 1997).

Recent U.S. data on DLCs in breast milk are limited. A 1986 study analyzed breast-milk samples for polychlorinated biphenyls (PCBs) and p,p'-dichloro-diphenyldichloroethylene in more than 800 predominantly white, educated women living in North Carolina (mean age of 27 years). The median PCB con-centration in breast milk was 1.8 ppm (fat basis) at birth, declining by almost 50 percent to 1.0 ppm by the infant's first birthday (Rogan et al., 1986). This study indicated that exposure to DLCs in breastfed infants was many times higher than exposures that would be expected for adults, but that these exposures declined after weaning, most likely as a result of lower DLC intakes from foods and the infants' expanding total body-fat pool. Feeley and Brouwer (2000) reported similar observations for PCBs, polychlorinated dibenzo-*p*-dioxins (PCDDs), and polychlorinated dibenzofurans (PCDFs) in U.S. and European cohorts in the early 1990s, and Furst and colleagues (1989) also observed that DLC concentrations were highest during the first weeks of lactation. A nursing infant's exposure to DLCs per unit of body weight will always be greater than his or her mother's exposure because DLCs are concentrated in the breast-milk fat. After weaning, since other environmental exposures are lower than the exposure to DLCs through breast milk, DLC levels in the infant's body fat will decline to adult levels as the adipose tissue pool size increases.

A study of congener-specific DLCs in nonpooled breast-milk samples from Belgium showed that the major DLC congeners in these samples were 2,3,7,8-tetrachlorodibenzo-*p*-dioxin (TCDD), 1,2,3,7,8-pentachlorodibenzo-*p*-dioxin (PeCDD), 2,3,4,7,8-pentachlorodibenzo-*p*-furan (PeCDF), and PCB-126. The mean value of DLCs present in breast-milk samples was 29.4 pg TEQ/g of fat. The estimated average intake by breastfeeding infants was 76 TEQ pg/kg/d, which is 20-fold higher than the World Health Organization's tolerable daily intake value for adults (Focant et al., 2002). A similar study on Korean women and infants showed that the predominant congeners in breast-milk samples were 2,3,4,7,8-PeCDD and PCB-126, and that the mean daily DLC intake for breastfeeding infants was 85 TEQ pg/kg/d (Yang et al., 2002).

DLC Exposure of Infants Fed Formula versus Breast Milk. In general, infant formulas have been found to contain significantly lower levels of DLCs than breast milk, due largely to the use of plant oils in place of animal fats in the formula (Ramos et al., 1998; Schecter et al., 1994). Bajanowski and colleagues (2002) compared tissue levels of PCDD/PCDF congeners in both breastfed and formula-fed infants from Germany that died unexpectedly in the years 1991 to 1992 and 1996 to 1997. Among infants of similar age, the mean concentration of DLCs in fatty tissue from breastfed infants was greater than from formula-fed infants, and concentrations were higher in infants breastfed for longer compared with shorter time periods. The average weekly dioxin intake (I-TEQ ng/kg of fat) was 10- to 25-fold higher in breastfed than in formula-fed infants. When cohorts from 1991 to 1992 were compared with those from 1996 to 1997, total DLC levels were lower in the latter, independent of length of breastfeeding, reflecting the trend toward lower DLC exposure seen in the population at large.

Altogether, DLC intake exposure to infants through breastfeeding is significantly greater than through formula feeding. Evidence indicates that, for the general population, the benefits of breastfeeding outweigh existing evidence for the detrimental effects to infants resulting from exposure to DLCs through breast milk (Work Group on Breastfeeding, 1997). Therefore, recommendations limiting breastfeeding as a means of reducing DLC exposure for the general population would result in unnecessary loss of the physiological advantages, psychological health, and societal benefits of breastfeeding. There are, however, documented incidences of unintentional high exposure due to unintended industrial releases in which breastfeeding may not be advisable; these situations should be addressed apart from recommendations to the general population.

Young Children. The main route of DLC intake exposure in growing children is through consumption of whole milk and other full-fat dairy products. Furthermore, as suggested by the previously discussed intake exposure scenario, consumption of added fats from all animal sources increases exposure to DLCs. Discretionary fat intake, particularly from animal foods, can be reduced in the diets of young children so that DLC intake exposure can be decreased without compromising the overall nutritional quality of the diet.

Additionally, using low-fat or skim milk rather than whole milk will reduce DLC exposure without compromising this important source of calcium for children. (FDA defines milk that contains 1 percent or less fat as low fat, and 2 percent or less fat as reduced fat.) Results from the 1994–1996 CSFII indicate that 47 percent of children ages 1 to 5 years who consumed milk had whole milk, while 37 percent had low-fat milk.

The consumption of milk and other dairy products by children offers a leverage point for reducing dietary exposure to DLCs for two reasons. First, substitution of low-fat and skim milk and other products for whole-milk counterparts carries no risk of compromising the nutritional quality of the diet for children over the age of 2 years. In fact, children would benefit from the standpoint of reduced long-term risk for chronic disease. Second, federal child nutrition programs, including WIC, the National School Lunch and School Breakfast Programs, and the Child and Adult Care Food Program, reach very large numbers of children and account for a large proportion of their milk consumption.

Adolescents 12 to 19 Years of Age. Avoidance of dietary exposure to DLCs by girls before they enter their childbearing years is of importance because of future DLC exposure to a developing fetus through the mother's total body burden. However, many popular food choices among adolescents are high in fat, including animal fat. Lin and colleagues (1996, 1999) reported that consumption of foods from fast-food restaurants represents approximately one-third of the average adolescent's meals eaten outside the home. French and colleagues (2001) recently described a strong association between high frequency of fast-food consumption and greater intakes of total energy and percentage of energy from fat and a lower intake of milk.

Whether these trends translate into increased exposure to DLCs from foods has not been determined. Bearing in mind the long half-life of DLCs in humans, and that adolescents will soon be entering their child-bearing years, it may be wise to monitor and evaluate potential DLC exposure of this group through the consumption of foods high in animal fats.

American Indian, Alaska Native, and Northern Dwelling Fish Eaters. The native peoples of America differ markedly in culture, language, lifestyle, and diet. Some of those who embrace traditional customs and foodways have in common a dependence on hunting, fishing, and gathering for foodstuffs, whereas others embrace agricultural traditions. Importantly, for a subset of native peoples, a major portion of their diet and cultural practices involves fish and marine mammals, and these groups are at particular risk of exposure to DLCs. In addition, for some agriculturally-based American Indian populations, the lands used for crops and grazing have been previously contaminated with herbicides containing DLCs (Personal communication, T. Dawes, Inter Tribal Council of Arizona, February 19, 2002). Their exposure is a function of both the quantity of fish, marine mammals, and agricultural products consumed, and the levels of contaminants in the foods.

There is limited information on the extent of fish consumption among various American Indian people. Fitzgerald and colleagues (1995) and Hwang and colleagues (1996) studied exposure to PCBs among the Mohawks in northern New York. The traditional Mohawk fishing grounds are immediately downstream of three aluminum foundries that have contaminated the water with PCBs, resulting in high levels of these compounds in the fish (Sloan and Jock, 1990). Gerstenberger and colleagues (1997, 2000) studied fish consumption in an Ojibwa population in Wisconsin. Both of these investigations found that, on average, the frequency of fish consumption was not different from the general population, although some individuals within the populations may have been at higher risk than others.

Native peoples living in the far north have fewer local sources of food compared with those living in more temperate climates, and a number of tribal populations are dependent upon fish or marine mammals to varying degrees. In the Americas, most studies have been conducted in northern Canada, particularly with the Inuit people. Dewailly and colleagues (1994) compared an Inuit population's dietary exposures to DLCs through fish consumption with the exposures of a group of Caucasian fishermen on the north shore of the Gulf of St. Lawrence. The Inuit ate many times greater amounts of fish than the Caucasians did, and they had significantly higher (eight to ten times) plasma levels of PCBs (Dewailly et al., 1994). The predominant source of DLCs in the diets of other Northern Dwellers is sea birds, sea bird eggs, fish, seal, and whale.

Muckle and colleagues (2001) report that Inuit women of child-bearing age living in Nunavik (northern Quebec) ate, on average, 3.3 fish meals per week, and that almost 80 percent ate fish at least once per week. In addition, they consumed beluga whale fat, muktuk (0.5 meals/wk), and meat (0.1 meals/wk), as well as seal fat (0.3 meals/wk), meat (0.4 meals/wk) and liver (0.1 meals/wk). These researchers also found that during pregnancy, 42 percent of the women increased their consumption of traditional foods. Ayotte and colleagues (1997) reported that Inuit living in Nunavik (Arctic Quebec) had seven times greater mean total DLC TEQs in breast milk than did women living in southern Quebec.

Scrudato and colleagues have an ongoing study of the Yupik Eskimos from St. Lawrence Island, Alaska (Carpenter et al., in press). This island is at the Arctic Circle, 40 miles from the Russian coast. The diet of the Yupiks is primarily whale, seal, and walrus, with berries and greens in the summer. Because of the high-fat content of their food and the likelihood that DLCs are present in the fat, this population is at greater than average risk of exposure to DLCs in their food supply. Although specific DLCs were not tested, initial studies of serum in Yupiks have shown a mean total PCB concentration of 7.1 ppb (range 1.2–19.4), higher than the values reported by ATSDR (2000) for unexposed populations (0.9–1.5 ppb).

Subsistence Fishers. There are few studies of non-Native subsistence fishermen, partly because they are difficult to identify and often do not purchase fishing licenses. Frequent consumption of game fish is of concern because many of these fish have particularly high levels of DLCs (Anderson et al., 1998). Subsistence fishing is a common cultural tradition in many African-American families and Southeast-Asian immigrants. The Executive Director of the Mississippi Rural Development Council (Rawls, 2001) stated that many rural African-American individuals eat fish two to three times per week and, in some rural areas, as often as five times per week. Waller and colleagues (1996) investigated Lake Michigan fish consumption by urban, pregnant, and poor African-American women in Chicago. Ninety percent said that they ate fish regularly, and of these, approximately 10 percent ate sport-caught fish on a regular basis.

Hutchison and Kraft (1994) documented fishing activities and fish consumption among Asian Hmong residents of Green Bay, Wisconsin. They found that 60 percent of Hmong families fished and one-fourth of these families ate fish at least once a week, including fish from locations covered by fishing advisories. Wong (1997) surveyed 228 people fishing in the San Francisco Bay, of which 70 percent were non-Caucasian, especially Asian. The average fish consumption by this population was 32 g (1.2 oz) per day, or about four meals per month.

Sport Fishers. Because DLCs are lipophilic and bioconcentrate in the food chain, there are significant levels of these compounds in predatory fish that live in contaminated waters (Kuehl et al., 1989; Schmitt et al., 1999). For example, total PCB levels in Chinook salmon from Lake Huron (338 ppb) and Lake Ontario (835 ppb) (Jordon and Feeley, 1999) exceeded the U.S. Environmental Protection Agency guidelines for unlimited consumption of 50 ppb. Frequent consumption of these contaminated fish may result in undesirable exposure levels.

A number of studies have been conducted with sport fishers who fished from the Great Lakes. A study of DLCs in breast milk from 242 women who reported moderate consumption of Lake Michigan fish over an average span of 16.1 years, compared with 71 women who reported no consumption, found that fish consumption correlated with levels of PCBs in maternal serum and breast milk (Schwartz et al., 1983). Of another 1,820 women who held Great Lakes fishing licenses (ages 18–40 years), 979 reported that they never ate Great Lakes fish and 408 had not eaten them recently (Mendola et al., 1995). Among consumers, the average consumption in 1990 to 1991 was 4.9 kg, and the estimated mean lifetime sport-fish consumption was 57.4 kg.

Anderson and colleagues (1998) compared levels of dioxins, furans, and PCBs in 32 sport-fish consumers from three of the Great Lakes. On average, these individuals ate 49 sport-fish meals per year for a mean of 33 years. Dioxin concentrations in the sport-fish consumers were 1.8 times, furans 2.4 times, and PCBs 9.7 times greater than background dioxin concentrations in the general population. Another 100 Great Lakes fish eaters reported an average of 43 sport-fish meals in the past year. Their dioxin TEQs averaged 9.6 (range 3.0–41.2), furan TEQs 7.4 (range 2.7–25.6), and coplanar PCB TEQs 4.3 (range 0.1–66.2) (Falk et al., 1999). In two Canadian communities, Great Lakes sport-fish consumption averaged 21 g/d among consumers (Kearney et al., 1999). The measured PCB concentrations in these populations were 4.0 µg/L in nonfish eaters, and 6.1 µg/L in fish eaters. A recent study noted that the total quantity of sport fish consumed by Lake Michigan anglers declined significantly over the period 1979 to 1993 (He et al., 2001), presumably as a result of an increased knowledge of advisories.

Exposure to DLCs from fish consumption is not limited to fresh-water fish. Fatty fish in the Baltic Sea, such as salmon and herring, have levels of PCDDs and PCDFs (Rappe et al., 1989) comparable to those of fish from the Great Lakes (Kuehl et al., 1989). One study of Swedish men who ate significant amounts of Baltic Sea salmon and herring found that fish consumers had on average 8.0 pg/g of serum lipid TCDD, compared with 1.8 pg/g of serum lipid for nonconsumers (Svensson et al., 1991).

In summary, persons who consume fatty fish are at risk of greater than average exposure to DLCs if it is a major part of their diet and if they fish from contaminated waters. Certain groups of American Indians and Alaska Natives, for whom fish is a central part of their traditional diet, may be at greatest risk from exposure. Sport fishers may also have high DLC exposure, depending on the levels of DLCs in the fish they consume.

Factors Affecting Food Consumption

Dietary-consumption patterns are influenced by many different factors, from economic to cultural. Between 1970 and 1997, the consumption of animal-based foods, including red meat, fish, poultry, cheese, fats, and oils, increased in the

United States, whereas the consumption of milk and eggs decreased (Putnam and Allshouse, 1999).

Among the factors identified as contributing to overall changes in food-consumption patterns among the general population were changes in price and disposable income, increased food availability to the poor, more convenience foods and food imports, an increase in eating away from home, changes in advertising strategies, and changes in food-enrichment standards and fortification policies. Additionally, changes in sociodemographics have contributed to current food-consumption patterns. In the United States, there are trends toward smaller household size, greater numbers of households with dual incomes, more single-parent heads of household, and an older, more ethnically diverse population (Putnam and Allshouse, 1999).

The availability and content of consumer information also influence food choices. Many messages and media for messages about food choices and health effects exist. Food choices are most likely to be influenced by this type of information when the content is meaningful, consistent, and delivered through media accessible to the consumer. Consistency refers to consumers learning the same message, associated with different health effects, from different information sources. For example, educational programs may focus on the benefits of low-fat foods for cardiovascular health benefits, as well as on the reduction in DLC exposure.

DLC-reducing behavioral advice is more likely to be effective if it supports, and is consistent with, other food choice behaviors that are recommended for other reasons and that are important to the consumer. Food choices are likely to be unaffected if information efforts do not recognize the cultural importance associated with preparing and eating food. Advice about reducing exposure to DLCs by modifying food-consumption practices will be more successful if the behaviors suggested (e.g., food preparation, choice of food types) are compatible with existing social, cultural, and familial norms.

Ethnic Variation in Food-Consumption Patterns

Consumption of whole milk and full-fat dairy products represents the major source of DLC intake exposure among children in the United States. The U.S. Dietary Guidelines recommend reducing total fat and saturated fat intake for Americans over 2 years of age (USDA/HHS, 2000). For children over 2 years of age, whole milk may be a major source of fat that can be reduced to decrease DLC exposure and, at the same time, comply with current dietary recommendations. A study of milk consumption patterns of non-Hispanic white, African-American, and Hispanic New York State WIC participants, ages 1–5 years, showed that fewer African-American and white children drank whole milk with increasing educational level of the adult caregiver. However, Hispanic children tended to consume whole milk equally, regardless of the caregiver's level of education (Dennison et al., 2001). The authors suggested that cultural and language barriers present in the Hispanic population may have contributed to the study findings.

Klesges and colleagues (1999) investigated the milk-consumption patterns of young adult Air Force recruits of various ethnicities. Among 17- to 24-year-old male and female recruits, milk consumption was highest among younger non-Hispanic whites, followed by Hispanics, then Asians, and lowest among African Americans. Within each ethnic group, milk consumption was greater among men than women and was negatively associated with age and education level.

Changing Dietary Patterns Among Northern Dwellers

A change in food-consumption patterns among northern-dwelling Inuit populations away from traditional foods to more marketed foods common to a Western diet has become a unique nutritional dilemma. Inuit women traditionally rely on caribou, seal, and narwhal mattak for protein, iron, and zinc. Similar foods, such as liver, mattak, kelp, and fish skin have, in the past, been this population's source of calcium and vitamin A (Kuhnlein, 1991). As the use of traditional Inuit foods has declined and been substituted with marketed foods, the risk for inadequate intake of important nutrients has increased in the population. The dilemma, however, is that the native species consumed by the Inuits are contributors of important nutrients but are also sources of DLCs (Kuhnlein, 1991). Although the impact of the consumption of these compounds at the levels present in the Inuit food supply is not fully understood, the substitution of the traditional diet with a more Western diet appears to be associated with other nutritional risks.

Vegetarian Food Patterns

Given the relatively lower contribution to DLC intakes from fruit, vegetables, and grains than from animal products, a lower body burden in vegetarians might be anticipated as compared with carnivores. Welge and colleagues (1993) found no difference in the mean blood levels of several different PCDD/PCDF congeners in 24 vegetarians compared with 24 nonvegetarian volunteers. Importantly, however, this study did not distinguish between lacto-vegetarian and vegan cohorts and lacto-vegetarian cohorts who consumed some foods that contained animal fats.

In a group of strict vegetarians, Schecter and Papke (1998) found that their mean blood DLC concentrations were lower than those of the general population. In another study, Noren (1983) analyzed breast milk for pesticides and PCBs from lacto-vegetarian and nonvegetarian women who frequently consumed fatty fish from the Baltic sea. The mean concentrations of four of the six compounds tested, adjusted for age and parity, were lowest in the milk of the lacto-vegetarian mothers compared with that of the nonvegetarian and fish-consuming mothers. These studies point out the contribution of DLCs in animal-based foods to total DLC intake exposures. Furthermore, the long half-life of DLCs in the human body suggests that changes in body burdens that may occur from dietary modification will likely take many years to realize.

Cooking Methods and Preparation

It is possible to reduce DLC exposure through the selection, handling, preparation, and processing of foods at the consumer level. These specific practices can form the basis for targeted consumer education. For dairy products, the selection of low-fat or skim milk (for individuals over 2 years of age) and low-fat dairy products in preference to full-fat products is an effective way to preserve the nutritional contribution of these foods yet minimize DLC exposure. This choice is consistent with the trend over the last three decades toward more use of low-fat and skim milks. However, available data show that approximately 47 percent of U.S. children still consume only or mostly whole milk (Enns et al., 1997). Cheese is also a significant source of animal fat for most of the population. While low-fat versions of cheeses are available, they are not widely consumed. For children under 2 years of age, the committee concluded that concern for their total caloric requirements and their developmental immaturity outweighs risk, as currently known, from exposure to DLCs.

The stability of DLCs and their persistence in the lipid fraction of animal foods make this route of exposure a concern. Several studies have shown that trimming prior to or discarding fat after broiling, pan-frying, grilling, roasting, and pressure-cooking meats decreases DLC levels in the foods by 50 percent or more (Petroske et al., 1998; Rose et al., 2001; Schecter et al., 1998). Rose and colleagues (2001) evaluated the effect of various cooking methods on dioxin and furan levels in cuts of beef from a single cow that had been dosed with five different congeners. Several different cooking methods were used: frying, grilling, and barbecuing ground beef; conventional and microwave oven-roasting beef; and open-pan and pressure cooking stew meat. The cooking process, per se, did not significantly change the concentrations of dioxins and furans. However, losses of fat in preparation and cooking did lower the total amount of the various congeners in the finished product.

Schecter and colleagues (1998) determined total pg TEQ and TEQ/kg wet weight for DLCs before and after broiling market basket samples of beef, bacon, and catfish. For all of the samples, total pg TEQ was reduced by an average of 50 percent from broiling. However, the concentrations, as measured by wet weight, were more ambiguous. The beef patty showed no change, the bacon sample greatly increased, and the catfish sample modestly decreased in TEQ/kg wet weight in cooked versus raw product (Schecter et al., 1998).

A study comparing pan-fried hamburger with raw hamburger showed reduced DLC levels in the cooked product by up to 50 percent. The loss of DLCs correlated with the loss of lipid collected from the pan juices after cooking (Petroske et al., 1998). Clearly, food preparation methods that retain, recycle, or add animal fat into products will increase the potential DLC exposure through those products. Examples are gravies made from meat fat and juices and using lard, bacon grease, or butter for frying, sautéing, or adding to mixed dishes.

Selecting lean cuts of meat and removing visible fat is clearly important. For fish and poultry, discarding the skin and trimming visible fat will also reduce exposure to DLCs. One study showed that trimming the lateral line and belly flap from fish and cooking skin-off filets decreased PCB levels in the cooked product by 33 percent (Zabik and Zabik, 1999). The cooking process neither forms nor degrades DLCs, but substantial decreases in the DLC levels of the prepared product can be achieved by cooking methods that result in the loss of fat. For example, one study found that by increasing the surface area and internal temperature of restructured carp surimi products, DLCs from the finished product were 55 to 65 percent lower than in the raw product, depending on the surface area of the fillet (Zabik and Zabik, 1999).

With the exception of members of the squash family (which also have a waxy lipophilic coat and can extract vapor phase or particulate-bound DLCs from the air), most contamination of plants by DLCs appears to be a result of surface deposition of contaminated soils or particulates (Lovett et al., 1997; Muller et al., 1994). One study (Hülster et al., 1994) showed that PCDD/PCDF detected in squashes (zucchini and pumpkin) came from root uptake, apart from the DLCs that adhered to the plant surfaces. An analysis of carrots, lettuce, and peas grown in highly contaminated soils compared with those grown in background soils revealed that DLCs in the whole vegetable grown in the contaminated soil was greatest for carrots, and that more than 75 percent of the contamination was in the peel (Muller et al., 1994). Likewise, DLC contamination of peas, although far less than in carrots, was concentrated in the pods. This analysis supports the observation that DLC levels in plants are related to the DLC content of the soils in which they are grown, but that peeling carrots and removing the outer pods from peas are effective ways to reduce DLC intakes from these vegetables.

Tsutsumi and colleagues (2000) and Hori and colleagues (2001) assessed the effect of washing and boiling on levels of PCDD/PCDFs and dioxin-like PCBs in samples of green vegetables from supermarket and home-garden sources. These reports showed that washing and cooking lowered the levels of detected DLCs, particularly for the home-garden samples. In addition, Hori and colleagues (2001) suggested that washing was most effective for reducing the levels of dioxins and furans in the samples, but was less effective for reducing the levels of dioxin-like PCBs.

In summary, food selection, handling, and preparation practices can provide the consumer with some tools to reduce DLC exposure that are essentially cost-free and easy to implement. Selecting low-fat products; trimming and discarding visible fat from meat, fish, and poultry; discarding skin from fish and poultry; avoiding practices that add or retain animal fat; and washing vegetables and peeling root and waxy-coated vegetables are recommended.

Food and Nutrition Assistance Programs

USDA administers several food assistance programs that provide a variety of benefits to target recipients. The four primary programs are the Food Stamp Program, the National School Lunch Program, the School Breakfast Program, and WIC. The nationwide average monthly participation rate for the Food Stamp Program was just over 19 million in fiscal year 2002 (FNS, 2003b). More than 28 million children participated in the National School Lunch Program in fiscal year 2002, and more than 8.1 million participated in the School Breakfast Program (FNS, 2003c, 2003d). Statistics for the WIC program show that in fiscal year 2002 nearly 7.5 million people participated in the program (FNS, 2003e). The target recipients for the National School Lunch and School Breakfast Programs and WIC participants are of particular interest to this report.

Although surveys of participants in the National School Lunch Program showed that they had greater intakes of energy, protein, and several vitamins and minerals compared with nonparticipants, they also consumed greater amounts of total fat and saturated fat per day than nonparticipants (Devaney et al., 1993). Data from the School Nutrition Dietary Assessment Study-II show that in the survey years 1998–1999, at least 70 percent of elementary school lunches served met the program standard for calories (one-third of the Recommended Dietary Allowance), but exceeded the program standards for levels of fat and saturated fat. Results also showed that, on average, elementary school lunches provided 33 percent of calories from total fat and 12 percent of calories from saturated fat. At the same time, calories provided from carbohydrate were below the recommended 55 percent (Fox et al., 2001).

WIC providers encourage pregnant women to breastfeed their babies and the program supplies a greater variety and quantity of foods and offers longer participation to breastfeeding participants (FNS, 2003a). Fifty-six percent of WIC participants initiate breastfeeding in the hospital; however, only 45 percent are still breastfeeding at discharge, and only 16 percent continue breastfeeding to 5 months of age. WIC participants who breastfeed and are most likely to continue breastfeeding are those who receive the special food package for breastfeeding mothers along with information and advice on breastfeeding (IOM, 2002b).

WIC participants receive a food package that includes milk, cheese, and eggs. The recent report, *Dietary Risk Assessment in the WIC Program* (IOM, 2002b), identified adherence to Food Guide Pyramid recommendations, including consumption of fat, saturated fat, and cholesterol, as an appropriate criterion for reducing dietary risk, recognizing that essentially low-income women and children will meet this criterion. However, whether WIC participants are at risk for greater exposure to DLCs than the general population has not been determined.

DIETARY GUIDANCE

The ultimate goal of dietary guidance and consumer education with regard to DLC exposure is to reduce body burdens across the whole population, and particularly for women prior to pregnancy and breastfeeding. Empowering individuals, through information, to make dietary changes that reduce DLC exposure is theoretically possible but requires a long-term view. The stability of these compounds in the body means that reduction in risk through dietary choice is ineffective in the short run, but it is likely very important over years and decades, particularly if begun in childhood.

There are several sources of dietary recommendations and associated educational tools for the general population that are consistent with the reduction of DLC exposure through the reduction of animal fat intake. HHS and USDA jointly developed the Dietary Guidelines for Americans (USDA/HHS, 2000), which provide recommendations to the general population, based on current scientific evidence, about the relationship between diet and risk for chronic disease, and which serve as guides for how a healthy diet may improve nutritional status. By statute, these guidelines now form the basis of federal food, nutrition education, and information programs. The Food Guide Pyramid is designed as an educational tool to translate these recommendations for the public (USDA, 1996).

The U.S. Dietary Guidelines recommend that Americans over the age of 2 years "choose a diet that is low in saturated fat and cholesterol and moderate in total fat," and specifically that individuals should keep their "intake of saturated fat at less than 10 percent of total calories" and should "aim for a total fat intake of no more than 30 percent of total calories" (USDA/HHS, 2000). In this context, the U.S. Dietary Guidelines recommend choosing "vegetable oils rather than solid fats (meat, dairy fats, shortening)" and fat-free or low-fat milks, yogurts, and cheeses, and limiting intakes of high-fat processed meats and organ meats (including liver) and sauces made with cream.

Similar to the U.S. Dietary Guidelines, the American Heart Association's (AHA) 2000 dietary guidelines "advocates a population-wide saturated fat intake of less than 10 percent energy, which can be achieved by limiting intake of foods rich in saturated fatty acids (e.g., full-fat dairy products and fatty meats)" (Krauss et al., 2000). Another report, *Dietary Reference Intakes for Energy, Carbohydrate, Fiber, Fat, Fatty Acids, Cholesterol, Protein, and Amino Acids* (IOM, 2002a), is consistent with the Dietary Guidelines and the AHA recommendations in that it identifies the association between the excessive intake of saturated fats with the increased incidence of heart disease and the development of certain cancers. Further, the reports cited above all recommend that individuals strive to achieve and maintain a healthy body weight, that is, to avoid gaining excess weight because of the known adverse health consequences of obesity, independent of the increased body burden of DLCs due to the accumulation of total body fat. The committee did not, however, find evidence to support a reduction in DLC body burden through weight loss. Thus, preventing obesity is preferred as a means of reducing the potential for DLC accumulation.

The committee identified two areas in which changes in individual behavior could reduce DLC exposure, but which are not advised because of countervailing risks. These are reductions in fatty-fish consumption below current recommendations of 1 to 2 fish meals per week and in breastfeeding. Each of these are discussed briefly below.

Fish

A recommendation that emerged very recently and does not appear specifically in the current U.S. Dietary Guidelines recognizes the potential health benefits of fish consumption. Unlike the saturated fat in dairy products and meats, the fatty acids in fish have been shown in epidemiological studies to be beneficial to health. The strongest evidence for the importance of omega-3 fatty acids is a reduction in risk of sudden death from cardiac arrest: epidemiological studies show a lower risk with greater fish consumption among both men and women (Daviglus, 1997; Guallar, 1995; Hu et al., 2002).

Two long-chained, highly polyunsaturated fatty acids, docosahexaenoic acid and eicosapentaenoic acid, are thought to be responsible for these health benefits by providing strong antiarrhythmic action on the heart, by serving as precursors to prostaglandins, and by providing anti-inflammatory and anti-thrombotic actions. There is a large body of epidemiological evidence to indicate that consumption of omega-3 polyunsaturated fatty acids, which are present in fatty fish, seafood, and certain vegetable oils, is beneficial in preventing heart disease (Conner, 2000; Hu et al., 2002; Siscovick et al., 1995). Potential adverse effects on immune function and increased clotting time from high intakes of omega-3 fatty acids have also been identified, but there is a need for greater clinical evidence to support the epidemiological findings (IOM, 2002a). Research is underway and will help address these issues.

Based on current evidence in support of this decrease in risk for heart disease associated with the consumption of omega-3 fatty acids, AHA recommends eating fish two times per week, including fatty fish such as mackerel, lake trout, tuna, and salmon (Krauss et al., 2000). While AHA acknowledges the reported effects of omega-3 fatty-acid supplements in patients with heart disease, it concludes that "further studies are needed to establish optimal doses of omega-3 fatty acids...for both primary and secondary prevention of coronary disease..." and that "consumption of 1 fatty fish meal per day (or alternatively, a fish oil supplement) could result in an omega-3 fatty acid intake...of approximately 900 mg/

day, an amount shown to beneficially affect coronary heart disease mortality rates in patients with coronary disease" (Krauss et al., 2000).

In addition to heart disease, some studies suggest that there are benefits of omega-3 fatty acids against a variety of health conditions (Conner, 2000). These include inflammatory diseases such as rheumatoid arthritis (Calder et al., 2002), epileptic seizure (Schlanger et al., 2002), endometrial cancer (Terry et al., 2002), age-related macular degeneration (Seddon et al., 2001), prostate cancer (Terry et al., 2001), and premature birth (Allen and Harris, 2001).

The increasing evidence of health benefits from the consumption of fish argues that efforts to reduce the DLC content in fish should be promoted, rather than restricting intake in order to reduce DLC exposure. One caution regarding fish consumption exists with regard to women who are pregnant or who may become pregnant: some species of fish, particularly long-lived, larger predatory fish may contain high levels of methylmercury, a developmental neurotoxin that may harm a fetus's developing nervous system if ingested at high levels.

FDA has issued guidance to women of childbearing age, advising them to limit their consumption of fish to an average of not more than 12 oz/wk of cooked fish (approximately two to four servings). FDA also recommended that pregnant or potentially pregnant women eat a variety of fish; avoid eating shark, swordfish, king mackerel, and tilefish; and check with their state or local health department to see if there are special advisories on fish caught from freshwater lakes and streams in their local area (CFSAN, 2001).

The degree of contamination of farmed and wild-caught fish varies geographically and with changes in aquaculture practices (Gerstenberger et al., 1997; He et al., 2001). However, as sources of fats used in fish feeds change, so do the levels of DLCs that are introduced through this route. Thus, accurate data on regional sources that contribute to geographic variability are needed. Additionally, the committee recognizes that there are comigrating contaminants that will be found with DLCs in many types of fish, but an examination of these other contaminants was beyond the committee's charge and were not considered in this report.

In summary, fish are important sources of nutrients and potentially beneficial omega-3 fatty acids, and they also have cultural importance in the traditional diets of many population groups. But, unlike meats and poultry, fish cannot easily be trimmed to reduce their fat content. Therefore, fish consumption should be encouraged at currently recommended levels (1 to 2 fish meals per week), except for fish caught where known DLC contamination has occurred and fish advisories are in place.

Breastfeeding

As previously mentioned, AAP recommends exclusive breastfeeding until the age of 6 months and continued breastfeeding with supplementary foods until at least the age of 1 year. The benefits of breastfeeding have been clearly and positively identified, at the level of the infant, the mother, the family, and society (Work Group on Breastfeeding, 1997), and it would be inappropriate to interfere with the current upward trend in breastfeeding rates in the United States because of as yet only theoretical detrimental effects of the DLC levels currently found in breast milk. It is important to emphasize that changing the mother's diet during pregnancy and lactation will not have an impact on reducing DLC levels in her breast milk, since the accumulated body burden is what determines DLC exposure to the fetus in utero and to the nursing infant. Rather, the focus on reducing DLC exposure should begin after weaning and throughout childhood, so that the next generation of women will enter their reproductive years with lower body burdens of DLCs. This is a long-term agenda and emphasizes the value of reducing animal-fat consumption for children, with particular benefits for girls. For children, the general dietary guidance that applies to adults is appropriate for children over the age of 2 years, including the use of low-fat or skim milk rather than whole milk (AAP Committee on Nutrition, 1983; AHA, 1983).

SUMMARY

The major sources of DLCs in the food supply are animal fats, although small amounts appear in most foods. Exposure can be reduced through the reduced intake of animal fat by selecting lean cuts of meat, poultry, and fish; trimming visible fat and removing skin, as appropriate; and by selecting low-fat dairy products. DLCs on fruits and vegetables can be reduced through washing and through peeling root and waxy-coated vegetables.

In general, these precautions are consistent with current dietary advice, with two major exceptions. Although breast milk is a relatively concentrated source of DLCs, particularly in first lactations, and contributes significantly to body burdens, the evidence of the benefits of human breast milk overall outweighs the potential risks from this source. The focus should be on reducing lifetime body burdens through other means so that future generations of breastfeeding mothers will increasingly transmit lower levels through this important source of infant nutriture. The second exception is the conflict between recommendations to consume more fatty fish and the fact that some of the highest concentrations of DLCs in the food supply come from these fish, both farmed and from the sea. Some subsets of the population are likely to be at greater than average risk from this source and should be monitored. The accumulating evidence of health benefit from fatty fish (e.g., mackerel and salmon) suggest that rather than avoiding this food, the focus must again be on reducing the content of DLCs in future stocks.

REFERENCES

AAP (American Academy of Pediatrics) Committee on Nutrition. 1983. Towards a prudent diet for children. *Pediatrics* 71:78–80.

AHA (American Heart Association). 1983. AHA committee report. Diet in the healthy child. *Circulation* 67:1411A–1414A.

Allen KG, Harris MA. 2001. The role of *n*-3 fatty acids in gestation and parturition. *Exp Biol Med (Maywood)* 226:498–506.

Anderson HA, Falk C, Hanrahan L, Olson J, Burse VV, Needham L, Paschal D, Patterson D Jr, Hill RH Jr, The Great Lakes Consortium. 1998. Profiles of Great Lakes critical pollutants: A sentinel analysis human blood and urine. *Environ Health Perspect* 106:279–289.

ATSDR (Agency for Toxic Substances and Disease Registry). 2000. *Toxicological Profile for Polychlorinated Biphenyls (PCBs)*. Atlanta: ATSDR.

Ayotte P, Dewailly E, Ryan JJ, Bruneau S, Lebel G. 1997. PCBs and dioxin-like compounds in plasma of adult Inuit living in Nunavik (Arctic Quebec). *Chemosphere* 34:1459–1468.

Bajanowski T, Furst P, Wilmers K, Beike J, Kohler H, Brinkmann B. 2002. Dioxin in infants—An environmental hazard? *Int J Legal Med* 116:27–32.

Bernt KM, Walker WA. 1999. Human milk as a carrier of biochemical messages. *Acta Paediatr Suppl* 88:27–41.

Calder PC, Yaqoob P, Thies F, Wallace FA, Miles EA. 2002. Fatty acids and lymphocyte functions. *Br J Nutr* 87:S31–S48.

Carpenter DO, DeCaprio AP, O'Hehir D, Scrudato RJ, Apatiki L, Kava J, Gologergen J, Miller PK, Eckstein L. In press. Organochlorine contaminants in serum of Siberian Yupik people from St. Lawrence Island, Alaska. Submitted to *Environ Health Perspec*.

CFSAN (Center for Food Safety and Applied Nutrition). 2001. *An Important Message for Pregnant Women and Women of Childbearing Age Who May Become Pregnant About the Risks of Mercury in Fish*. Online. U.S. Food and Drug Administration. Available at http://www.cfsan.fda.gov/%20~dms/admehg.html. Accessed June 4, 2003.

Conner WE. 2000. Importance of *n*-3 fatty acids in health and disease. *Am J Clin Nutr* 71:171S–175S.

Daviglus ML, Stamler J, Greenland P, Dyer AR, Liu K. 1997. Fish consumption and risk of coronary heart disease. What does the evidence show? *Eur Heart J* 18:1841–1842.

Dennison BA, Rockwell HL, Nichols MJ. 2001. Use of low-fat milk by children in the New York State WIC varies with parental characteristics. *J Am Diet Assoc* 101:464–466.

Devaney B, Gordon A, Burghardt J. 1993. *The School Nutrition Dietary Assessment Study: Dietary Intakes of Program Participants and Nonparticipants*. Alexandria, VA: Food and Nutrition Service, U.S. Department of Agriculture (USDA).

Dewailly E, Ryan JJ, Laliberte C, Bruneau S, Weber JP, Gingras S, Carrier G. 1994. Exposure of remote maritime populations to coplanar PCBs. *Environ Health Perspect* 102:205–209.

Dewey KG. 2003. Is breastfeeding protective against childhood obesity? *J Hum Lact* 19:9–18.

Dewey KG, Peerson JM, Brown KH, Krebs NF, Michaelsen KF, Persson LA, Salmenpera L, Whitehead RG, Yeung DL. 1995. Growth of breast-fed infants deviates from current reference data: A pooled analysis of US, Canadian and European data sets. *Pediatrics* 96:495–510.

Douglass JS, Murphy MM. 2002. *Estimated Exposure to Dioxins in the Food Supply*. Paper prepared for the Committee on the Implications of Dioxin in the Food Supply. Institute of Medicine, Washington, DC.

Enns CW, Goldman JD, Cook A. 1997. Trends in food and nutrient intakes by adults: NFCS 1977–78, CSFII 1989–91, and CSFII 1994–95. *Fam Econ Nutr Rev* 10:2–15.

Falk C, Hanrahan L, Anderson HA, Kanarek MS, Draheim L, Needham L, Patterson D Jr. 1999. Body burden levels of dioxin, furans, and PCBs among frequent consumers of Great Lakes sport fish. The Great Lakes Consortium. *Environ Res* 80:S19–S25.

Feeley M, Brouwer A. 2000. Health risks to infants from exposure to PCBs, PCDDs and PCDFs. *Food Addit Contam* 17:325–333.

Fitzgerald EF, Hwang SA, Brix KA, Bush B, Cook K, Worswick P. 1995. Fish PCB concentrations and consumption patterns among Mohawk women at Akwesasne. *J Expo Anal Environ Epidemiol* 5:1–19.

FNS (Food and Nutrition Service). 2003a. *Breastfeeding Promotion and Support in WIC.* Online. USDA. Available at http://www.fns.usda.gov/wic/Breastfeeding/breastfeedingmainpage.htm. Accessed January 10, 2003.

FNS. 2003b. *Food Stamp Program: Average Monthly Participation (Persons).* Online. USDA. Available at http://www.fns.usda.gov/pd/fsfypart.htm. Accessed June 24, 2003.

FNS. 2003c. *National School Lunch Program: Participation and Lunches Served.* Online. USDA. Available at http://www.fns.usda.gov/pd/slsummar.htm. Accessed June 24, 2003.

FNS. 2003d. *School Breakfast Program Participation and Meals Served.* Online. USDA. Available at http://www.fns.uda.gov/pd/sbsummar.htm. Accessed June 24, 2003.

FNS. 2003e. *WIC Program: Total Participation.* Online. USDA. Available at http://www.fns.usda.gov/pd/wifypart.htm. Accessed June 24, 2003.

Focant JF, Pirard C, Thielen C, De Pauw E. 2002. Levels and profiles of PCDDs, PCDFs and PCBs in Belgian breast milk. Estimation of infant intake. *Chemosphere* 48:763–770.

Food Surveys Research Group. 1997. *Data Tables: Results from USDA's 1994–96 Continuing Survey of Food Intakes by Individuals and 1994–96 Diet and Knowledge Survey. Table Set 10.* Online. Agricultural Research Service (ARS), USDA. Available at http://www.barc.usda.gov/bhnrc/foodsurvey. Accessed December 9, 2002.

Fox MK, Crepinsek MK, Conner P, Battaglia M. 2001. *School Nutrition Dietary Assessment Study-II Summary of Findings.* Report prepared for the USDA. Cambridge, MA: Abt Associates.

French SA, Story M, Neumark-Sztainer D, Fulkerson JA, Hannan P. 2001. Fast food restaurant use among adolescents: Associations with nutrient intake, food choices and behavioral and psychosocial variables. *Int J Obes Relat Metab Disord* 25:1823–1833.

Furst P, Kruger C, Meemken HA, Groebel W. 1989. PCDD and PCDF levels in human milk— Dependence on the period of lactation. *Chemosphere* 18:439–444.

Gerstenberger SL, Tavris DR, Hansen LK, Pratt-Shelley J, Dellinger JA. 1997. Concentrations of blood and hair mercury and serum PCBs in an Ojibwa population that consumes Great Lakes region fish. *J Toxicol Clin Toxicol* 35:377–386.

Gerstenberger SL, Dellinger JA, Hansen LG. 2000. Concentrations and frequencies of polychlorinated biphenyl congeners in a Native American population that consumes Great Lakes fish. *J Toxicol Clin Toxicol* 38:729–746.

Guallar E, Hennekens CH, Sacks FM, Willett WC, Stampfer MJ. 1995. A prospective study of plasma fish oil levels and incidence of myocardial infarction in U.S. male physicians. *J Am Coll Cardiol* 25:387–394.

He JP, Stein AD, Humphrey HE, Paneth N, Courval JM. 2001. Time trends in sport-caught Great Lakes fish consumption and serum polychlorinated biphenyl levels among Michigan anglers, 1973–1993. *Environ Sci Technol* 35:435–440.

HHS (U.S. Department of Health and Human Services). 2000a. *Healthy People 2010: Understanding and Improving Health*, 2nd ed. Washington, DC: U.S. Government Printing Office.

HHS. 2000b. *HHS Blueprint for Action on Breastfeeding.* Washington, DC: Office on Women's Health, HHS.

Holladay SD. 1999. Prenatal immunotoxicant exposure and postnatal autoimmune disease. *Environ Health Perspect* 107:687–691.

Hori H, Nakagawa R, Tobiishi K, Iida T, Tsutsumi T, Sasaki K, Toyoda M. 2001. Effects of cooking on concentrations of polychlorinated dibenzo-*p*-dioxins and related compounds in green leafy vegetables 'Komatsuna'. *J Food Hyg Soc Japan* 42:339–342.

Hu FB, Bronner L, Willett WC, Stampfer MJ, Rexrode KM, Albert CM, Hunter D, Manson JE. 2002. Fish and omega-3 fatty acid intake and risk of coronary heart disease in women. *J Am Med Assoc* 287:1815–1821.

Huisman M, Koopman-Esseboom C, Fidler V, Hadders-Algra M, van der Paauw CG, Tuinstra LG, Weisglas-Kuperus N, Sauer PJ, Touwen BC, Boersma ER. 1995. Perinatal exposure to polychlorinated biphenyls and dioxins and its effect on neonatal neurological development. *Early Hum Dev* 41:111–127.

Hülster A, Müller JF, Marschner H. 1994. Soil-plant transfer of polychlorinated dibenzo-*p*-dioxins and dibenzofurans to vegetables of the cucumber family (*Cucurbitaceae*). *Environ Sci Technol* 28:1110–1115.

Hutchison R, Kraft CE. 1994. Hmong fishing activity and fish consumption. *J Great Lakes Res* 20:471–478.

Hwang SA, Fitzgerald EF, Bush B, Cook KSE. 1996. Exposure to PCBs from hazardous waste among Mohawk women and infants at Akwasasne. *J Franklin Inst* 333A:17–23.

IOM (Institute of Medicine). 2002a. *Dietary Reference Intakes for Energy, Carbohydrate, Fiber, Fat, Fatty Acids, Cholesterol, Protein, and Amino Acids.* Washington, DC: National Academy Press.

IOM. 2002b. *Dietary Risk Assessment in the WIC Program.* Washington, DC: National Academy Press.

Jacobson JL, Fein GG, Jacobson SW, Schwartz PM, Dowler JK. 1984. The transfer of polychlorinated biphenyls (PCBs) and polybrominated biphenyls (PBBs) across the human placenta and into maternal milk. *Am J Public Health* 74:378–379.

Jacobson JL, Jacobson SW. 2002. Association of prenatal exposure to an environmental contaminant with intellectual function in childhood. *J Toxicol Clin Toxicol* 40:467–475.

Jacobson SW, Chiodo LM, Jacobson JL. 1999. Breastfeeding effects on intelligence quotient in 4- and 11-year old children. *Pediatrics* 103:e71.

Jordan SA, Feeley MM. 1999. PCB congener patterns in rats consuming diets containing Great Lakes salmon: Analysis of fish, diets, and adipose tissue. *Environ Res* 80:S207–S212.

Kearney JP, Cole DC, Ferron LA, Weber J-P. 1999. Blood PCB, *p,p*?-DDE, and mirex levels in Great Lakes fish and waterfowl consumers in two Ontario communities. *Environ Res* 80:S138–S149.

Kennedy ET, Bowman SA, Powell R. 1999. Dietary-fat intake in the US population. *J Am Coll Nutr* 18:207–212.

Klesges RC, Harmon-Clayton K, Ward KD, Kaufman EM, Haddock CK, Talcott GW, Lando HA. 1999. Predictors of milk consumption in a population of 17- to 35-year-old military personnel. *J Am Diet Assoc* 99:821–826.

Koopman-Esseboom C, Huisman M, Weislas-Kuperus N, Van der Paauw CG, Tuinstra LGM, Boersma ER, Sauer PJJ. 1994a. PCB and dioxin levels in plasma and human milk of 418 Dutch women and their infants. Predictive value of PCB congener levels in maternal plasma for fetal and infant's exposure to PCB's and dioxins. *Chemosphere* 28:1721–1732.

Koopman-Esseboom C, Morse DC, Weisglas-Kuperus N, Lutkeschipholt IJ, Van der Paauw CG, Tuinstra LG, Brouwer A, Sauer PJ. 1994b. Effects of dioxins and polychlorinated biphenyls on thyroid hormone status of pregnant women and their infants. *Pediatr Res* 36:468–473.

Koopman-Esseboom C, Weisglas-Kuperus N, de Ridder MA, van der Paauw CG, Tuinstra LG, Sauer PJ. 1996. Effects of polychlorinated biphenyl/dioxin exposure and feeding type on infants' mental and psychomotor development. *Pediatrics* 97:700–706.

Krauss RM, Eckel RH, Howard B, Appel LJ, Daniels SR, Deckelbam RJ, Erdman JW Jr, Kris-Etherton P, Goldberg IJ, Kotchen TA, Lichtenstein AH, Mitch WE, Mullis R, Robinson K, Wylie-Rosett J, St. Jeor S, Suttie J, Tribble DL, Bazzarre TL. 2000. AHA dietary guidelines: Revision 2000: A statement for healthcare professionals from the Nutrition Committee of the American Heart Association. *Circulation* 102:2284–2299.

Kreuzer PE, Csanady GA, Baur C, Kessler W, Papke O, Greim H, Filser JG. 1997. 2,3,7,8-Tetra-chlorodibenzo-p-dioxin (TCDD) and congeners in infants. A toxicokinetic model of human lifetime body burden by TCDD with special emphasis on its uptake by nutrition. *Arch Toxicol* 71:383–400.

Kuehl DW, Butterworth BC, McBride A, Kroner S, Bahnick D. 1989. Contamination of fish by 2,3,7,8-tetrachlorodibenzo-p-dioxin: A survey of fish from major watersheds in the United States. *Chemosphere* 18:1997–2014.

Kuhnlein HV. 1991. Nutrition of the Inuit: A brief overview. *Arctic Med Res* S:728–730.

Lin B-H, Guthrie J, Blaylock JR. 1996. *The Diets of America's Children: Influences of Dining Out, Household Characteristics, and Nutrition Knowledge.* Report No. AER-746. Washington, DC: Economic Research Service, USDA.

Lin B-H, Guthrie J, Frazao E. 1999. Quality of children's diets at and away from home: 1994–1996. *Food Rev* 22:2–10.

Lovett AA, Foxall CD, Creaser CS, Chewe D. 1997. PCB and PCDD/DF congeners in locally grown fruit and vegetable samples in Wales and England. *Chemosphere* 34:1421–1436.

Lucas A, Morley R, Cole TJ, Lister G, Leeson-Payne C. 1992. Breastmilk and subsequent intelligence quotient in children born preterm. *Lancet* 339:261–264.

Mendola P, Buck GM, Vena JE, Zielezny M, Sever LE. 1995. Consumption of PCB-contaminated sport fish and risk of spontaneous fetal death. *Environ Health Perspect* 103:498–502.

Muckle G, Ayotte P, Dewailly E, Jacobson SW, Jacobson JL. 2001. Determinants of polychlorinated biphenyls and methylmercury exposure in Inuit women of childbearing age. *Environ Health Perspect* 109:957–963.

Muller JF, Hulster AA, Papke OC, Ball MC, Marschner H. 1994. Transfer of PCDD/PCDF from contaminated soils into carrots, lettuce and peas. *Chemosphere* 29:2175–2181.

NCHS (National Center for Health Statistics). 2003. *National Health and Nutrition Examination Survey.* Online. Centers for Disease Control and Prevention. Available at http://www.cdc.gov/nchs/nhanes.htm. Accessed January 30, 2003.

Noren K. 1983. Levels of organochlorine contaminants in human milk in relation to the dietary habits of the mothers. *Acta Paediatr Scand* 72:811–816.

Nutrient Data Laboratory. 2002. *USDA National Nutrient Database for Standard Reference, Release 15.* Online. ARS, USDA. Available at http://www.nal.usda.gov/fnic/foodcomp. Accessed December 9, 2002.

Petroske E, Zaylskie RG, Feil VJ. 1998. Reduction in polychlorinated dibenzodioxin and dibenzo-furan residues in hamburger meat during cooking. *J Agric Food Chem* 46:3280–3284.

Playford RJ, Macdonald CE, Johnson WS. 2000. Colostrum and milk-derived peptide growth factors for the treatment of gastrointestinal disorders. *Am J Clin Nutr* 72:5–14.

Putnam JJ, Allshouse JE. 1993. *Food Consumption, Prices, and Expenditures, 1970–92.* Statistical Bulletin No. 867. Washington, DC: USDA.

Putnam JJ, Allshouse JE. 1999. *Food Consumption, Prices and Expenditures, 1970–97.* Statistical Bulletin No. 965. Washington, DC: USDA.

Ramos L, Torre M, Laborda F, Marina ML. 1998. Determination of polychlorinated biphenyls in soybean infant formulas by gas chromatography. *J Chromatogr A* 823:365–372.

Rappe C, Bergqvist PA, Kjeller LO. 1989. Levels, trends and patterns of PCDDs and PCDFs in Scandinavian environmental samples. *Chemosphere* 18:651–658.

Rawls O. 2001. Perceptions of fish safety: Voices from the community. In: *National Risk Communication Conference. Proceedings Document.* Research Triangle Park, NC: RTI International. Pp. II-17–II-19.

Rogan WJ, Gladen BC, McKinney JD, Carreras N, Hardy P, Thullen J, Tingelstad J, Tully M. 1986. Polychlorinated biphenyls (PCBs) and dichlorodiphenyl dichloroethene (DDE) in human milk: Effects of maternal factors and previous lactation. *Am J Public Health* 76:172–177.

Rose M, Thorpe S, Kelly M, Harrison N, Startin J. 2001. Changes in concentration of five PCDD/F congeners after cooking beef from treated cattle. *Chemosphere* 43:861–868.

Ryan AS. 1997. The resurgence of breastfeeding in the United States. *Pediatrics* 9:E12.

Schecter A, Papke O. 1998. Comparisons of blood dioxin, dibenzofuran, and coplanar PCB levels in strict vegetarians (vegans) and the general United States population. *Organohalogen Compd* 38:179–182.

Schecter A, Startin J, Wright C, Kelly M, Papke O, Lis A, Ball M, Olson JR. 1994. Congener-specific levels of dioxins and dibenzofurans in U.S. food and estimated daily dioxin toxic equivalent intake. *Environ Health Perspect* 102:962–966.

Schecter A, Dellarco M, Papke O, Olson J. 1998. A comparison of dioxins, dibenzofurans and coplanar PCBs in uncooked and broiled ground beef, catfish and bacon. *Chemosphere* 37:1723–1730.

Schlanger S, Shinitzky M, Yam D. 2002. Diet enriched with omega-3 fatty acids alleviates convulsion symptoms in epilepsy patients. *Epilepsia* 43:103–104.

Schmitt CJ, Zajicek JL, May TW, Cowman DF. 1999. Organochlorine residues and elemental contaminants in U.S. freshwater fish, 1976–1986: National Contaminant Biomonitoring Program. *Rev Environ Contam Toxicol* 162:43–104.

Schwartz PM, Jacobson SW, Fein G, Jacobson JL, Price HA. 1983. Lake Michigan fish consumption as a source of polychlorinated biphenyls in human cord serum, maternal serum, and milk. *Am J Public Health* 73:293–296.

Seddon JM, Rosner B, Sperduto RD, Yannuzzi L, Haller JA, Blair NP, Willett W. 2001. Dietary fat and risk for advanced age-related macular degeneration. *Arch Ophthalmol* 119:1191–1199.

Siscovick DS, Raghunathan TE, King I, Weinmann S, Wicklund KG, Albright J, Bovbjerg V, Arbogast P, Smith H, Kushi LH, Cobb LA, Copass MK, Psaty BM, Lemaitre R, Retzlaff B, Childs M, Knopp RH. 1995. Dietary intake and cell membrane levels of long-chain *n*-3 polyunsaturated fatty acids and the risk of primary cardiac arrest. *J Am Med Assoc* 274:1363–1367

Sloan R, Jock K. 1990. *Chemical Contaminants in Fish from the St. Lawrence River Drainage on Lands of the Mohawk Nation at Akwesasne and Near the General Motors Corporation/Central Foundry Division, Massena, NY, Plant.* Tech. Doc. 90-1 (BEP). Albany, NY: New York State Department of Environmental Conservation.

Smiciklas-Wright H, Mitchell DC, Mickle SJ, Cook AJ, Goldman JD. 2002. *Foods Commonly Eaten in the United States: Quantities Consumed per Eating Occasion and in a Day, 1994–1996.* Online. USDA. Available at http://www.barc.usda.gov/bhnrc/foodsurvey/Products9496.html. Accessed June 16, 2003.

Svensson BG, Nilsson A, Hansson M, Rappe C, Akesson B, Skerfving S. 1991. Exposure to dioxins and dibenzofurans through the consumption of fish. *N Engl J Med* 324:8–12.

Terry P, Lichtenstein P, Feychting M, Ahlbom A, Wolk A. 2001. Fatty fish consumption and risk of prostate cancer. *Lancet* 357:1764–1766.

Terry P, Wolk A, Vainio H, Weiderpass E. 2002. Fatty fish consumption lowers the risk of endometrial cancer: A nationwide case-control study in Sweden. *Cancer Epidemiol Biomarkers Prev* 11:143–145.

Tsutsumi T, Iida T, Hori T, Yanagi T, Kono Y, Uchibe H, Toyoda M. 2000. Levels of PCDDs, PCDFs and co-PCBs in fresh and cooked leafy vegetables in Japan. *Organohalogen Compd* 47:296–299.

USDA (U.S. Department of Agriculture). 1996. *The Food Guide Pyramid.* Home and Garden Bulletin No. 252. Washington, DC: USDA.

USDA/HHS. 2000. *Nutrition and Your Health: Dietary Guidelines for Americans,* 5th ed. Home and Garden Bulletin No. 232. Washington, DC: U.S. Government Printing Office.

Waller DP, Presperin C, Drum ML, Negrusz A, Larsen AK, van der Ven H, Hibbard J. 1996. Great Lakes fish as a source of maternal and fetal exposure chlorinated hydrocarbons. *Toxicol Ind Health* 12:335–345.

Welge P, Wittsiepe J, Schrey P, Ewers U, Exner M, Selenka F. 1993. PCDD/F-levels in human blood of vegetarians compared to those of non-vegetarians. *Organohalogen Compd* 13:13–17.

Wold AE, Adlerberth I. 2000. Breast feeding and the intestinal microflora of the infant—Implications for protection against infectious diseases. *Adv Exp Med Biol* 478:77–93.

Wong K. 1997. *Fishing for Food in San Francisco Bay.* Oakland, CA: Save San Francisco Bay Association.

Work Group on Breastfeeding. 1997. American Academy of Pediatrics: Breastfeeding and the use of human milk. *Pediatrics* 100:1035–1039.

Yang J, Shin D, Park S, Chang Y, Kim D, Ikonomou MG. 2002. PCDDs, PCDFs, and PCBs concentrations in breast milk from two areas in Korea: Body burden of mothers and implications for feeding infants. *Chemosphere* 46:419–428.

Zabik ME, Zabik MJ. 1999. Polychlorinated biphenyls, polybrominated biphenyls, and dioxin reduction during processing/cooking food. In: Jackson LS, Knize MG, Morgan JN, eds. *Impact of Processing on Food Safety.* New York: Kluwer Academic/Plenum Publishers. Pp. 213–231.

6

Framework for the Development of Policy Options to Reduce Exposure to Dioxins and Dioxin-like Compounds

A central element of the committee's charge was to identify, evaluate, and recommend policy options for reducing human exposure through diet to dioxin and dioxin-like compounds (referred to collectively as DLCs). In considering DLC exposure reduction options, the committee was charged with taking into account both possible reductions in human health risk associated with reducing DLC intake and the possible nutritional consequences of dietary changes that might result from the selected measures. In response to this charge, the committee has identified a number of possible exposure reduction (or risk-management) options, which are described in Chapter 7; in Chapter 8, the committee makes risk-management recommendations.

One of the committee's primary conclusions, however, is that the data required to evaluate and recommend a comprehensive set of risk-management interventions for DLCs are lacking. There remains substantial uncertainty about the magnitude of the risks posed by low levels of dietary exposure to DLCs and corresponding uncertainty about the magnitude of the risk reductions that are achievable by reducing such exposures. This uncertainty will be a substantial limiting factor for the government as it considers risk-management strategies, especially to the extent it considers the use of its traditional food safety regulatory tools, as discussed below. The study's sponsors, however, asked the committee not to evaluate or repeat the extensive risk assessment work that has been done on DLCs (see AEA Technology, 1999; ATSDR, 1998; EPA, 2000; Fiedler et al., 2000; IARC, 1997; and Scientific Committee on Food, 2000, 2001). In keeping with its charge, the committee has proceeded with its identification of risk-management options on the premise that reductions in DLC exposure are desirable for health reasons in light of: (1) the extensive animal data demonstrating the toxicity of DLCs at very low exposure levels, (2) the cumulative effect even low levels of exposure can have on an individual's total DLC body burden, and (3) the existence of vulnerable subpopulations, such as fetuses and breastfeeding infants and individuals (e.g., subsistence fishers) whose traditional diets, for cultural or economic reasons, may result in higher levels of DLC exposure.

Even with the assumption that reducing dietary exposure to DLCs is desirable, insufficient data are available to assess and support at least some of the potential risk-management options for DLCs. For example, the data on DLCs in the food supply, including the incidence, levels, and geographical distribution of DLC contamination in various animal production systems and in raw, processed, and cooked foods, are extremely limited, as discussed in Chapters 4 and 5. Such data would be required to most effectively target and prioritize risk-management interventions and support the establishment of regulatory limits on DLCs in food, if the government considered that an appropriate option for reducing DLC exposure. The committee also found very little data that would help to assess the feasibility and calculate the costs and other impacts of the wide range of risk-management options that are potentially available to reduce DLC exposure through the food supply. This paucity of data is particularly evident in the animal production arena, where the ecology and epidemiology of these compounds under normal production conditions are not available. Further examples of data limitations that affect the consideration of risk-management options are provided later in this chapter and in Chapter 7.

Because of the large gaps in the data required to devise an optimal, comprehensive risk-management strategy for DLCs and to assess specific options, the committee was unable to reach conclusions on the likely effectiveness of a number of the possible risk-management options it identified, or on a relative ranking of these options. It thus focused its efforts on devising a framework for the development and evaluation of options for reducing DLC exposure. The proposed framework is described in this chapter.

The purpose of the framework is to provide a systematic basis to identify risk-management options and to collect and analyze the data required to evaluate the options. As discussed below, the framework divides the universe of options into three categories: (1) options to reduce DLCs in animal forage and feeds, (2) options to reduce DLCs in the human food supply, and (3) options to change food-consumption patterns to reduce DLC exposure. The framework is suitable for evaluating a wide array of risk-management options, including regulatory interventions, voluntary and collaborative interventions, and nonregulatory incentives. In identifying possible options, the committee assumed that the government would be operating within current statutory authorities; the key food safety statutory provisions and other legal tools applicable to DLCs are thus summarized in this chapter.

Finally, the framework addresses the committee's charge to consider the possible nutritional consequences of dietary changes that might occur as a result of measures taken to reduce exposure to DLCs. The committee recommends for this purpose an approach called "risk-relationship analysis," which takes a comprehensive approach to assessing how measures taken to prevent or mitigate a particular risk can affect other risks to individuals or populations, whether by causing a countervailing increase in risk or by causing ancillary risk reductions. This chapter concludes with a discussion of risk-relationship analysis and how it applies to the evaluation of risk-management options for DLCs.

CATEGORIZING POSSIBLE RISK-MANAGEMENT OPTIONS

In developing an inventory of possible risk-management options to reduce exposure to DLCs, the committee concluded that opportunities to reduce dietary exposure fall into three major categories, which correspond to the three distinct stages of food production, processing, and consumption within the farm-retail chain. The committee's framework organizes the possible options into these three categories, each of which has its own distinct attributes and its own features that affect the feasibility and effectiveness of possible interventions.

Animal Production Systems

As discussed in Chapter 4, the forage and feeds consumed by food-producing animals, including fish, is one of the most critical pathways for the entry of DLCs into human food and for human exposure to these compounds. It is also the most promising pathway that provides an opportunity to interdict and reduce DLC contamination of the human food supply.

One major source of DLC contamination is air deposition of DLCs onto plants that are used for forage and feeds. When animals consume plants, their by-products, sediment-contaminated waters, or soils directly, DLCs are ingested and stored in the animals' fat. When humans consume the animals' fat as part of their diets, they are exposed to DLCs.

The importance of animal feeds as sources of DLCs in human diets is enhanced by the practice at abattoirs of rendering fats and animal proteins from fallen (recently dead) animals and offal not acceptable for human consumption, and using the salvaged fat in animal feeds, which is then distributed nationally in the animal feed system. This practice has the effect of recycling and potentially concentrating the DLCs that are present in animal fat because of environmental exposures, and of increasing DLC exposure to humans through the food supply. Interrupting this cycle of DLC contamination of the food supply via the animal feed pathway is, in the committee's judgment, a high-priority risk-management option.

The advantage of forage and feeds as a focal point for measures to reduce DLC exposure is that steps taken at this stage can prevent DLCs from entering human food. Such steps include limiting access to forage in areas known to be high in DLC contamination and limiting the use of recycled animal fat in animal feeds or setting direct limits on the levels of DLCs permitted in animal feeds. Measures of this kind impose costs (e.g., loss of use of land for grazing in highly contaminated areas, increased production costs) and may have other consequences that the committee believes should be considered under the proposed framework. However, any reductions in DLC contamination that are achieved at the forage and feed stage have the important benefit of directly reducing the reservoir of DLCs in food to which humans are potentially exposed.

Human Food

The second major category of opportunities to reduce DLC exposure identified by the committee involves reducing DLC levels in the foods people consume. Once DLCs are present in human food or in the fat of food-producing animals, risk-management options are available, but they may be more limited in their potential to reduce DLC exposure than interventions at the forage and feed stage. In some cases, such as when DLCs are present on the exterior surfaces of fruits and vegetables through direct airborne deposition onto leaves or the presence of DLC-contaminated soil that clings to peels, washing leafy surfaces or removing the outer peels from produce can reduce DLC exposure. DLCs can also be reduced in human food by trimming excess fat from cuts of meat or otherwise reducing the amount of animal fat in food. At present, processing methods to directly remove DLCs from animal foods do not exist.

When the removal or reduction of DLCs through washing or other processing interventions is not possible, reduction of DLC exposure at the human-food stage can be achieved by limiting the levels of DLCs permitted in the foods. This would require removing foods from the food supply if they contain DLCs above some permissible level. Because human foods or food components generally have greater economic value than animal feeds or feed components, the cost issues that would have to be considered under the food safety laws or in the committee's analytical framework could be greater. However, this would depend on a wide range of cost factors associated with intervention strategies at different points in the farm-retail supply chain.

Food-Consumption Patterns

Finally, the committee identified changes in individual food-consumption patterns as a vehicle for reducing DLC exposure. This involves reducing the consumption of foods or food components known or expected to be high in DLCs and substituting foods that are likely to be lower in DLCs. Because individual food-consumption patterns are generally beyond direct regulatory control, the risk-management options in this category are likely to involve education, expansion of food choices, incentives, and other interventions to encourage voluntary changes in what people eat.

The options for reducing DLC exposure through changes in food-consumption patterns are likely to vary in their feasibility and effectiveness depending on the circumstances of the individual. For example, changing consumption patterns may be very difficult for indigenous populations highly dependent on wild-caught fish and other game from areas with high environmental contamination levels. It may be relatively easy for school-age children who can reduce DLC intake by replacing whole milk with low-fat (1 percent milk fat) or skim milk. The health benefit of changing consumption patterns to reduce DLC exposure may also vary considerably depending on age and other factors. For example, adolescent girls and their future offspring might benefit substantially by reducing the animal fat content of their diets and thus their DLC intake, while the benefits for adults may not be as great. These factors should be considered in designing risk-management interventions and analyzing them in the committee's proposed framework.

THE RANGE OF POSSIBLE RISK-MANAGEMENT OPTIONS

As suggested by the preceding discussion and presented in more detail in Chapter 7, there is a wide range of possible risk-management options for reducing DLC exposure from food. The range is defined not only by the three categories or stages in the food chain that are outlined above, but also by the type of intervention. These include: (1) the use of traditional food safety regulatory tools, (2) collaborative and voluntary interventions involving food safety agencies and the food industry, (3) public policy and regulatory interventions beyond traditional food safety interventions, including subsidies, economic incentives, and other measures to reduce DLCs and expand dietary options, and (4) information and education interventions. All of these have been considered by the committee in identifying possible risk-management options, and they are addressed by the committee's proposed framework.

Traditional Food Safety Regulatory Tools

The most common risk-management intervention by the federal government with respect to food contaminants such as DLCs is the enforcement of the food adulteration provisions in the Federal Food, Drug, and Cosmetic Act (FDCA) of 1938 and other federal food safety laws that deal specifically with meat and poultry. Jurisdiction over food safety is shared between the U.S. Food and Drug Administration (FDA), the U.S. Department of Agriculture (USDA), and the U.S.

Environmental Protection Agency (EPA). As discussed in more detail below, current laws prohibit the presence in food of substances that are potentially harmful to health, and they authorize FDA to act to prevent contamination of the food supply with such substances, either by setting and enforcing standards for the amount of such substances that can be present, or by bringing case-by-case enforcement actions against the products that the government believes are contaminated at harmful or potentially harmful levels. FDA has used these tools to restrict the presence of mercury, lead, aflatoxin, and other naturally occurring and man-made contaminants in foods. The central theme of these and other traditional food safety regulatory tools is that they involve the use of the government's legal powers to prohibit harmful contamination or practices that could result in such contamination.

Collaborative and Voluntary Interventions

As an alternative to mandatory regulatory interventions, the food safety agencies have the option to initiate collaborative programs with the food industry dependent on the willingness of the industry to act voluntarily to reduce a hazard. FDA could, for example, work with animal producers to develop guidance or voluntary codes of practice that producers could follow to reduce DLCs in animal forage or feeds, or the government could provide technical or other support for industry-sponsored DLC testing programs. Such collaborative and voluntary measures assume that the industry has its own interest or incentive for taking action to reduce DLC levels in foods. In such cases, they may be viable risk-management options, especially if the data required to support the use of traditional regulatory tools are unavailable.

Subsidies, Economic Incentives, and Other Measures to Reduce DLCs and Expand Dietary Options

Beyond the traditional regulatory or cooperative approaches to reducing DLCs, there are a number of other tools of public policy that, though rarely used to address food safety hazards, are at least theoretically available to the government to foster reductions of DLCs. These could include, for example, subsidies or other economic incentives, such as taxes, to induce reductions in the use of animal fat in animal feeds; development of alternative uses for animal fats other than feeds (e.g., biofuels); or the reduction of the fat content of processed foods. The government could also act to expand the information and choices available to consumers who might want to reduce animal fat in their diets, such as by expanding nutrition labeling of meat or mandating the availability of low-fat or skim milk in school lunch and other government feeding programs.

Information and Education Interventions

The final category of possible risk-management interventions is aimed primarily at consumers and involves the use of information and educational campaigns to encourage the changes in dietary patterns that would be expected to reduce DLC exposure. Such campaigns might be aimed at indigenous populations whose traditional diets are drawn from contaminated areas, or at mainstream consumers who might achieve reductions in DLC exposure by conforming their diets to the government's Dietary Guidelines or to the Food Guide Pyramid.

LEGAL FRAMEWORK FOR THE IMPLEMENTATION OF INTERVENTIONS

As noted earlier, the committee assumed for purposes of identifying and analyzing risk-management options that the government would continue to operate under current statutes, although legislation to address a particular hazard is always possible. The following summary focuses on these options because they traditionally have been the primary means through which the government has acted to reduce dietary intakes of environmental contaminants such as DLCs. There are, however, constraints in the form of substantive standards and burdens of proof that have an important effect on what the government can do to address a particular hazard. These constraints are among the reasons the committee has considered a range of nonregulatory, as well as regulatory, options for reducing dietary intakes of DLCs.

Food Safety Regulatory Tools

Among the regulatory agencies in the federal government, FDA has the primary food safety regulatory jurisdiction over DLCs in animal feeds and human food categories other than meat and poultry. For meat and poultry, USDA's Food Safety and Inspection Service (FSIS) plays the enforcement role. FDA is the federal government's food safety standard-setting agency for environmental contaminants, other than pesticides. EPA considers food safety and acceptable levels of DLCs in fish by setting acceptable air and water emission levels for DLCs.

The FDCA provides FDA with the broad authority to control or prohibit the presence in food of contaminants such as DLCs. The authority is broad in the sense that FDA can act to control DLCs in both animal feeds and human food; can intervene at virtually any stage of food production, processing, and marketing; can set mandatory standards applicable to broad categories of DLC-containing feeds and foods; and can take case-by-case enforcement action to remove DLC-contaminated food from commerce. FDA's authority is constrained, however, by the fact that its exercise is conditioned on the agency being able to make certain factual showings concerning DLCs and the risks they pose.

Statutory Adulteration Provisions

FSIS has similar adulteration authorities for meat and poultry. Under the Poultry Products Inspection Act and the Federal Meat Inspection Act, FSIS may take action against meat or poultry that contains an added poisonous or deleterious substance that may render the product injurious to health. Also, FSIS has authority to act against a product that is for any reason unhealthful, and against product that has been prepared under insanitary conditions whereby it may have been rendered injurious to health.

Under these provisions, DLCs generally would qualify as poisonous or deleterious substances. The "may render" adulteration standard has been interpreted by the courts to mean that there must be a "reasonable possibility" that the substance will cause harm at the levels at which it is present in food. This standard does not require proof of actual harm to health, but it does require scientific evidence demonstrating that there is a reasonable possibility of harm to at least some who consume the food. Likewise, the adulteration provision that deems added but avoidable poisonous or deleterious substances to be adulterants requires factual evidence that the particular level of the substance in food is avoidable using GMPs. The practical meaning of this avoidability standard of adulteration has not been clearly defined by FDA or the courts. As persistent and pervasive environmental contaminants, DLCs would likely be considered unavoidable at some low concentration, but determining the line between avoidability and unavoidability for DLCs poses complex factual and policy issues, and that line has not been drawn.

To enforce the first two adulteration provisions noted above, FDA could choose to proceed on the basis of case-by-case enforcement action, through informal or formal standard-setting, or a combination of these two. FDA has the option, if it can prove that a particular lot of food contains an environmental contaminant at a level that violates one of the adulteration provisions, to seek court action to remove that lot from commerce and prevent its future distribution. This can be done without any prior notice or setting of standards concerning the level of the contaminant the agency considers to be in violation of one of the adulteration provisions. FDA rarely proceeds on this basis, and it would be an inefficient way to address a pervasive environmental contaminant such as DLCs.

FSIS has the authority to withhold the mark of inspection from product, which is necessary for the product to enter commerce, if it is not able to find that the product is not adulterated. Once the product is in commerce, FSIS's authority is similar to FDA's.

Action Levels

Instead, prior to taking court enforcement action, FDA typically announces in advance the level of an environmental contaminant against which it is prepared to act. FDA calls these levels *action levels*. They are not regulations and do not have any binding legal effect on FDA or on private parties. Technically, they are guidelines for use by FDA enforcement officials in deciding when to bring court enforcement action against a contaminated food or its purveyor under one of the adulteration provisions. In such an enforcement action, FDA still has to prove to a court's satisfaction that an adulteration provision has been violated. As a practical matter, however, FDA's action levels operate as guidance to the food industry and are recognized by many food producers and processors as informal standards with which they seek to comply. To date, FDA has not set action levels for DLCs as a class, though it has established an action level for polychlorinated biphenyls (PCBs) in meat.

Tolerances

In the case of added substances that are at least to some extent unavoidably present in food, as in the case of environmental contaminants like DLCs, FDA has statutory authority to establish binding legal limits on the level of the substance that may lawfully be present in food. These limits are called *tolerances*. They are set at the level FDA considers necessary to protect public health, taking into account the extent to which the presence of the substance in food is unavoidable. Tolerances are regulations that the FDCA requires be established under "formal" rulemaking procedures, which involve public notice and comment and formal trial-like hearings. These rulemaking procedures are costly and time consuming. The advantage of tolerances, from an enforcement perspective, is that they are legally binding and directly enforceable through court action without any further showing by FDA that the food is unsafe or adulterated. The only tolerances FDA has established to date are for PCBs in certain human foods (milk, dairy products, poultry, eggs, fish, infant and junior foods) and in animal feeds and feed components for food-producing animals.

Regulatory Limits

Due to the cumbersomeness of the formal rulemaking procedures required to establish tolerances, FDA has created an "informal" rulemaking process for establishing regulatory limits for added poisonous or deleterious substances. Although FDA has yet to use this procedure to establish any regulatory limits, it is available in cases in which FDA is not prepared to establish a tolerance (because, for example, the levels of a contaminant may be in flux), but is prepared to define the level at which it deems the contaminant to adulterate food because it "may render [the food] injurious to health" (Merrill and Schewel, 1980). Regulatory limits would be established following public notice and comment, but without the formal hearings required for tolerances. It is uncertain how much deference courts would accord FDA's regulatory limits in an enforcement case.

Good Manufacturing Practices

The provision that declares food to be adulterated based on the conditions under which it has been prepared has been used by FDA to promulgate regulations establishing GMPs and other requirements for food production that are intended to prevent potentially harmful contamination. FDA has established GMP regulations for animal feeds and human food, but these currently address general practices for safe and sanitary food production; they do not establish any standards or requirements for preventing or minimizing contamination with specific contaminants such as DLCs. FDA could attempt to use its GMP authority for this purpose, but the agency would have to support such regulations with evidence that establishes a link between any specific GMPs for DLCs and the prevention of situations in which the food may be injurious to health. This evidence would be difficult to provide given the lack of data on these linkages.

Practical Constraints on Regulatory Options

While FDA has successfully used these regulatory tools to address a wide range of food safety issues, including hazards posed by environmental contaminants such as lead, mercury, and aflatoxin, the agency would face significant practical constraints in applying them to DLCs in feeds and food. As discussed in Chapter 2, there continues to be scientific debate and uncertainty about the level of human intake of DLCs required to cause adverse health effects in the population at large and in specific subpopulations, especially in light of the fact that the risks of DLCs are generally thought to be associated with their accumulation in human tissue over time. This uncertainty does not necessarily preclude, but would certainly complicate, any effort by FDA to establish tolerances or other regulatory standards through rulemaking or to enforce adulteration provisions through court proceedings. In such rulemaking and court enforcement proceedings, FDA would bear the burden of proving the potential of DLCs to cause harm to human health at relatively low intake levels.

In addition, effective compliance monitoring and enforcing regulatory standards both require the availability of practical analytical methods and a credible level of testing to detect violations. The expense of available testing methods creates a barrier to enforcing regulatory standards for DLCs. The difficulty of meeting this legal burden, in light of the uncertainty about risk at low exposure levels and the paucity of data on actual levels of DLCs in food, is one of the reasons the committee is not recommending any immediate action using these traditional food safety regulatory tools. The committee is aware of the recent initiative of the European Union (EU) to establish quantitative standards for DLCs in feeds and food, as discussed in Chapter 2, but notes that EU regulatory policymakers are not subject to the same legal constraints as U.S. regulators since, in the EU system of government, the adoption of the DLC policy was itself a legislative act.

Voluntary Guidance, Cooperative Programs, and Subsidies to Reduce DLCs

Regulatory agencies, including FDA and FSIS, commonly supplement their regulatory standard setting and enforcement programs with collaborative programs to encourage and support voluntary efforts by food producers and processors to address health, safety, and environmental problems. Such efforts can be productive when private entities are provided incentives to address issues when regulatory mandates are absent, and they are often pursued in circumstances in which there are legal, scientific, resource, or other practical constraints that make regulatory intervention difficult or impracticable. For example, both FDA and FSIS have begun to work with food producers to foster voluntary good agricultural practices (GAPs) and other measures at the food production level to address food safety hazards posed by microbial pathogens.

The committee has identified a number of voluntary efforts as options to reduce DLC levels in feeds and food. They include setting nonbinding targets for reducing DLC levels in feeds and food, working with the industry to establish and encourage the adoption of voluntary GAPs and GMPs, and providing data on the current levels and sources of DLCs in feeds and food to support voluntary reduction efforts. There are no legal constraints on such efforts.

The committee has also identified various economic subsidies to food producers and processors as possible options for reducing DLC levels in feeds and food. These options may be more novel as a policy matter and require additional resources that are not provided for in the budgets of the food safety regulatory agencies. In addition, the regulatory agencies likely lack legal authority for such subsidies. These options would thus require congressional action to implement.

Reducing Dioxin Intake by Changing Food-Consumption Patterns

Perhaps the most direct way for an individual or a population to reduce dietary intake of DLCs is to reduce their consumption of dietary fat, especially from animal sources that are known to contain higher levels of these compounds. The federal government has little or no legal authority to directly control what people eat, but the committee has identified a number of options to encourage changes in food-consumption patterns that, if achieved, would reduce dietary intake of DLCs. These include the provision of information to consumers about DLC levels in various foods, public education programs, and changes in government feeding programs to facilitate and encourage the consumption of foods less likely to be contaminated with DLCs. Most of these efforts could be pursued within the current statutory authorities of the food safety and other federal agencies. In some cases, however, changes in existing regulations would be required.

ANALYTICAL TOOLS FOR HEALTH RISK DECISION-MAKING

In recent years, policymakers and regulators have developed the *risk analysis* framework as a means of organizing the information needed to make decisions about controlling health and environmental risks. The framework starts with *risk assessment*, which is a method for identifying the possible harms that may result from a specified activity and estimating the probability that such harms will occur and the number of people that may be affected. It continues with *risk management*, which is the process of weighing policy alternatives and, if required, selecting and implementing appropriate control options. The final element of the framework is *risk communication*, which is the interactive exchange of information and opinions concerning risk among risk assessors, risk managers, consumers, and other interested parties.

As noted previously, the sponsors of this study did not charge the committee with conducting a risk assessment for DLCs in the food supply, but rather asked it to rely on already existing information. The committee was charged to develop and assess risk-management options for reducing DLC exposure from foods. In considering these options, it was also asked to incorporate *risk-risk* or *risk-tradeoff analysis*. This approach is a method for evaluating regulatory and other options to reduce risk by highlighting the risks that may be created by these actions. It provides a means for considering multiple countervailing risks that may indicate that it is riskier to remediate a problem than to take no action. As discussed in detail below, the committee chose to employ a broader framework, termed *risk-relationship analysis*, which takes into consideration both the countervailing risks and the ancillary benefits that may result from a particular risk-management option. The committee also considered several aspects of risk communication, particularly for management options that involve voluntary actions by consumers or industry.

The objective of the committee's approach, as presented in the framework, is to point out the types of information needed for comprehensive risk-management decisions and to explain the range of risk-management options that were considered by the committee in making its final recommendations.

The choice of efficient risk-management options may make use of *cost-effectiveness* or *benefit-cost analysis*. Both are methods of economic-welfare analysis that attempt to comprehensively evaluate the beneficial and adverse effects of a policy. Cost-effectiveness analysis measures the primary effects (i.e., the benefits) in some natural units (e.g., cases of illness prevented, Quality Adjusted Life Years) and other consequences (i.e., the costs) in monetary units. Benefit-cost analysis measures all outcomes in monetary units. The essential idea is to choose risk-management options that provide the most risk reduction for a given cost (cost-effectiveness analysis) or offer the largest surplus of benefits over costs (benefit-cost analysis).

While some elements of the application of these economic approaches to the reduction of DLC exposure from food are discussed below and in Chapters 7 and 8, the committee was not able to conduct a comprehensive analysis of this type that would allow evaluation across DLC management options because it would require resources (e.g., time, data, and money) to conduct such an analysis that was far beyond the ability of the committee. Further, the committee recognizes the importance of using *comparative-risk analysis* as a method of identifying government priorities among diverse health-related risks and allocating resources across those risks. Such between-risk comparisons were beyond the scope of this study.

RISK-RELATIONSHIP ANALYSIS

Past government efforts to reduce risks posed by environmental contaminants in food generally have focused on the health risk posed by the contaminant. They have relied primarily on the traditional food safety regulatory tools outlined earlier, sometimes coupled with public advisories, such as ones issued to encourage reductions in the consumption of certain fish due to high levels of mercury. Other essentially economic considerations have also played a role in regulatory choices. For example, the FDA tolerances for PCBs in fish were established on the basis of a consideration of how much fish would have to be removed from the food supply to comply with the ultimately enacted tolerance of 2 ppm. This consideration ultimately served as a practical way of assessing what levels of contamination would be deemed avoidable (above 2 ppm) and unavoidable (below 2 ppm).

In contrast, the charge to the committee was to look beyond these traditional considerations in food safety regulation, both to consider a broader array of possible risk-management options and to consider a broader array of factors in evaluating possible options. Specifically, the study sponsors asked the committee to consider the possible nutritional consequences of dietary changes that might result from measures taken to reduce DLC exposure. The hypothesis underlying this charge was that such dietary changes, which might stem, for example, from increases in food prices due to costs imposed by new regulatory requirements or by the shifting of food-consumption patterns away from animal fat, might create countervailing health risks by depriving some individuals of needed nutrients.

While the consideration of countervailing health risks is not common in food safety regulation, it is an appropriate approach in any public health context where interventions that are being considered to reduce a target risk could plausibly increase some other health risk. Consideration of such countervailing risks is sometimes called risk-risk or risk-tradeoff analysis. In the case of DLCs in food, however, the committee considered it at least equally plausible that dietary changes to reduce DLC exposure could have ancillary health benefits, in addition to whatever health benefit was achieved by reducing DLC intake. In particular, reductions in the consumption of animal fat to reduce DLC exposure may also be associated with a number of other health benefits, including reductions in cardiovascular disease and cancer. The committee thus decided to adopt a framework for analysis of DLC risk-management options that is broader than a risk-risk or risk-tradeoff framework. It involves risk-relationship analysis.

In the context of diet-related issues, risk-relationship analysis recognizes that most human foods provide a combination of desirable and undesirable components. In addition to attractive taste, texture, and other aesthetic attributes, desirable components include macronutrients (protein, carbohydrate, certain fats, and fiber) and micronutrients (vitamins and minerals) that are required to support health and maintain adequate nutritional status. Other food components that may be undesirable and pose health concerns for consumers include excess levels of saturated fats and cholesterol and both naturally occurring and man-made foodborne toxins. Components that should not be present in foods include microbial pathogens, pesticides, and other hazardous contaminants. At a fundamental level, optimizing dietary choice can be viewed as choosing a mixture of foods that provides adequate and balanced qualities of desirable components, while minimizing health risks associated with the consumption of undesirable components or contaminants.

Defining a "healthy diet" remains a matter of controversy, precisely because food is compositionally so complex, dietary patterns are so varied, and there remains considerable scientific uncertainty about the interactions of dietary components and the impact of changes in dietary patterns. Nevertheless, the U.S. Dietary Guidelines for Americans (USDA/HHS, 2000), which establishes the policy, and the Food Guide Pyramid (USDA, 1996), which provides the educational component, reflect consensus recommendations about desirable intakes of micro- and macronutrients. The committee concludes that, in considering risk-management options to reduce DLC exposure, it is appropriate to conduct a risk-relationship analysis that considers whether a proposed intervention would affect dietary patterns, for better or worse, in relation to current dietary guidelines.

Ideally, a risk-relationship analysis would provide a quantitative basis for considering and comparing the risk reduction likely to be achieved by the proposed intervention, any countervailing health risks the intervention might cause, and any expected ancillary health benefits (see Box 6-1). Proposed interventions that have positive net health impacts, based on an aggregation of such quantitative assessments, would be candidates for consideration, subject to whatever

---

BOX 6-1 Risk-Relationship Analysis

Risk-relationship analysis is an analytical method that can be used to evaluate the countervailing risks and ancillary benefits of a policy option, and then to compare them with the target risk reduction. Risk-relationship analysis includes the following steps:

1. Identify and determine the magnitude of the target risk reduction expected to be achieved by a risk-reduction option.
2. Identify possible countervailing risks and ancillary health benefits of the risk-reduction option. This requires information on the likely or possible actions that food producers, consumers, and others will take in the event of a specific policy.
3. Evaluate changes in exposure to the agents directly responsible for the countervailing risks or ancillary benefits. The magnitude of a change in a countervailing risk or ancillary benefit typically depends on the magnitude of exposure to specific agents, such as nutrients or contaminants, and how the magnitude of the response varies in relation to changes in exposure (commonly referred to as the exposure-response relationship).
4. Characterize the exposure-response relationship for each agent over the relevant range of exposures. For many agents, it may be sufficient to describe the general direction of the relationship over the relevant range; for others, there may be thresholds of exposure below which the response does not occur or other significant nonlinearities that must be described.
5. Aggregate the changes in target risk, countervailing risks, and ancillary health benefits. It is necessary to aggregate all the significant risk changes, whether positive or negative, to determine the net risk effect of a policy.

---

legal standards and other impact assessments that might be relevant to the selection and implementation of a risk-management option.

As is made clear in the discussion of specific risk-management options in Chapter 7, it may not be possible to conduct quantitative risk-relationship analyses, and even the qualitative outcome (i.e., are net health benefits likely to be positive or negative) of a risk-management option may be unclear. This is the case for several reasons: (1) the models and data required to predict changes in dietary patterns are extremely limited, (2) a given intervention might push diets in more than one direction; for example, reductions in the consumption of full-fat dairy products could not only reduce overall intake of undesirable animal fat, but it could also reduce the overall intake of desirable calcium if the low-fat product is not consumed, and (3) the health consequences of a given dietary change will vary among demographic groups (e.g., by age and sex) and among individuals (depending on the individual's health status and the nature of the diet prior to the intervention).

Despite these complexities and uncertainties, which inevitably make the answers produced by risk-relationship analysis uncertain, questions about countervailing health risks and ancillary health benefits are worth asking. In at least some cases, they can help affirm a proposed course of action or point policy-makers in a different direction than originally proposed.

OVERVIEW OF THE PROPOSED FRAMEWORK

With the foregoing considerations in mind, the committee adopted a framework for identifying and analyzing risk-management options for reducing DLC exposure. In Chapter 7, these options are presented in three matrices corresponding to the three major categories of intervention options:

- Matrix 1: Intervention Options Considered to Reduce DLC Exposure Through Pathway I: Animal Production Systems
- Matrix 2: Intervention Options Considered to Reduce DLC Exposure Through Pathway II: Human Foods
- Matrix 3: Intervention Options Considered to Reduce DLC Exposure Through Pathway III: Food-Consumption Patterns

Within these matrices, each considered option is presented in the format shown in Figure 6-1. This format captures the series of questions the committee believes government agencies should consider in identifying and analyzing risk-manage-

Option	
Alternate/Interim Actions	
Current Barriers to Implementation	
DLC Exposure Reduction	Exposure Reduction:
Risk-Relationship Analysis	Ancillary Benefits:
	Countervailing Risks:

FIGURE 6-1 Format for analyzing intervention options considered to reduce DLC exposure.

ment options for DLCs in food: (1) possible alternate or interim actions that could be taken in the short term if the proposed intervention is not immediately feasible, (2) the current barriers to implementation of a proposed option and its practical feasibility, (3) the potential target DLC exposure reduction achievable through the proposed intervention, and (4) the existence and size of ancillary benefits and countervailing risks associated with the option.

The proposed framework and matrices respond directly to the committee's charge to consider options for reducing DLC exposure, taking into account the possible nutritional consequences of dietary changes that might result as a consequence of the measures taken. The charge to the committee and the proposed framework address the health consequences of efforts to reduce DLC exposure, not the economic consequences. The committee considers the practical feasibility of a proposed intervention to be a valid threshold question because no health benefit will accrue from an intervention that cannot feasibly be implemented. While the committee considers the economic feasibility of the options to be a key issue, it has not included benefit-cost analyses of the specific options proposed in the framework. This is primarily because consideration of the economic costs of DLC reduction options is beyond the scope of the committee's charge, except to the extent that economic costs imposed by an intervention affect dietary patterns and health. In addition, the traditional food safety regulatory tools do not provide for the consideration of benefit-cost analyses in setting food safety standards.

Nevertheless, the committee recognizes that major government regulatory decisions, even if not controlled by cost-effectiveness, benefit-cost, or other economic analyses under their authorizing legislation, are subject to regulatory impact assessments under presidential Executive Orders administered by the Office of Management and Budget in the Executive Office of the President. Moreover, such economic analyses are likely to play a role in decisions on whether to mount significant new initiatives to reduce DLCs. These analyses raise their own complicated issues, from both methodological and policy perspectives, and these issues are beyond the scope of this report.

The elements included in the committee's proposed analytical framework are described in more detail in the following sections.

Alternate or Interim Actions

Closely related to the issue of feasibility is the need to consider, for each proposed intervention, whether there are alternate or interim actions that could be taken in the short term to begin making progress in reducing DLCs or to make necessary preparations for interventions that may not be feasible at present. The committee finds that, while substantial public investments have been made to collect the information required for risk assessment with respect to DLCs (e.g., toxicity and epidemiological and exposure data), relatively little public investment has been made to collect the data and develop the tools required for risk management, such as rapid and inexpensive test methods. For example, there are very limited data on the actual distribution of DLCs in the production chain and food supply at the level of detail required to establish binding regulatory limits on DLCs in feeds or food, or for mandating significant changes in current production practices that affect the presence of DLCs in food. Thus, it is appropriate with respect to a number of the proposed interventions discussed in Chapter 7 to consider alternate or interim steps, such as developing improved test methods, expanding data collection programs, and fostering voluntary efforts to reduce DLCs.

Current Barriers to Implementation

The committee considered at length how DLCs enter the food supply, their pervasiveness, and some of the obstacles to reducing DLC exposure from food. No one desires the presence of DLCs or directly controls their entry into the food chain. DLC contamination is, for many participants in the food chain, an unavoidable phenomenon. The widespread presence of DLCs in the environment means that DLC contamination is, to some extent, embedded in the way food is produced, beginning on the farm or in waterways. This is why, in the committee's view, continued efforts to reduce environmental emissions must remain a key element of any DLC exposure reduction strategy.

The fact of widespread DLC contamination of the environment does not mean there are no feasible interventions to reduce DLC exposure through food. The committee has identified, and discusses in Chapter 7, a number of risk-management options that are more or less feasible, depending on the point in the food chain at which the option seeks to reduce DLCs and on the magnitude of the reduction being sought. Some options are clearly feasible technically, such as replacing the animal fat with high DLC levels that is contained in animal feeds with other energy sources, but they may require such significant change in current practices or may raise other issues (such as finding alternative means to use or dispose of the fat) that their practical feasibility may be legitimately questioned. Other options may simply not be technically feasible, at least at present, such as the removal of DLCs from feeds or food components through processing. In any event, technical feasibility is an appropriate first question to ask in considering DLC exposure reduction options.

An evaluation of the economic feasibility of different risk options would require a much more extensive analysis than could be carried out by the committee. Experience with evaluating other risk-reduction programs does indicate a range of private and public costs that might be associated with reducing exposure to DLCs in food. In the private sector, risk-reduction options may require increased expenditures in the supply chain, such as costs of monitoring, certification, substitution of higher priced inputs, or other changes in standard operations. Consumers may also be called upon to do more to protect themselves from DLC exposure. A significant lesson from prior research is that both short- and long-term costs should be evaluated, since innovation and new approaches may cause long-term costs to diverge from short-term costs. In the public sector, risk-management options incur costs related to monitoring, establishing programs, and setting and enforcing standards. Again, while not fully analyzed by the committee, economic feasibility remains an important consideration in evaluating risk-management options.

DLC Exposure Reduction

In the committee's proposed framework, DLC exposure reduction is the declared goal, but it should be understood to be a surrogate for risk reduction. The reason to reduce DLCs in the food supply and in diets is to reduce the potential for adverse health effects resulting from exposure and accumulation of DLCs in human tissue. The estimation of the risk reduction associated with any increment of exposure reduction is, however, fraught with uncertainty, due both to the inherent difficulties of extrapolating from test conditions in which potential hazards have been identified and to the complexity and diversity of conditions under which humans are exposed.

Numerous health effects have been ascribed to exposure to DLCs, including dermatological damage, cancer, noninsulin-dependent diabetes in adults, neurological and immune system impairments in infants, and endocrine system disruption (see Chapter 2). Many of these effects were identified in individuals subjected to high levels of exposure (e.g., through unintended industrial releases); others have been described in specific population groups such as the Ranch Hands cohort. However, information on adverse health effects caused by low-level exposure to DLCs through foods is limited.

Another source of information on the toxicity of DLCs is research on animal models. These studies, however, have inherent limitations that affect their use in human risk assessment. The dose-response effect of DLC exposure varies widely among animal species and between animal models and humans. This variability makes it difficult to establish an accurate threshold dose and leads to increased uncertainty, particularly when extrapolating toxicity findings from animal models to humans. These studies are further limited by the need to use single compounds for exposure assessments, whereas in human exposures, multiple compounds and other confounding variables are present. Estimations of risk using animal models must be approached carefully and in conjunction with other resources, including epidemiological evidence.

In addition to the uncertainties and variability inherent in the DLC health-effects data, there is wide variability in the vulnerability of human subpopulations. For example, fetuses and breastfeeding infants and young children may be especially vulnerable to the hazards posed by DLCs. This compounds the difficulty of assessing the health importance of any projected decrease in DLC exposure and underscores the need in evaluating proposed interventions to consider how the anticipated reductions in exposure will be distributed.

Despite these scientific uncertainties and complexities, the assumption underlying the charge to the committee was that it is desirable to reduce DLC exposure through food. The committee embraces this assumption as an expression of sound public health policy. The committee recognizes, however, that it will be difficult to predict the magnitude of the exposure reduction likely to result from any proposed intervention or the health benefit of the reduction, and that the benefits, whether measured in terms of direct health outcomes or the monetary value of those outcomes, will vary among individuals and subpopulations. The committee did not conduct a detailed analysis of this kind for any of the options it identified. The framework identifies exposure reduction—whether, to what extent, and for whom it occurs—to be an important question in evaluating risk-management options.

Risk-Relationship Analysis

As discussed earlier, risk-relationship analysis goes beyond consideration of the target risk reduction likely to be achieved by a proposed intervention—it also considers *ancillary benefits* and *countervailing risks* potentially caused by the intervention. In the case of options for reducing DLC exposure from food, the central source of ancillary benefits and countervailing risks are potential changes in dietary patterns that may occur as a result of implementing a risk-reduction strategy.

Increases in countervailing risks and ancillary benefits could occur at the same time and to the same or different people. For example, one approach to reduce exposure is to reduce the consumption of meat, poultry, fish, egg, and milk products that may contain undesirable levels of DLCs. However, these foods are sources of protein, iron, calcium, and vitamin D and, in the case of fish, omega-3 fatty acids. Therefore, decreases in intake may increase the risk for nutritional deficits and for some diseases. Alternatively or at the same time, there may be corollary benefits related to the decreased consumption of some animal foods. For example, a decreased intake of saturated fats and cholesterol is associated with a decreased risk for heart disease and possibly some cancers.

The committee concluded that it is appropriate to consider both the potential ancillary health benefits and countervailing health risks associated with different risk-management options, recognizing that the potential health impact of any dietary change would vary among population subgroups and individuals depending on their nutritional needs and nutritional status at the time of the dietary change. As discussed in Chapter 5, the committee utilized two data resources to help predict the outcomes of certain scenarios of dietary modification that could result from efforts to reduce DLC exposure through foods. The Continuing Survey of Food Intakes by Individuals (a food-consumption survey conducted by

USDA) was mapped to FDA's Total Diet Study to estimate probable DLC exposure by food type for various age and gender groups. The results of the analysis, summarized in Chapter 5, suggest that reductions in the consumption of animal fats can reduce DLC exposure levels to varying degrees; approximately 10 percent in the case of this analysis. Whether the reduction is significant for the general population is not known. However, it could be a way to reduce exposure to sensitive population groups in such a manner that negative nutritional consequences could be avoided.

The committee also considered the possibility that regulatory interventions to reduce DLC exposure in the diet could impose costs on the food production system that would be reflected in higher consumer prices for food, which could in turn affect dietary choices and nutrient intake levels to an extent that could have negative health consequences. The impact of the price of food on nutrient availability may be significant for some low-income, sensitive, or highly exposed groups.

The importance of such an effect depends on three factors. The first is how much food purchases and resulting nutrient intakes respond to changes in food prices. The second is whether any resulting changes in nutrient intakes are nutritionally significant, that is, are they large enough to have possible health consequences. The third is whether the changes are adverse or beneficial. Research by Huang addresses these points.

For the general population, Huang (1996) developed a methodology to measure how a change in the price of a food-product category (e.g., beef) would impact the consumption of that product category (the own-price elasticity effect) and the consumption of other food-product categories (the cross-price elasticity effect). Huang then linked these changes in consumption to changes in nutrient intakes through the use of a database containing nutrient values of different food products. He found that a 1 percent increase in the price of beef would reduce per capita food energy by 0.027 percent, protein by 0.091 percent, and fat by 0.025 percent, but vitamin A would increase by 0.064 percent, with the latter change being particularly affected by cross-product effects. Thus, while a price change affects nutrient intake, the impact in percentage terms is much smaller than the change in price due to the ability of consumers to attain nutrients from substitute products. The effects of changes in nutrient intake on health depend on whether current intakes are above or below optimal levels. Moreover, while some of the changes in nutrient intake may adversely effect health (e.g., a reduction in protein), others may be beneficial (e.g., a reduction in fat, an increase in vitamin A).

Huang (1997) went on to assess the nutritional significance of changes in nutrient intakes resulting from price changes by evaluating them relative to FDA daily values for those nutrients. In this case, he evaluated the impact on nutrient intake of a much larger change in price of 10 percent. Even at this higher level, Huang concluded "a comparison of the nutrient quantity changes with the daily values shows that the magnitudes of nutrient quantity changes are relatively small and, therefore, unlikely to be of nutritional significance" (p. 18). He cautioned, however, that the nutrient effects could exacerbate existing nutritional problems if they occurred over prolonged periods of time.

A risk-management option to reduce DLC exposure might affect the prices of more than one category of food products at the same time, potentially increasing the impact of the option on nutrient intakes. Nevertheless, it is the committee's judgment that the potential countervailing risk stemming from a causal link between increased production costs, increased food prices, changed dietary intakes, and poorer health is not likely to be an important consideration in evaluating risk-management options for reducing DLC exposure. Indeed, some of the changes in dietary intake are likely to be beneficial for most consumers (e.g., a reduction in fat). Unless price increases were quite large (e.g., greater than 10 percent) and widespread across food product categories, it is likely that the general population would be able to attain sufficient nutrients by substituting other food products for those that became more expensive. However, the impact of price changes for the nutritionally vulnerable could require closer scrutiny.

The committee also discussed a separate strand of economic analysis, health-health analysis, which focuses on expenditure tradeoffs made by consumers (Chapman and Hariharan, 1994, 1996; Keeney, 1990, 1997; Lutter et al., 1999; Viscusi, 1994). This approach argues that when risk-reduction options instituted by the government cause consumers' costs to increase, consumers must spend less on other goods, some of which promote health and safety. This decrease in spending on health and safety could result in increases in other risks that could offset any decrease in risk resulting from the government policy. The committee had diverse views on whether this countervailing risk scenario was likely to be an important factor in analyzing options to reduce DLC exposure from foods and did not include further discussion of it in Chapter 7.

RESEARCH NEEDS

The framework for discussing intervention options ends with research needs related to each option. The committee found that these needs were numerous and diverse, such as testing methodologies and techniques, sampling for DLC presence in the food-supply chain, and analyses of the impacts of dietary changes on DLC exposure. In discussing each option in Chapter 7, the committee lists key research needs related to the possible implementation and assessment of the option.

CONCLUSION

The proposed framework summarized in this chapter essentially poses a series of questions that the committee considers relevant in identifying and evaluating possible DLC exposure reduction options. In-depth analysis of the kind that would likely be necessary to support government action would require both a detailed description of the proposed intervention and its implementation, as well as good data and information to answer the questions posed by the framework, especially on the specific intervention's feasibility and likely contribution to exposure reduction. The committee did not attempt to develop and analyze options at this level of detail.

The framework can also be used, however, to identify and test options preliminarily to consider whether they are plausible and deserving of further refinement and evaluation. In Chapter 7, how the framework can be used for this purpose by applying it to some possible options is illustrated, and some of the issues and data needs that would have to be addressed to more fully develop and evaluate these options are discussed.

REFERENCES

AEA Technology. 1999. *Compilation of EU Dioxin Exposure and Health Data.* Prepared for the European Commission DG Environment. Oxfordshire, England: AEA Technology.

ATSDR (Agency for Toxic Substances and Disease Registry). 1998. *Toxicological Profile for Chlorinated Dibenzo-p-dioxins.* Atlanta, GA: ATSDR.

Chapman KS, Hariharan G. 1994. Controlling for causality in the link from income to mortality. *J Risk Uncertain* 8:85–94.

Chapman KS, Hariharan G. 1996. Do poor people have a stronger relationship between income and mortality than the rich? Implications of panel data for health-health analysis. *J Risk Uncertain* 12:51–63.

EPA (U.S. Environmental Protection Agency). 2000. *Exposure and Human Health Reassessment of 2,3,7,8-Tetrachlorodibenzo-p-Dioxin (TCDD) and Related Compounds.* Draft Final Report. Washington, DC: EPA.

Fiedler H, Hutzinger O, Welsch-Pausch K, Schmiedinger A. 2000. *Evaluation of the Occurrence of PCDD/PCDF and POPs in Wastes and Their Potential to Enter the Foodchain.* Prepared for the European Commission DG Environment. Bayreuth, Germany: University of Bayreuth.

Huang KS. 1996. Nutrient elasticities in a complete food demand system. *Am J Agr Econ* 78:12–19.

Huang KS. 1997. *How Economic Factors Influence the Nutrient Content of Diets.* Technical Bulletin No. 1864. Washington, DC: U.S. Department of Agriculture, Economic Research Service.

IARC (International Agency for Research on Cancer). 1997. *IARC Monographs on the Evaluation of Carcinogenic Risks to Humans. Volume 69: Polychlorinated Dibenzo-para-Dioxins and Polychlorinated Dibenzofurans.* Lyon, France: World Health Organization.

Keeney R. 1990. Mortality risks induced by economic expenditures. *Risk Anal* 10:147–159

Keeney R. 1997. Estimating fatalities induced by the economic costs of regulations. *J Risk Uncertain* 14:5–23.

Lutter R, Morrall JF III, Viscusi WK. 1999. The cost-per-life-saved cutoff in safety-enhancing regulations. *Econ Inq* 37:599–608.

Merrill RA, Schewel M. 1980. FDA regulation of environmental contaminants of food. *Va Law Rev* 66:1357.

Scientific Committee on Food. 2000. *Opinion of the Scientific Committee on Food on the Risk Assessment of Dioxins and Dioxin-like PCBs in Food.* Brussels: European Commission.

Scientific Committee on Food. 2001. *Opinion of the Scientific Committee on Food on the Risk Assessment of Dioxins and Dioxin-like PCBs in Food, Update.* Brussels: European Commission.

USDA (U.S. Department of Agriculture). 1996. *The Food Guide Pyramid.* Home and Garden Bulletin No. 252. Washington, DC: USDA.

USDA/HHS (U.S. Department of Health and Human Services). 2000. *Nutrition and Your Health: Dietary Guidelines for Americans*, 5th ed. Home and Garden Bulletin No. 232. Washington, DC: U.S. Government Printing Office.

Viscusi WK. 1994. Mortality effects of regulatory costs and policy evaluation criteria. *Rand J Econ* 225:94–109.

7

Policy Options to Reduce Exposure to Dioxins and Dioxin-like Compounds

INTRODUCTION AND BACKGROUND

The central task assigned to the committee was to identify risk-management options to reduce exposure to dioxins and dioxin-like compounds (referred to collectively as DLCs) through the food supply and to recommend ways to minimize current exposure levels. The committee was not asked to do risk assessment, thus the focus was on evaluating risk relationships and risk-management options.

In the course of its deliberations, the committee considered many potential intervention scenarios within the framework outlined in Chapter 6. These interventions can be initiated at a number of points along the animal production, human food, and food-consumption pathways (Figures 7-1 through 7-3). The generated range of risk-management options took into account current statutory food safety standards and other regulatory policies, procedures, and practices that frame and constrain the adoption of regulatory interventions. The committee also considered risk-management options that involve encouraging voluntary actions that could be taken by food system participants to reduce DLC exposure. As explained in Chapter 6, the committee considered these options within an overall framework of a risk relationship analysis.

The committee's approach was to deliberate and present a broadly representative list of risk-management options and some guides toward the evaluation of the desirability of these options. All of the options presented have the intention of reducing DLC exposure through food, working from the starting point that reductions in exposure are desirable. However, a central conclusion of the committee is that the desirability of particular risk-management options in many instances

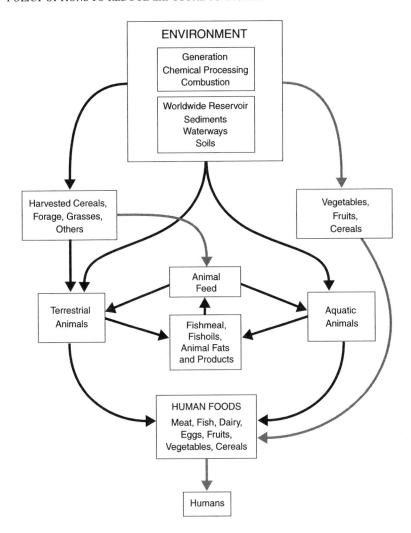

FIGURE 7-1 Pathways leading to exposure to dioxin and dioxin-like compounds through the food supply. Boxes depict point sources in the pathways. Dark arrows refer to pathways with a greater relative DLC contribution than the pathways with light arrows.

cannot be adequately analyzed due to insufficient information. Regulatory agencies have invested relatively little toward generating the data required to support risk-management decisions related to options to reduce DLC exposure. For example, little data are available that would permit more refined estimates of current DLC levels in feeds, forage, and human foods; the current distribution and

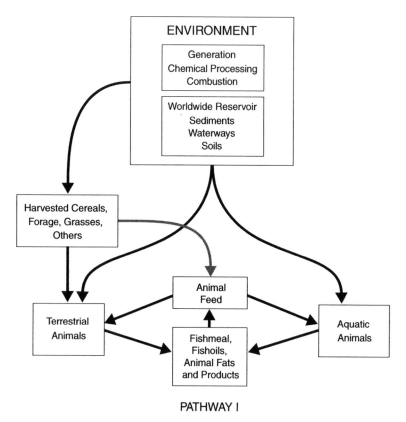

PATHWAY I

FIGURE 7-2 Pathway I, animal production systems. Boxes depict point sources in the pathways. Dark arrows refer to pathways with a greater relative DLC contribution than the pathway with a light arrow.

levels of human exposure; and the amount of exposure reduction that could be achieved through various interventions.

The risk-management options discussed by the committee encompassed a wide range of regulatory and nonregulatory options, which are presented in this chapter. The options are presented in outline form in matrices that correspond to each of the three pathways (see Chapter 6) and are discussed in detail. Not all of the options considered were selected as recommendations for immediate action; specific recommendations and research needs are presented in Chapter 8.

The committee analyzed only the likely outcomes or the ultimate desirability of the most significant risk-management options considered. Thus, the matrices contain an array of options greater than those that the committee considered in detail. The listings of current barriers to implementation, expected DLC exposure

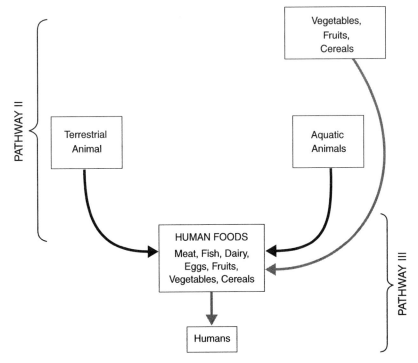

FIGURE 7-3 Pathway II, human foods, and Pathway III, food-consumption patterns. Boxes depict point sources in the pathways. Dark arrows refer to pathways with a greater relative DLC contribution than the pathways with light arrows.

reduction, and risk relationships (ancillary benefits and countervailing risks) should be viewed as suggestive—not definitive. Additionally, due to gaps in data and high analytical costs, many options are not feasible for immediate implementation, although they may be used in the future.

The size and direction of these impacts must be the subject of further analysis. For example, a potential countervailing risk of several options could occur through the following chain of events: higher costs of production cause higher prices for some food products (e.g., meat and dairy products), and consumers shift away from the higher-priced foods to the lower-priced foods, which results in less access to important nutrients and a negative impact on nutritional status. The committee's evaluation of the available economic research suggests that this chain of events is unlikely unless price changes were large (i.e., greater than 10 percent) and widespread. Moreover, some of the effects of shifts in diet are likely to provide ancillary health benefits, either through reduction in intake of dietary constituents that are currently consumed in excess, or through increased intake of constituents that are consumed at suboptimal levels. However, inclusion of this possible risk relationship in matrices suggests the need for further evaluation.

While the committee did not fully analyze the available risk-management options, it did identify two areas that appear to be the most promising leverage points for affecting DLC exposure from food. One of these is at the animal production stage, where DLCs enter the food supply through forage and feeds and are subsequently recycled back through the system by practices such as the reuse of animal fats (where DLCs accumulate) as ingredients in animal feeds. The second leverage point is in food-consumption patterns, where consumption of foods with higher levels of animal fat, particularly by children, contributes to DLC exposure and life-long DLC body burdens. These two areas are where further analysis of risk-management options should begin.

ENVIRONMENTAL CONSIDERATIONS

The committee was clearly instructed by the sponsors to address ways to reduce DLCs in the food supply, but not the environment. However, the committee felt it was necessary to take into account the sources of the DLCs that enter feeds and foods. Therefore, the committee addressed reservoir and high-exposure sources of DLCs as contributors to the overall exposure through foods.

Dramatic reductions have occurred in the production of DLCs from combustion sources in the United States as a result of enforcement of the provisions of the Clean Air Act. For example, DLC emissions from municipal waste incinerators fell from 8,877 g $TEQ_{DF-WHO98}$ in 1987, to 1,250 g $TEQ_{DF-WHO98}$ in 1995, and are projected to fall to 12 g $TEQ_{DF-WHO98}$ in 2002 to 2004 (Winters, 2001). There have been similar dramatic declines in emissions from medical waste incinerators, smelters, and cement kilns. The largest current source of DLC production in the United States is now backyard barrel burning (estimated at 628 g $TEQ_{DF-WHO98}$/y) and other poorly characterized sources such as sewage sludge application, residential wood burning, and coal-fired utilities.

Actions should be taken at all possible levels to monitor and reduce emissions from these poorly characterized sources. The net effect of restricting the emission of DLCs into the atmosphere is a long-term continual decline in DLC levels in pastureland, soil, and sediment and, thus, reduced entry of these contaminants into the pathways that lead to human exposure through the food supply.

PATHWAY I: ANIMAL PRODUCTION SYSTEMS

Effective, long-term reduction of DLC exposure through the human food supply requires interrupting the introduction of DLCs into pathways leading to human foods. Thus, a range of possible measures to reduce DLC levels in forage and feeds were identified and are presented in Matrix 1. Due to global environmental contamination, it is expected that all forages and many grain sources will have detectable, although widely variable, DLC levels.

The gaps in current information on the quantitation of risks due to DLCs in forage and feeds make it more difficult to develop uniform policy. However, research studies currently underway will improve future decision-making options for the management of these issues.

Option: Require Testing for DLC Levels in Forage, Feeds, and Feed Ingredients

This option would require feed manufacturers and animal producers to test samples of forage, feeds, and feed ingredients used in animal production systems to determine DLC levels in these products. Data would be collected for samples of ingredients, with the sampling and tracking of samples being performed under close supervision. The data obtained from testing and surveys would be combined with data from other sources to form a nationwide database on the levels and distribution of DLCs in forage, feeds, and human food. The database would be developed and maintained as a private/public cooperative project, with participation from meat and produce commodity groups, feed manufacturers, meat and vegetable/fruit processors and retail markets, state departments of agriculture and public health, and the U.S. Department of Agriculture (which would act as the lead). An interagency coordination group should develop sampling priorities to identify and remediate deficits in DLC contamination data for all feeds, feed ingredients, forage, and animal production environments associated with production agriculture and for the human food supply. In addition, the interagency coordination group should develop uniform sampling techniques, analytic processes and data reporting, and storage algorithms to enable direct cross-industry and regional comparisons.

To supplement the mandated testing, government agencies, commodity groups, and other food industry suppliers, including importers, would be encouraged to supply analytic and descriptive product data to further expand the available information. All data collected should be maintained with source anonymity and be collected for informational purposes only. As data accumulate and are evaluated, modifications in collection procedures may be required to maximize the resources available to the industry and the government.

The justification for mandated testing, as discussed in Chapter 4, would be the need to build the base of data required to support a coherent risk-management strategy to identify and remove DLCs sources that enter animal production systems and, thus, interrupt the cycle that allows DLCs to accumulate in the human food supply. At present, DLC data are collected piecemeal and are insufficient for use in tracking regional or national trends. Under a mandated data collection and distribution system, disparate public and private information resources would

Matrix 1. Intervention Options Considered to Reduce DLC Exposure Through Pathway I: Animal Production Systems

A. Require testing for DLC levels in forage, feeds, and feed ingredients

Alternate/Interim Actions	Establish a national voluntary monitoring and information system for DLCs in forage, feeds, and feed ingredients: 1. Develop new, less costly methods for DLC analysis 2. Encourage cooperative reporting of DLC results among feed manufacturers and other industry suppliers 3. Continue or expand government DLC monitoring programs 4. Make monitoring data available to industry and to the public
Current Barriers to Implementation	Current food safety laws are not applicable to DLCs The current cost of analysis creates a barrier to required testing
DLC Exposure Reduction	Results of testing provide supportive data to establish acceptable levels of DLCs in forage, feeds, and feed ingredients and augment monitoring for unintentional contaminants
Risk-Relationship Analysis	*Ancillary Benefits:* An expanded database to track unintentional contaminants in forage, feeds, and feed ingredients that could be used in conjunction with forthcoming analyses of the human food supply
	Countervailing Risks: Possible increased cost of some foods, which may decrease the availability of these foods and impact the nutritional status of at-risk populations

B. Establish tolerance levels for DLCs in forage, feeds, and feed ingredients

Alternate/Interim Actions	1. Implement a voluntary program to reduce DLC-contaminated fat in animal feeds 2. Establish nonbinding targets for DLCs in forage, feeds, and feed ingredients
Current Barriers to Implementation	Data are insufficient to establish specific evidence-based DLC levels, and there is a low-priority resource status for implementing industry- or government-sponsored testing programs
DLC Exposure Reduction	Reduced levels of DLCs in animals and farmed fish that enter the food supply
Risk-Relationship Analysis	*Ancillary Benefits:* Long-term reduced levels of other undesirable lipophilic contaminants that have been found with DLCs in fats from animal-based foods
	Countervailing Risks: Possible increased cost of some foods, which may decrease the availability of these foods and impact the nutritional status of at-risk populations

C. Restrict the use of animal products and forage, feeds, and feed ingredients that originate from specific areas that are considered to be contaminated

Alternate/Interim Actions	1. Identify and target for further analyses those geographic areas with high potential for contamination and determine appropriate reduction strategies 2. Implement educational strategies to teach consumers who reside in contaminated areas how to modify cooking and food preparation to minimize exposure
Current Barriers to Implementation	The uncertainty of risk and insufficient forage, feed, or feed-ingredient data available to assess local, regional, or national contamination levels and identify high-risk contaminant classifications
DLC Exposure Reduction	Reduced levels of DLCs in animal feeds that contain fats from animal sources
Risk-Relationship Analysis	*Ancillary Benefits:* Decreased opportunity for other undesirable contaminants that have been found with DLCs in lipid-based ingredients to enter into food animals and farmed fish via feeds
	Countervailing Risks: Possible increased costs of some foods, which may decrease the availability of these foods and impact the nutritional status of at-risk populations

D. Restrict the reuse of the animal by-products in agriculture, animal husbandry, and manufacturing processes

Alternate/Interim Actions	1. Work with industry to develop voluntary good agricultural, animal husbandry, manufacturing, and transportation practices to achieve reduction in DLCs: a) Monitor building materials and bedding for DLC contamination b) Eliminate growing foods and forages in high-exposure areas or do not use first-cut forages from high exposure areas c) Increase intensive growing practices for livestock d) Reduce or eliminate the use of animal fats and oils, which may be high in DLCs, as ingredients in animal feeds 2. Provide subsidies to encourage the adoption of practices that reduce DLCs in forage, feeds, and feed ingredients
Current Barriers to Implementation	The use of vegetable fats may lead to increased spoilage and off-flavors in some products; disposal of unused animal fats will create additional environmental problems
DLC Exposure Reduction	Reduced reintroduction of DLCs into animal production systems, which will reduce DLC levels in human foods in the long-term
Risk-Relationship Analysis	*Ancillary Benefits:* The entry of other undesirable lipophilic contaminants into lipid-based agricultural products also is reduced as a result of revised production practices
	Countervailing Risks: Possible increased cost of some foods, due to costs of expanded testing and reporting, which may decrease the availability of these foods and impact the nutritional status of at-risk populations

combine, over time, into a body of information that would be useful for an ongoing analysis of risk and the development of risk-management strategies that reflect the evolution of environmental DLC levels.

Alternate/Interim Actions

An alternative to mandated DLC testing is a voluntary national monitoring and information system, developed to contain reported DLC levels in forage, animal feeds, feed ingredients, and animal production environments as described above. This option was developed in consideration of available regulatory tools and the constraints inherent in using them.

Current Barriers to Implementation

Current food safety laws do not provide regulatory agencies with the legal authority to mandate testing except possibly as part of a regulatory intervention that the agencies could justify as necessary to prevent illegal "adulteration" of a product. As discussed in Chapter 6, the evidence required for this appears not to be currently available. Thus, new legislation would likely be required to adopt the mandatory data collection option. The committee assumes that, under current law, the agencies could collaborate with the private sector in a voluntary data collection effort. Beyond legal authority, the committee strongly believes that a major factor in the paucity of DLC contamination data for all feeds and food classes is the high cost and complexity of available, validated analytic procedures. New and less costly methods for DLC analysis in forage, feeds, feed ingredients, and animal production environments, as well as in a wide range of food commodities and processed foods, should be encouraged. Screening tests that detect baseline contamination levels at costs that are significantly below current levels, as well as simplified instrumentation and detection processes, are needed to move forward. Funding for developmental research in DLC detection methods should be encouraged as part of the national research agenda.

DLC Exposure Reduction

Identifying "hot spots" where DLC contamination is particularly high would help narrow the focus for short-term or immediate interventions to reduce DLC exposures to food animals through forage and feeds. Such actions would also help minimize problems that could arise from the intermingling of food animals raised in areas where a contamination event has occurred with those raised in uncontaminated areas. Long-term testing would further allow for trend monitoring to map the rate of decrease of DLCs in exposure pathways and would be a useful adjunct to monitoring for human exposures and health outcomes that chronologically coincide with trends in DLC levels.

Risk-Relationship Analysis

Ancillary Benefits. The committee did not identify a specific ancillary benefit to mandated or voluntary DLC testing and establishment of a nationwide database other than the possibility that the proposed system may, in the future, be useful in identifying other unintended or unwanted contaminants (similar to DLCs) in animal production systems.

Countervailing Risks. A potential countervailing risk could occur if implementation of this option increased the cost and prices of some foods. However, the size of the production-cost increases resulting from expanded testing and reporting, and the resultant consumer-level food price increases, would have to be significant for this to be a strong likelihood. The committee did not view this as probable.

Option: Establish Tolerance Levels for DLCs in Forage, Feeds, and Feed Ingredients

The option to regulate DLC levels in forage, feeds, and feed ingredients would likely take the form of tolerances or action levels (see Chapter 6). These levels would be set, based on current DLC levels in forage, feeds, and feed ingredients and projected target levels, to achieve exposure reduction. Setting tolerances or action levels requires that the levels be set at a maximum limit necessary to protect public health (see Chapter 6). Tolerances, but not action levels, are legally binding and enforceable through court action. The purpose of setting tolerances or action levels for DLCs is to achieve a specific reduction goal that is consistent with a maximum level of exposure considered to be safe. The justification for setting tolerances or action levels is based on the premise that DLC exposure reduction is necessary for public health and safety.

Alternate/Interim Actions

As an alternative to the binding limitations established by tolerances, it may be feasible to set nonbinding goals aimed at reaching lower DLC levels in forage and feeds that would minimize the amounts of these compounds available to food animals through these pathways. Furthermore, by maintaining flexible and adjustable goals, these nonbinding target levels could conform to changing trends in environmental DLC loads as new data are gathered.

Current Barriers to Implementation

The limited amount of data available on DLC levels in forage and feeds creates a barrier to establishing enforceable tolerance levels. However, even if tolerance levels were in place, foods such as wild-catch fish and imported foods would be missed by established feed regulations. An additional barrier to implementing enforceable tolerance levels is that testing is expensive.

Establishing nonbinding goals to reduce DLC levels in animal forage and feeds is limited by the expense and difficulty of the analytical procedures currently in place. There is also the question of whether nonbinding goals would be followed and whether they would achieve a reduction in DLC levels in these products.

DLC Exposure Reduction

Implementing measures to ensure that DLCs are reduced in animal forage, feeds, and feed ingredients is an important point of intervention in reducing DLCs in human food because the contamination of animal feeds is directly linked to the presence of contaminants in human food. By establishing maximum limits for DLCs in animal feeds, products contaminated at levels above the defined limits would be legally excluded from the food supply.

Risk-Relationship Analysis

Ancillary Benefits. If taking action to reduce the source of DLCs in the food chain results in a decrease in other hazardous contaminants, this could produce an ancillary risk-reduction benefit.

Countervailing Risks. A potential countervailing risk could occur if implementation of this option increased the cost and prices of some foods. However, the size of the production-cost increases resulting from expanded testing and reporting, and the resultant consumer-level food price increases, would have to be significant for this to be a strong likelihood. The committee did not view this as probable.

Option: Restrict the Use of Animal Products and Animal Forage, Feeds, and Feed Ingredients That Originate from Specific Areas That Are Considered to Be Contaminated

This option mandates that forage, feeds, and feed ingredients obtained from geographic regions known to be contaminated with greater than background levels of DLCs be removed from animal production systems, and that the use of animal food products obtained from contaminated areas be restricted.

This option may also be a means to protect potentially highly exposed populations from DLC sources in their immediate areas. Within specific geographic regions, certain populations may have higher daily intake levels of DLCs than the general population, due to food-consumption patterns, use of locally obtained foods such as marine and wild animals, and culture-based behaviors (see Chapter 5).

As discussed in Chapter 4, the sources of DLCs in various geographic regions include point-source contamination (including some industrial manufacturing), uncontrolled waste incineration, naturally occurring contamination leached from clays or emitted by forest fires, and long-range atmospheric deposition onto a watershed. Thus, fish and shellfish from impacted waters will generally have higher DLC levels than similar species from less-impacted areas.

This option would highlight the need to reduce environmental generation of DLCs and minimize future contamination of marine and wild animals, animal feedstuffs, and aquatic environments.

Alternate/Interim Actions

A possible interim action is to identify specific geographic regions that have a high potential for contamination and institute testing in these targeted areas. An area could be identified as having high potential for contamination if it were within an air transport pattern known to carry DLCs or if it were located in an area of high industrial activity that has previously produced DLC contaminants or wastes that have entered into the surrounding environment. This interim action would initiate testing that would provide information on DLC levels, which are needed to implement reduction strategies in areas of high potential DLC exposure. Because it would occur in targeted areas, it could be used to identify areas of high priority for early intervention. Such testing would be carried out on a smaller scale than in the action above, which would reduce the need for expensive DLC analyses.

Another possible interim action to protect specific populations that reside in contaminated geographic areas and consume locally obtained foods is to implement educational strategies to teach consumers how to modify cooking and food preparation to minimize exposure (see Chapter 5).

Current Barriers to Implementation

Under the statutory standards governing food safety interventions, the uncertainty of risk reduction at low doses and the lack of sufficient data on DLC levels in animal production systems would be a constraint to setting mandatory tolerance levels, enforcing action levels, or mandating restrictions on the use of animal products, forage, and animal feeds that originate from specific regions. Additionally, even though some specific geographic regions have already been investigated, information on DLC levels across broad geographic areas is limited since current data are insufficient to identify high-risk sites for contaminated soil, sediment, forage, and surface waters. As a result, it is not always possible to accurately predict, outside of certain previously identified regions of contamination, the origin of unexpectedly high DLC levels in animal feeds. However, geographic areas that may be considered to have high potential of DLC contamination due to the presence of known sources can be targeted for testing and analysis.

DLC Exposure Reduction

Removing highly contaminated feeds and feed ingredients will reduce the introduction of new DLCs into food animals and farmed fish, particularly since the ingredients for animal feeds come from many different sources and regions. Ultimately, the total DLC level will decrease in the finished food or product consumed and risks to consumers who use these products will decline.

Risk-Relationship Analysis

Ancillary Benefits. The soils that provide forage for animals and the sediments where fish feeds are known to contain many contaminants at varying concentrations. Some of these contaminants, such as dichlorodiphenyltrichloroethane, dichlorodiphenyldichloroethylene, dieldrin, and hexachlorocyclohexane, like DLCs, accumulate in the fatty tissues of exposed animals and fish (see Chapter 5). Whether these contaminants concentrate in the same geographic regions where DLC levels are high has not been established, except in some specific regions where unintended industrial or contaminant releases have occurred. There is an increased probability of reduced exposure to other lipophilic contaminants when the use of DLC-contaminated areas is restricted. However, the committee did not have access to the data needed to assess the extent to which this might occur.

Countervailing Risks. A potential countervailing risk of this option is increases in the cost and prices of some foods. However, the size of the production-cost increases resulting from expanded testing and reporting, and the resultant consumer-level food price increases, would have to be significant for this to be a strong likelihood. The committee did not view this as probable.

Option: Restrict the Reuse of Animal By-Products in Agriculture, Animal Husbandry, and Manufacturing Processes

This option requires that the by-products of animal production systems, particularly animal fats, be restricted from re-entering the pathways that lead to the human food supply via animal feed ingredients. Removing animal fats from animal diets will, over time, reduce DLC levels in animal food products and interrupt the cycle of introducing DLCs into the food supply.

Alternate/Interim Actions

An alternative to direct restriction of animal products, forage, and feed ingredients from specific geographic areas would be a voluntary program to develop good agricultural, animal husbandry, manufacturing, and transportation practices to achieve reductions in DLCs. These could focus on monitoring DLC levels, changing feed production practices, and changing animal feeding practices. This approach could also involve subsidies to encourage the adoption of new practices.

Current Barriers to Implementation

Animal fats are unavoidable by-products of animal-based food production. If the reuse of animal fats as components of animal feeds is restricted, this will create a problem with disposal of the unused fat (see Chapter 4). Incineration, as a disposal alternative, may release DLCs into the air, adding to the current environmental burden. Because there currently are no viable and economical alternative uses for these by-products of animal production systems, this creates a barrier to the implementation of regulatory restrictions that would apply to the reuse of animal fats in animal feeds. Alternative uses for these animal by-products would need to be found if reuse of these products is restricted from entering the animal feed supply.

Another barrier to restricting the use of animal fats in animal feeds is that fat sources can have a significant impact on the quality of certain animal products, specifically swine and poultry meat. A greater volume of vegetable oil would be required to meet the same caloric demand as a solid animal fat in the feeds of swine and poultry, and would therefore create technical difficulties in the mixing and delivery of the feeds (see Chapter 4). Additionally, as noted in two National Research Council reports (see Chapter 4), when vegetable fats are fed to these animals, the quality of their body fat is affected; that is, it is more unsaturated, which increases spoilage and causes off-flavors in the final food product, thus reducing consumer acceptance.

DLC Exposure Reduction

The long-term effect of implementing restrictions on the reuse of animal fats in the animal feed supply is that the reintroduction of DLCs into the human food supply through the animal feed pathway will be reduced, ultimately resulting in lower DLC exposure to the humans who consume animal food products.

Risk-Relationship Analysis

Ancillary Benefits. The committee did not identify specific ancillary benefits to restricting the use of animal by-products from agriculture, animal husbandry, and manufacturing processes. However, if other lipid-based contaminants are present in DLC-containing animal fats, their reintroduction into the food supply would also be interrupted.

Countervailing Risks. A potential countervailing risk could occur if implementation of this option increased the cost and prices of some foods. However, the size of the production-cost increases resulting from expanded testing and reporting, and the resultant consumer-level food price increases, would have to be significant for this to be a strong likelihood. The committee did not view this as probable.

PATHWAY II: HUMAN FOODS

Most human foods contain DLCs at some level. However, the fat component of animal-based foods appears to have the greatest DLC concentrations. Efforts to reduce DLC levels in foods must take into consideration that many of the foods containing these contaminants are also sources of important nutrients. The committee's consideration of various intervention options to reduce DLC exposure from foods, therefore, also took into account the need to promote good nutrition and health (see Matrix 2). Thus, reducing DLC exposure by not consuming foods that may be sources of important nutrients is appropriate only if nutritionally equivalent foods are substituted.

Option: Require Testing and Publishing of Data on DLC Levels in the Human Food Supply, Including Food Products, Dietary Supplements, and Breast Milk, to Use in Establishing Tolerance Levels in Foods

This option would require food manufacturers and processors of food products and dietary supplements to determine current levels of DLCs in their products. As with the testing option for agricultural products, required testing for DLCs in the human food supply would provide the baseline data needed to adequately assess the need for setting allowable DLC limits in foods. Furthermore, data from testing could be used to model potential exposure scenarios in a risk-management strategy to reduce exposure, particularly for sensitive and highly exposed population groups.

This option would be coupled with government-sponsored DLC testing and monitoring of breast milk as a food source for infants. There are breast-milk banks in the United States and Canada that carefully screen donors. Various studies have demonstrated that infants can receive levels of DLCs through breast milk that are, initially, many times in excess of the dietary exposure levels for adults (see Chapter 5). Infants are especially vulnerable to being exposed to higher levels of DLCs through breast milk due to their developmental immaturity (see Chapters 2 and 5).The Food and Drug Administration, as a component of its

Total Diet Study (TDS), conducts DLC analyses in foods and supplements. The committee recommends expansion of the current TDS data-collection program and increased systematic data collection for high-risk foods. Areas where systematic data collection could be improved include expanding the source and variation in foods, as well as geographic variation, since many foods (e.g., milk and fish) are produced in regions where DLC levels vary greatly. (See Chapters 4 and 5 for detailed discussions of regional variations in DLC levels.) The committee did not identify a lead agency for collection and analysis of breast milk.

Alternate/Interim Actions

An alternate to mandatory DLC testing would be to encourage voluntary analysis of DLC levels in foods, dietary supplements, and human breast milk. Voluntary DLC data collection and analysis would likely fall under the purview of the food industry. Individual food producers and processors are currently conducting some DLC analyses, and the committee encourages these industries to make information on DLC levels in foods available to regulatory agencies and to the public.

Current Barriers to Implementation

The government lacks authority under current food safety laws to require private sector DLC testing for the purposes proposed here. The high cost of DLC analysis is also an obstacle to widespread testing for DLC levels in all foods. The committee considers that data gathering for monitoring DLC levels in breast milk is more feasible because several breast-milk banks in the country could serve as sample sources. Thus, no expensive infrastructure would be required to acquire samples on an ongoing basis, although the cost of analysis remains high.

DLC Exposure Reduction

The outcome of the collection and analyses of DLC levels in foods, supplements, and breast milk is that a nationwide database would be established and made available to regulators, industry, and the public. This information could be used to develop an informational framework to assess risks from DLC exposure through foods. The availability of up-to-date information on DLC levels in foods would contribute to the body of evidence needed to establish targets for safe levels of these compounds in foods. Such information is particularly important in the case of sensitive populations, such as preadolescent girls, whose future children may be at risk during development in utero and through exposure to DLCs through breast milk.

Matrix 2. Intervention Options Considered to Reduce DLC Exposure Through Pathway II: Human Foods

A. Require testing and publishing of data on DLC levels in the human food supply, including food products, dietary supplements, and breast milk, to use in establishing tolerance levels in foods

Alternate/Interim Actions	Encourage voluntary analysis of the DLC content in human foods, dietary supplements, and human breast milk
Current Barriers to Implementation	No government authority to require private-sector testing; low feasibility at the current cost of testing and analysis
DLC Exposure Reduction	Establishment of a database that will allow the accurate assessment of DLC exposure through foods and provide a basis for setting tolerance or action levels in foods
Risk-Relationship Analysis	*Ancillary Benefits:* Changes in food-consumption patterns away from consumption of saturated fats from animal-based foods will decrease the risk for some chronic diseases
	Countervailing Risks: Possible increased costs of some foods, which may decrease the availability of these foods and impact the nutritional status of at-risk populations

B. Establish enforceable standards or allowable levels for DLC in processed foods and in food-product packaging that directly contacts foods

Alternate/Interim Actions	Modify Hazard Analysis and Critical Control Point (HACCP) standards to include practices that change processing and packaging to lower potential DLC exposure in foods

Risk-Relationship Analysis

Ancillary Benefits. Animal fats provide a large proportion of the saturated fats consumed in the United States. Therefore, by reducing animal fat consumption, consumers also will reduce their intake of saturated fats, producing the net benefit that such dietary change may produce a reduction in risk for DLC exposure and some chronic diseases, such as heart disease, that are associated with consuming high levels of saturated fats.

Countervailing Risks. A potential countervailing risk could occur if implementation of this option increased the cost and prices of some foods. However, the size of the production-cost increases resulting from expanded testing and reporting, and the resultant consumer-level food price increases, would have to be significant for this to be a strong likelihood. The committee did not view this as probable.

Current Barriers to Implementation	The lack of current data is a barrier to setting enforceable standards
DLC Exposure Reduction	Reduced levels of DLCs in human foods, although food processing and packaging appear to play minor roles in human exposure to DLCs
Risk-Relationship Analysis	*Ancillary Benefits:* The cost of a DLC monitoring program for food packaging materials could be spread over several suppliers within the industry, thus diminishing the cost to individual companies
	Countervailing Risks: Possible increased costs of some foods, which may decrease the availability of these foods and impact the nutritional status of at-risk populations

C. Require cleaning or washing practices for all vegetable, fruit, and grain crops that potentially had contact with soil

Alternate/Interim Actions	Recommend washing practices to consumers
Current Barriers to Implementation	Cost impact is assumed to be small
DLC Exposure Reduction	Reduced levels of DLCs in processed foods of nonanimal origin
Risk-Relationship Analysis	*Ancillary Benefits:* May reduce exposure to bacterial and other food contaminants, especially on fresh fruits, vegetables, and grains
	Countervailing Risks: Possible increased costs of some foods, which may decrease the availability of these foods and impact the nutritional status of at-risk populations

Option: Establish Enforceable Standards for DLC Levels in Processed Foods and in Food-Product Packaging That Directly Contacts Foods

This option mandates that, within the current regulatory framework, enforceable standards be set that limit the maximum allowable DLC levels in materials used for food packaging.

In the past, DLCs have been found to enter the food supply through the processing and packaging of foods (see Chapter 4). Fortunately, these instances were infrequent and corrective measures were taken. In addition, there are no materials currently in use in processing and packaging that are known sources of DLCs. Based on past experience, predicting inadvertent contamination events that may occur through processing, although not easily done, would be prudent as a means of preventing contamination incidents in the future.

Alternate/Interim Actions

An alternative option is for regulatory agencies to develop a proactive monitoring program that will identify DLC-contaminated food packaging materials for removal prior to their use. As an example of ways to reduce DLC levels by exploiting food safety intervention programs already in place, manufacturers could take a proactive role in monitoring product processing and packaging using the current HACCP program as a model.

Current Barriers to Implementation

As previously discussed in Matrix 1, the greatest barrier to implementing either enforceable standards or allowable levels for a detection program for DLCs in food-product packaging is a lack of legal authority, given the limitations in available data, which is due in large part to the high cost of testing and analysis. Lower-cost alternatives (see Chapter 2) may be implemented until DLC testing and analysis becomes more affordable. However, widespread testing and accurate information on current exposure levels are needed to establish enforceable guidelines.

DLC Exposure Reduction

Although food processing and packaging play a minor role in contributing to DLC exposure through foods, identifying the critical points at which the entry of DLCs into the food and food packaging process can be controlled. Setting action levels at these points, could further reduce the DLC level in many processed foods, which may be important if DLC exposure is found to be hazardous at background levels.

Risk-Relationship Analysis

Ancillary Benefits. A potential ancillary benefit to food producers and processors that may result from implementation of a DLC monitoring program in food packaging materials is that the cost of DLC analyses would not be borne by a single supplier, but rather would be spread over several suppliers within the industry, thus diminishing the cost to individual companies.

Countervailing Risks. A potential countervailing risk could occur if implementation of this option increased the cost and prices of some foods. However, the size of the production-cost increases resulting from expanded testing and reporting, and the resultant consumer-level food price increases, would have to be significant for this to be a strong likelihood. The committee did not view this as probable.

Option: Require Cleaning or Washing Practices for All Vegetable, Fruit, and Grain Crops That Potentially Had Contact with Soil

This option mandates that food processors and producers implement a uniform standard for washing practices for fresh vegetables, fruits, and grains used in food products. Soil containing DLCs may adhere to vegetables, fruits, and grains intended for human consumption. Although vegetables have not been shown to contribute large amounts of DLCs to overall intake levels, they can increase DLC exposure when contaminated soil is present on their surface. Thus, washing vegetable and fruit surfaces, leaves, and stems to remove soil and peeling root and waxy-coated vegetables are effective ways to reduce DLC exposure through these foods. Washing can be incorporated into industry production practices, and can be encouraged among consumers and food preparers.

Alternate/Interim Actions

An alternative and interim action to mandated washing practices within the food processing industry is to develop a public education and information program to disseminate the information to consumers that they can reduce their exposure to DLCs by washing fresh vegetables and fruits. This option should include information on cleaning foods obtained from home gardens.

Current Barriers to Implementation

Uniform standards for washing fresh vegetables, fruits, and food grains are not currently in place in the food-processing industry. The cost to industry of implementing washing procedures is unknown, and a cost-impact analysis would be needed to determine the cost feasibility of such procedures. However, washing has been shown to be effective in reducing DLC levels in vegetables, fruits, and grain crops (see Chapter 5), and the cost of washing may be smaller than the cost of discarding foods that would be considered unusable due to DLC contamination.

DLC Exposure Reduction

The committee regards washing procedures as an important intermediary step to reduce DLC exposure through vegetables, fruits, and grains. Even though such foods have not previously been considered major contributors to DLC exposure in humans, they may deserve more consideration than they have been afforded if DLC testing and analysis are implemented and levels are found to be high in some of these foods.

Risk-Relationship Analysis

Ancillary Benefits. DLCs adhere to the outside surfaces of plants, including the skins and peels of vegetables and fruits. These compounds are carried in soil, along with many other potential contaminants (e.g., bacteria, other organisms, and lead) that may pose some health risk, particularly to children. Implementing stringent cleaning and washing practices will reduce not only DLC levels on the surfaces of vegetables, fruits, and grains, but also any bacterial or other contaminants on these surfaces and thus reduce exposure, particularly of young children, to these potential hazards.

Countervailing Risks. A potential countervailing risk could occur if implementation of this option increased the cost and prices of some foods. However, the size of the production cost increases resulting from expanded cleaning and washing practices, and the resultant consumer-level food price increases, would have to be significant for this to be a strong likelihood. The committee did not view this as probable.

PATHWAY III: FOOD-CONSUMPTION PATTERNS

Given that the majority of human exposure to DLCs is from food sources, information on the human consumption patterns of foods that may contain DLCs is an important component of any assessment of body burden for these contaminants. Furthermore, an accurate analysis of food consumption by individuals in the general population and in sensitive or highly exposed subpopulations can serve as the basis for identifying certain foods, such as animal fats, that are major sources of DLC exposure.

Food-consumption patterns are dynamic, and consumers tend to make specific choices within the various food groups to meet their nutritional needs. Within a food group (e.g., the "meat and meat alternates" food group, which contains meat, poultry, fish, dry beans, eggs, and nuts), specific foods vary greatly in terms of their animal fat content and potential source of DLC exposure.

It is generally recognized that food-consumption patterns that include a variety of food choices are important for maintaining nutritional health. When the selection of a food is limited, whether by choice, accessibility, or custom, the potential to limit nutrient availability is increased, as is the probability for DLC exposure if the food chosen is a source of DLCs. The committee identified various dietary choices that may reduce exposure to DLCs (see Matrix 3), yet have a minimal impact on nutrient availability or intake (see Chapter 5).

Option: Increase the Availability of Low-Fat and Skim Milk in Federal Feeding Programs Targeted to Children by Amending the Current Act Favoring the Provision of Whole Milk in the National School Lunch Program

This option would increase the availability of low-fat and skim milk to participants in federal nutrition programs that are targeted to children, such as the National School Lunch and School Breakfast Programs, and the Child and Adult Care Food Program. The statute governing the National School Lunch Program (Richard B. Russell National School Lunch Act, amended Dec. 8, 2000, § 9(2)) states that "schools participating in the school lunch program under this Act shall offer students a variety of fluid milk consistent with prior year preferences unless the prior year preference for any such variety of fluid milk is less than 1 percent of the total milk consumed at the school," which currently favors the purchase of whole milk. Thus, this option requires the amendment of regulations that favor the provision of whole milk to participants in the School Lunch Program.

Whole milk and full-fat dairy foods have been identified as important contributors to the DLC intake of children, whereas low-fat and skim milk contain at least as much calcium and other essential nutrients as whole milk, but have lower levels of DLCs (see Chapter 5). Thus, choices of low-fat milk and dairy products can satisfy nutrient requirements and reduce exposure to DLCs.

Alternate/Interim Actions

An alternative to requiring that low-fat and skim milk be provided in federal nutrition programs is to encourage the inclusion and increase the marketing of low-fat and skim milk in these programs, and encourage the selection of low-fat milk and dairy products for children over 2 years of age in the Special Supplemental Nutritiom Program for Women, Infants and Children. In addition, public education and information efforts can be used to encourage parents to choose, when appropriate, low-fat and skim milk in place of whole milk in diets of their children above the age of 2 years.

Current Barriers to Implementation

The committee did not identify any significant direct cost barriers to program sponsors that would result from implementation of the option to reduce DLC exposure through foods provided in federal nutrition programs. However, the committee recognizes that the dairy and related industries face a significant economic issue with regard to the disposal of excess butterfat if it is removed from milk; interfacing with relevant federal agencies for assistance may help resolve this problem.

Matrix 3. Intervention Options Considered to Reduce DLC Exposure Through Pathway III: Food-Consumption Patterns

A. Increase the availability of low-fat and skim milk in federal feeding programs targeted to children by amending the current act favoring the provision of whole milk in the National School Lunch Program

Alternate/Interim Actions	the inclusion and marketing of low-fat and skim milk in federal feeding programs targeted to children
Current Barriers to Implementation	Highly feasible; possible cost barriers to the dairy industry
DLC Exposure Reduction	Reduced consumption of whole milk and increased consumption of low-fat and skim milk, which have lower levels of DLCs
Risk-Relationship Analysis	*Ancillary Benefits:* Improved long-term health benefits due to reduced consumption of saturated fats
	Countervailing Risks: The nutritional status of children who will not consume low-fat or skim milk may be at risk

B. Establish a maximum saturated fat content for meals served in schools that participate in federal child nutrition programs

Alternate/Interim Actions	Encourage following recommended dietary guidelines for the maximum saturated fat content for individual meals served in schools that participate in federal feeding programs, while determining the feasibility of setting maximum limits on saturated fats in individual meals

DLC Exposure Reduction

Based on current evidence for DLC exposure through whole milk and full-fat dairy products (see Chapter 5), the committee expects that reduced consumption of these products will decrease the DLC intake levels in children who consume them. The commissioned analysis of DLC exposure through foods further indicates that, especially in young children over 2 years of age and girls under 19 years of age, consuming low-fat or skim rather than whole milk and choosing low-fat versions of other dairy foods will reduce their DLC intake (see Chapter 5).

Risk-Relationship Analysis

Ancillary Benefits. There are no clear nutritional benefits from eating foods high in saturated fats that would not be obtained by eating lower-fat versions of

Current Barriers to Implementation	High feasibility if recommendations are accepted and implemented, but cost may be high, especially to food industries
DLC Exposure Reduction	Reduced exposure to DLCs present in animal fats
Risk-Relationship Analysis	*Ancillary Benefits:* Changed food-consumption patterns away from the consumption of saturated fats from animal-based foods will decrease the risk for some chronic diseases
	Countervailing Risks: The nutritional status of some children who will not consume foods low in saturated fats may be at risk

C. Promote changes in dietary-consumption patterns of the general population that more closely conform to recommendations from the Dietary Guidelines for Americans to reduce the consumption of foods high in saturated fats

Alternate/Interim Actions	Implement revisions to the federally sponsored dietary guidelines
Current Barriers to Implementation	High feasibility if recommendations are accepted and implemented by the public
DLC Exposure Reduction	Reduced exposure to DLCs through foods containing animal fats
Risk Relationship Analysis	*Ancillary Benefits:* Changed food-consumption patterns away from the consumption of saturated fats from animal-based foods will decrease the risk for some chronic diseases
	Countervailing Risks: Dietary recommendations to increase fish consumption are not conducive to reducing DLC intake

the same foods. If young children, both male and female, decrease their intake of whole milk and full-fat dairy products, they will consume lower levels of saturated fats, while consuming the same levels of calcium, vitamin D, and other important nutrients. This modification will likely have the net benefit of decreasing the risk for certain chronic diseases that are associated with consuming high levels of saturated fats (see Chapter 5).

Countervailing Risks. A possible countervailing risk that could arise from offering low-fat and skim milk in place of whole milk to children is that they could choose to not drink any milk, which could place them at nutritional risk. However, a risk analysis would have to be performed for this scenario in order to verify this outcome.

Option: Establish a Maximum Saturated Fat Content for Meals Served in Schools That Participate in Federal Child Nutrition Programs

This option would, within the current regulatory framework, establish a maximum allowable content for saturated fats in individual meals served through federal child nutrition programs, including the School Lunch and School Breakfast Programs and the Child and Adult Care Food Program.

There is currently no standard for saturated fat levels for individual meals in these programs. There is, however, a requirement that meals averaged over a week contain less than 10 percent of total calories from saturated fats (School Meals Initiative for Healthy Children, 7 C.F.R. 20). Since these standards were adopted, reductions in the saturated fat content of school meals have occurred, although the goal for saturated fat to be less than 10 percent of total calories has not been met.

Alternate/Interim Actions

An alternate or interim action is to encourage compliance with recommended dietary guidelines through public education and information programs that will inform parents of the benefits of providing foods low in saturated fats to their children, while determining the feasibility of setting maximum limits on saturated fat in school breakfast and lunch meals.

Current Barriers to Implementation

Children who are not accustomed to consuming foods low in saturated fats may not initially accept such foods offered through federal nutrition programs. If the meals served in a program are the primary source of nutrients for children, their nutritional intakes are affected by the food choices they make. There are, however, a wide variety of foods low in saturated fat that can be made available to children. Thus, although this option will change the variety of foods offered, it will not necessarily limit it. One barrier to implementation of this option is likely to be poor acceptance of foods lower in saturated fats, particularly if they are not presented in an attractive and appealing way.

Current studies suggest that implementing reductions in the saturated fat content of school meals can be accomplished at minimal cost to consumers. However, because the economic feasibility of reducing the saturated fat content of school meals, and the impact of other factors, such as reduced acceptance by children and increased costs to the food industry have not been established, this is another barrier to implementation.

DLC Exposure Reduction

DLCs are found primarily in animal fats. Additionally, these fats are also the primary source of saturated fats in the American diet. By reducing the availability of foods high in saturated fats, exposure to DLCs that may be present will also be reduced. Since this option is targeted to young children, their exposure to DLCs through these foods is expected to decrease. As previously stated, the committee believes that DLC exposure reduction is important in young children, particularly girls, so that they may enter their reproductive years with lower DLC body burdens.

Risk-Relationship Analysis

Ancillary Benefits. Research has shown that reducing the consumption of foods high in saturated fats is associated with the reduced risk for many chronic diseases, including heart disease and cancer. Therefore, long-term health benefits would be expected to result from reducing saturated fat intake beginning in childhood.

Countervailing Risks. Children who do not accept foods lower in saturated fats may not have adequate nutrient intakes and may be at increased nutritional risk.

Option: Promote Changes in Dietary-Consumption Patterns of the General Population That More Closely Conform to Recommendations from the Dietary Guidelines for Americans to Reduce the Consumption of Foods High in Saturated Fats

This option recommends that the federal sponsors of the U.S. Dietary Guidelines institute public education and information dissemination campaigns and actively promote compliance with the recommendations of the Guidelines, particularly those that advise the decreased intake of foods high in saturated fats. Although the U.S. Dietary Guidelines are available through publications and informational websites, current promotional efforts by federal sponsors are insufficient.

The dietary recommendations and associated educational tools for the general population that are consistent with reducing DLC exposure by reducing animal-fat intake include the Dietary Guidelines for Americans, last revised in 2000; the Food Guide Pyramid, based on the Dietary Guidelines; the dietary findings of the Institute of Medicine in the recently released report on *Dietary Reference for Intakes Energy, Carbohydrate, Fiber, Fat, Fatty Acids, Cholesterol, Protein, and Amino Acids*; and the recommendations of the American Heart Association (AHA). Although these documents do not provide specific guidance regarding foods high in DLCs, their recommendations to reduce intake of animal fat, the primary source of saturated fats, will result in a reduction in exposure to DLCs. As discussed in Chapter 5, dietary surveys, such as the Continuing Survey of Food Intakes by Individuals and the National Health and Nutrition Examination Survey have shown a trend of decreasing saturated fat intake in the general population, and the committee believes this trend should be encouraged to continue. In addition, the recommendation to reduce intake of saturated fats is in keeping with current national policies for good nutrition and health. Compliance with this recommendation does not require a reduction in protein intake; rather, it is a recommendation to trim visible fats from animal foods. Thus, the committee believes that compliance with this recommendation will not place consumers at increased nutritional risk and would likely reduce the risk for chronic disease.

Alternate/Interim Actions

An alternate action that may be taken is to revise the content of the U.S. Dietary Guidelines to include a food safety component that would recommend low-fat, but nutritionally equivalent, alternate choices to foods that are high in saturated fats as a way to reduce DLC exposure.

Current Barriers to Implementation

Since compliance with dietary recommendations is not mandatory, the success of such programs as a means to reduce DLC exposure depends entirely on their acceptance by the population. Educational campaigns and other promotional efforts can be undertaken to enhance public awareness and improve compliance with dietary recommendations. However, the committee recognizes there is uncertainty about the impact of educational programs on behavioral changes.

DLC Exposure Reduction

Adherence to dietary recommendations to reduce intake of saturated fats and animal fats is expected to, over a period of years, reduce DLC body burdens.

Risk-Relationship Analysis

Ancillary Benefits. In addition to reducing DLC exposures, there are concomitant ancillary benefits to overall health from compliance with recommendations to reduce total and saturated fat intake, such as the reduced risk for chronic disease, especially heart disease.

Countervailing Risks. The dietary recommendation in the U.S. Dietary Guidelines and the AHA recommendations that is not conducive to both reducing

DLC intake and improving health outcome is that of fish consumption. As discussed in Chapter 5, certain fish, including salmon and catfish, have been found in some cases to contain relatively high levels of DLCs, although these vary by source, location, and feeding exposure. As with other foods, the DLC content of fish closely parallels the fat content. Unfortunately, the fatty component of fish that is the source of beneficial omega-3 fatty acids is also the component of fish where DLCs accumulate.

To some extent, DLC concentrations in fish differ depending not only on the species, but also on the geographic region in which they are found, because some areas have greater environmental contamination than others. Nevertheless, overall wide variability is seen in analytical data reported for DLC levels in many foods, including fish, which increases the level of uncertainty and decreases the reliability of measures used for trend analysis. This level of uncertainty increases the difficulty of making reliable recommendations to reduce exposure to DLCs, particularly when it may be to the detriment of health benefits derived from certain foods.

Chapter 8 presents the committee's recommendations to reduce DLC exposure through foods from the array of options presented in this chapter.

8

Risk-Management Recommendations and Research Priorities

Chapters 6 and 7 describe a framework for developing policy options to reduce exposure of dioxins and dioxin-like compounds (referred to collectively as DLCs) through food, and they illustrate how the framework might be applied to a broad range of specific risk-management interventions. As the committee emphasized in those chapters, there are substantial gaps in the data needed to adopt many of the possible interventions that were identified, particularly regulatory interventions based on the government's food safety regulatory authority. Tolerances were not recommended not only because of the paucity of data on which to establish limits, but also because the committee recognized the significant ramifications of such actions without substantive evidence to support them.

The committee considered both the scientific uncertainties in risk at current levels of exposure and the concern within the general population about exposure to DLCs. It further recognized that there are substantial gaps in the data that have to be filled before many of the identified policy options can be adopted. Based on the analysis of current data and deliberations concerning the strategic options available to the government, the committee recommended some risk-management actions. The committee's recommendations are qualitative rather than quantitative in light of the paucity of data to support specific reduction goals, and they fall into four categories: (1) general strategic recommendations, (2) high-priority risk-management interventions, (3) other risk-management interventions that deserve consideration, and (4) research and technology development to support risk management.

GENERAL STRATEGIC RECOMMENDATIONS

DLCs have been the object of concern and activity on the part of federal regulatory agencies for many years, prompted by the large volume of evidence demonstrating the toxicity of DLCs at low levels of exposure in animals and suggesting the potential for DLCs to pose significant risks to humans. Important progress has been made in reducing new environmental discharges of DLCs. With respect to the problem of DLC exposures through food, however, most of the efforts have focused on assessment of the potential risks of DLCs, as discussed in Chapter 2. This has included substantial investments in toxicity testing and extended interagency efforts to refine and reach agreement on DLC risk assessments. This effort has been worthwhile. Risk assessment provides the essential starting point for risk management and, for any important problem, it is desirable to have the most definitive assessment of risk that available data and understanding will permit. Uncertainty is, however, inherent in risk assessment generally, and uncertainties in assessment of the risks posed by exposure to DLCs through food will persist for the foreseeable future. These uncertainties are not, however, an obstacle to sound and effective risk management.

Given that the risk assessments that have been conducted have raised concerns about the health impacts of DLCs, and that there is no benefit but possible harm from DLC exposure through foods, the committee considers it appropriate for the federal government to focus its efforts on exposure reduction strategies.

To move effectively toward reducing human exposure to DLCs through foods, the federal government should begin by pursuing the following strategic courses of action: (1) establish an integrated risk-management strategy and action plan, (2) foster collaboration between the government and the private sector to reduce DLCs, and (3) invest in the data required for effective risk management.

Develop an Integrated Risk-Management Strategy and Action Plan

Justification

DLC exposure through food is a widely shared problem in the food system. It results not from the actions of any one segment of the food system, but from the complex interaction of widespread environmental contamination and the established practices and behaviors of animal producers, food processors, and consumers. Its solution requires action across the system, based on consideration of all the factors that contribute to exposure and all the possibilities for reducing exposure.

At the federal level, regulatory responsibility for DLCs in food is shared by three agencies: the U.S. Department of Health and Human Services' Food and Drug Administration (FDA), the U.S. Department of Agriculture's (USDA) Food Safety and Inspection Service (FSIS), and the U.S. Environmental Protection Agency (EPA), and by multiple other program offices. For example, FDA's

Center for Food Safety and Applied Nutrition is responsible for setting regulatory limits on the levels of DLCs that may lawfully be present in human food and for regulating the practices of food processors. However, FDA's Center for Veterinary Medicine is responsible for regulating DLCs in animal feeds. A third organizational component of FDA, the Office of Regulatory Affairs, manages the field function of the agency, which involves conducting inspections, product sampling, and enforcement, and often requires balancing, in the face of scarce resources, the food-related functions of FDA with the agency's responsibilities for drugs and other medical products. FSIS has jurisdiction over the safety of meat and poultry, which would include enforcement of any regulatory restrictions that might be placed on the level of DLCs in these products. EPA sets limits on permissible discharges of DLCs into waterways based on assessments of the resulting impact of DLCs on the safety of fish. At the state and local level, multiple health, agriculture, and natural resource agencies can be involved in issuing and enforcing advisories and regulatory restrictions on the harvesting and consumption of fish from contaminated waters.

In addition to these regulatory agencies, a wider set of federal agencies would be involved in devising and implementing any DLC risk-management interventions that involve changing food-consumption patterns or conducting research. For example, multiple agencies within USDA and the U.S. Department of Health and Human Services collaborate in developing the government's Dietary Guidelines for Americans, and both departments have nutrition education programs. In addition, the Food and Nutrition Service within USDA sets policy for the National School Lunch Program and other government feeding programs, while USDA's Agricultural Marketing Service acts as the government's purchasing agent for these programs. Most of the regulatory agencies have food safety research programs that could contribute to the development of risk-management interventions, but so also do USDA's Agricultural Research Service and its Cooperative State, Research, Education and Extension Service, the Centers for Disease Control and Prevention, and other federal agencies.

Recommendation

The committee recommends that the sponsoring agencies empower an interagency coordination group with the authority and mandate to develop and to implement a single, integrated risk-management strategy and action plan. No one agency has the mandate, resources, expertise, or authority to address DLC exposure through food on a system-wide basis. With an integrated, comprehensive, and system-wide approach however, it will be possible to set achievable and widely shared goals for DLC exposure reduction, to identify optimal risk-management interventions, to set priorities, and to make the best use of available government resources. An integrated approach on the part of the federal government would also help to ensure effective interaction and collaboration between the public and private sectors and help to minimize private sector disruptions and costs.

The committee recognizes the difficulty of working across agency lines on a problem of this complexity. There will be a need to establish clearly defined leadership for the effort, accountability for results, and mechanisms for on-going collaboration and coordination.

Foster Collaboration Between the Government and the Private Sector to Reduce DLCs in the Food Supply

Justification

No one desires the presence of DLCs in food, and no one set of participants in the food system can fairly be assigned sole responsibility and accountability for their presence or for actions needed to reduce foodborne exposure to DLCs. DLC exposure through food is a shared problem requiring shared, collaborative solutions. This does not preclude the possibility that regulatory interventions will play a role at some point in reducing DLC exposure, but, as discussed in Chapters 6 and 7, the committee sees no immediate regulatory solutions. There is instead a need, among other things, to generate data and develop practices in the food-production and processing systems that will reduce DLC levels and resulting exposures over time. This will require active collaboration between the federal government and the private sector.

Recommendation

The committee recommends that, as part of the process of developing an integrated risk-management strategy and action plan, the federal government, through an interagency coordination group, create an atmosphere and program of collaboration with the private sector. This recommendation is premised on both the government and industry being willing to define and actively approach the problem of DLC exposure through food as a shared problem. If that willingness exists, the government should establish an organizational focal point and define processes through which government and the private sector could collaborate in developing and implementing the integrated strategy and action plan for reducing DLC exposure through food.

Invest in the Data Required for Effective Risk Management

Justification

As discussed in Chapters 6 and 7, there are significant gaps in the data required to devise, implement, and evaluate risk-management interventions to reduce DLC exposure through food. The lack of data is most glaring with respect to consideration of options to reduce DLC levels at the animal production stage and in human food. To target and prioritize efforts to reduce DLCs at these stages and to determine what reductions are desirable and feasible, the critical starting point is good information on the current levels of DLCs in forage, feeds, and feed components, across geographical regions, and in the array of human foods in which DLCs are found. As discussed in Chapters 4 and 5, however, the available data on these points are very limited. A reliable, reasonably complete picture of the current levels and distribution of DLCs in animal feeds and human food is lacking. There is also insufficient data on the levels and distribution of DLC body burdens in the general population and among sensitive and highly exposed subgroups. Body burden data are a prerequisite for evaluating the effectiveness of efforts to reduce DLC exposure through food.

Recommendation

The committee recommends that a commitment of resources by all sponsoring agencies for data collection be a central element of any risk-management strategy and action plan for reducing DLC exposure through food. The committee recognizes that DLC analysis is expensive and that expense has been a limiting factor in data-collection efforts to date. In the committee's judgment, however, a commitment to risk management to reduce DLC exposure necessitates a commitment to data collection. As discussed below, the committee recommends, as one of its research priorities, an effort to develop less costly analytical methods for DLCs in feeds and food.

HIGH-PRIORITY RISK-MANAGEMENT INTERVENTIONS

The committee has identified two areas that it believes deserve high-priority attention as part of any risk-management strategy for reducing DLC exposure through food. The committee recommends that the government focus its initial risk-management interventions on: (1) interrupting the cycling of DLCs through forage, animal feeds, and food-producing animals, and (2) reducing DLC exposure in girls and young women in order to protect fetuses and breastfeeding infants from exposure to DLCs.

Interrupt the Cycling of DLCs Through Forage, Animal Feeds, and Food-Producing Animals

Justification

As discussed in Chapter 4, animal forage and feeds are primary pathways for DLC contamination of the human food supply. This occurs as a result of the direct contamination of forage and plants used for animal feeds, typically by airborne deposition of DLCs. When animals consume contaminated forage and feeds, DLCs are stored in their fat and subsequently enter the human food supply. In addition, several billion pounds of rendered animal fat are used annually as a component of animal feeds, which results in the recycling of DLCs back through the feed and the possibility of increasing levels of DLCs in meat and other animal-derived food products (see Figure 4-1).

Recommendations

The committee recommends that the government's risk-management strategy for DLCs give high-priority attention to reducing the contamination of animal forage and feeds and interrupting the recycling of DLCs that results from the use of animal fat in animal feeds. The committee considers the animal feed and animal production stages of the food system to be key leverage points for reducing DLC exposure through food because it stops DLCs at their primary point of entry into the human food supply. This is the only available means the committee could identify for preventing DLC contamination of human food. At subsequent stages human exposure can be reduced only by discarding or otherwise avoiding consumption of DLC-contaminated food.

While FDA has the authority to set legally binding limits (i.e., tolerances) on the levels of DLCs in animal feeds, the committee is not recommending such regulatory action at this time. This is due to the current lack of data to support binding tolerances, the consequences of trying to establish limits without adequate supportive evidence, and the committee's belief that there is a need for collaboration between the government and the animal production industry to develop alternative feeding practices and to overcome other practical obstacles to reducing DLCs in animal forage and feeds that may obviate the need for regulatory action in the future.

As an initial step, the government, in collaboration with the animal production and feed industries, and directed by an interagency coordination group, should establish a nationwide data-collection effort and data repository on the levels of DLCs in animal forage and feeds, which should be accessible for both public and private use. An expanded data-collection effort by the government and industry and the pooling of the data would provide a better understanding of current DLC levels in forage and feeds, including their sources and geographic distribution. This would in turn provide a basis for devising and targeting specific interventions to reduce DLCs in forage and feeds.

The government and industry should also begin collaborating immediately to define voluntary guidelines for good animal feeding and production practices that would reduce DLC levels in forage and feeds and would minimize other potential sources of DLC exposure in animal production. Such guidelines could include avoiding forage or feeds obtained from areas known to have high levels of DLC contamination. They could also include criteria for the use of animal fat in animal feeds, which would result in reducing DLC levels in the finished feeds, and criteria for the use of non-DLC containing materials in animal enclosures and in feed packaging and transportation. The use of "cleaner" (i.e., uncontaminated) fats is not a practical alternative at the present time because of the high cost of analysis that would be needed to implement widespread testing.

The committee further recommends that the government, in collaboration with the animal production industry, identify means to achieve the reduction or elimination of DLC-containing animal fat as a component of animal feeds. This would require, among other things, the development of criteria for determining what constitutes an acceptable use of animal fat and economically feasible analytical methods for distinguishing acceptable from unacceptable animal fat. It would also require the development of alternative uses or acceptable disposal methods for the large quantity of animal fat now used in animal feeds. However, the committee recognizes that reducing or eliminating the use of animal fat as a component of animal feeds could have unintended negative consequences: (1) increased cost of food, (2) problems of animal waste disposal, (3) increased food spoilage, and (4) changes in the taste of food that consumers find unacceptable.

Only when more complete data are generated on DLC levels in forage and feeds, and a better understanding is developed of how DLC contamination can be avoided, should the government consider setting legally binding limits on DLCs in forage and feeds.

Reducing DLC Exposure in Girls and Young Women

Justification

As discussed in Chapter 2, fetuses and breastfeeding infants may be at particular risk from exposure to DLCs. This is due to the potential for DLCs to cause adverse neurodevelopmental, neurobehavioral, and immune system effects in developing systems, combined with the potential for in utero exposure of the fetus to DLCs and exposure of breastfeeding infants to relatively high levels of DLCs in breast milk. These exposures are a result of the body burden of DLCs that girls and young women accumulate during their childhood, adolescence, and young adult years. Data suggest that, due to the bioaccumulation phenomenon, reduction of DLC intake during pregnancy has no significant impact on the mother's body burden or on the baby's exposure in utero or through breastfeeding. The committee recommends the following options as ways to reach a broad number of individuals with the least potential undesired outcomes.

Recommendations

The committee recommends that the government place a high public health priority on reducing DLC intakes by girls and young women in the years well before pregnancy is likely to occur. Such reductions can be achieved by reducing the intake of animal fat among this population. **The committee therefore recommends, as an immediate intervention, that the government take steps to increase the availability of foods low in animal fat in government-sponsored school breakfast and lunch programs and in child- and adult-care food programs. Specifically, the committee recommends that the National School Lunch Program increase the availability of low-fat (1 percent) and skim milk and have the option of offering other milks.** Because the current law (Richard B. Russell National School Lunch Act, amended Dec. 8, 2000, §9(2)) favors the provision of whole milk, it should be amended to require that schools offer low-fat and skim milk and have the option of offering reduced-fat (2 percent fat) or whole (3.5 percent fat) milk. **The committee further recommends that participants in the Special Supplemental Nutrition Program for Women, Infants and Children be encouraged, except for children under 2 years of age, to choose low-fat or skim milk and low-fat versions of other animal-derived foods in their food packages.**

In addition, to reduce other sources of animal fat, the committee recommends that USDA's Economic Research Service undertake detailed analyses of the feasibility of and barriers to setting limits on the amount of saturated fat that can be present in individual meals in the National School Lunch Program. There are insufficient data to establish limits on saturated fat in school lunch and breakfast meals to reduce DLC exposure. Detailed analyses are needed to determine an appropriate level of saturated fat to reduce DLC exposure and to promote and maintain good nutrition habits.

Although data are insufficient to set limits on saturated fat intake to reduce DLC exposure, there is a strong body of evidence to support the benefit of reducing saturated fat intake to reduce the risk of chronic disease. The Dietary Guidelines for Americans recommend that less than 10 percent of calories should be derived from saturated fat. School lunches are currently required to provide less than this level when the fat content of the meals is averaged over a week. There is currently no saturated fat guideline for individual meals. If a limit were established for saturated fat in individual meals served in the National School Lunch Program, more lean meat, poultry, and seafood products would be used in place of high-fat versions of these foods, which would lower levels of animal fat (and thus DLCs) in school lunches. While potential changes in the school lunch program are being considered, there should be ongoing educational efforts aimed at reducing the consumption of animal fat by girls and young women.

As discussed in Chapter 5, substituting low-fat or skim milk for whole milk, especially when coupled with other substitutions of foods lower in animal fat such as selecting lean cuts of meat, poultry, low-fat fish, and low-fat cheeses, could significantly reduce DLC intakes and resulting body burdens of DLCs in girls and young women in the crucial years preceding pregnancy. Reducing DLC intakes in this way is the only practicable intervention the committee could identify that would, in the relatively near term, reduce DLC exposure to fetuses and breastfeeding infants, which appear to be the populations most vulnerable to the toxicity of DLCs.

OTHER RISK-MANAGEMENT INTERVENTIONS THAT DESERVE CONSIDERATION

Although more data are needed, there are several other specific interventions that could be considered as part of an integrated risk-management strategy and action plan for reducing DLC exposure through food. These include: (1) reducing DLC discharge sources in animal production areas, (2) removing DLC residues from foods during processing, (3) providing advisories and education to highly exposed populations, and (4) educating the general population about strategies for DLC exposure reduction.

Reducing DLC-Discharge Sources in Animal Production Areas

Justification

As discussed in Chapter 3, the largest quantifiable source of new DLC formation in the United States is the backyard burning of yard, home, and farm waste. To the extent this occurs in animal production areas or in areas where animal feed is produced, it is potentially an important pathway for DLC contamination of food and resulting human exposure.

Recommendation

The committee recommends that the government consider, as part of an integrated risk-management strategy, a focused effort to reduce unregulated (e.g., backyard) burning, especially in animal production areas. This could be pursued initially as an element of the collaborative effort with the animal production industry to reduce DLC contamination of animal forage and feeds. The committee has not examined whether there are practicable regulatory interventions to address this potential source of DLCs in the environment.

Removing DLC Residues from Foods During Processing

Justification

DLCs can directly contaminate human foods through airborne deposition on food plants and soils. As discussed in Chapter 4, as much as 25 percent of DLC exposure through foods may come from this source. In such cases, the DLCs and DLC-contaminated soils typically adhere to the external surfaces of the plant and its edible portion, including vegetables, fruits, and grains. These DLCs can, to a significant extent, be prevented from entering the food supply at the food-processing stage through readily available cleaning and processing measures and by peeling root and waxy-coated vegetables. In addition, the physical removal of fat from meat products through trimming prevents DLCs from entering the human food supply.

Recommendation

The committee recommends that the government explore, with the food industry, practicable steps to reduce, during processing, the DLC contamination of food. This effort could result in the development of voluntary good manufacturing practices for reducing DLCs in food. Consideration and adoption of regulatory measures in this area would require more data than currently exist on the magnitude of the DLC reductions that could be achieved through processing, as well as on the feasibility and cost of achieving these reductions.

Providing Advisories and Education to Highly Exposed Populations

Justification

As discussed in Chapter 5, there are population groups that for economic, cultural, or other reasons consume large amounts of fish and marine mammals that tend to be high in DLCs. These highly exposed groups, and in particular sensitive members of these groups (such as developing fetuses and breastfeeding infants), may be at higher risk of adverse health effects from DLCs than the general population. Currently, EPA and some state and local agencies issue advisories on the consumption of fish caught in highly contaminated areas.

Recommendation

The committee recommends that the government continue collaborating with state and local officials to provide up-to-date fishing advisories on waters that are highly contaminated with DLCs. In addition, the committee recommends that the government work with highly exposed populations to develop information and education programs about the potential risks of DLCs and offer practical ideas for reducing DLC exposure, taking into account each group's economic and cultural situation.

Educating the General Population About Strategies for Reducing Exposure to DLCs

Justification

Within the general population there is much that could be done at the individual and household levels to reduce exposure to DLCs through food. The intervention that could be the most immediate and could have the most impact is for individuals to reduce their intake of animal fat. This could be done by reducing the consumption of fat from meat, poultry, and fish (e.g., by trimming and discarding their excess fat) and by choosing lower-fat versions of these foods. Individuals could reduce DLC intakes by washing and, as appropriate, peeling root and waxy-coated vegetables prior to consumption.

Achieving changes in dietary patterns and food preparation practices is difficult. The government has been communicating the fat reduction message, based on the U.S. Dietary Guidelines, for a number of years. While fat intake has declined as a percentage of calories, it has remained fairly constant in absolute terms while total calorie intake has risen. The communication of the fat-reduction message is further complicated in the case of DLCs since one of the major sources of DLCs in the diet—fish—has nutritional advantages as a source of protein and potentially "healthy" fats, such as the omega-3 fatty acids. Furthermore, while it may seem that the provision of information about DLCs on food labels would be an alternative educational approach, the limited range of foods tested and the complexities of the toxicity equivalents measurement system and its interpretation could mislead consumers about the DLC content in food.

Recommendations

The committee recommends that the government continue (and explore ways to enhance) its promotion of the U.S. Dietary Guidelines, including the message to limit intake of saturated fat. This message has the potential to produce health benefits that go well beyond the reduction of DLC exposure, including reductions in the risk of cardiovascular disease and cancer. **The committee further recommends that the government consider linking, in its Dietary Guidelines and associated information campaigns, the saturated fat reduction message based on cardiovascular disease and cancer with the message that reducing saturated fat has the added benefit of reducing exposure to DLCs and other lipophilic contaminants.** The committee recognizes that communicating clear and effective messages on this complex subject is difficult, and that great care must be exercised in crafting messages that do not mislead or confuse consumers by making the messages ineffective for their intended purposes or by causing dietary changes that could be detrimental. As discussed below, behavioral research may be needed to craft effective messages.

RESEARCH AND TECHNOLOGY DEVELOPMENT TO SUPPORT RISK MANAGEMENT

As one of its general strategic recommendations, the committee recommends above that the government focus its efforts on exposure-reduction strategies and that it invest in data collection to support risk management. There is also a need, in the committee's judgment, for a broader research and technology development agenda to support risk management. Among the many possible subjects for such efforts, the committee recommends that the government consider placing a priority on the following: (1) low-cost analytical methods development and toxicity equivalents review, (2) research to support removal of DLCs from animal feeds, (3) expansion of the National Health and Nutrition Examination Survey's (NHANES) data collection on DLC body burdens, (4) research on the effects of dietary DLCs on fetuses and breastfeeding infants, (5) behavioral research on achieving dietary change, and (6) predictive modeling studies on DLCs in the food supply.

Analytical Methods Development and Toxicity Equivalents Review

Justification

As discussed in Chapter 2, the analysis of forage, feeds, and food commodities for DLC contamination is expensive, and this expense has been a limiting factor in the collection of data to design and evaluate risk-management interventions. In addition, accuracy and reproducibility in DLC measurements are often compromised by analytical variances and measurement bias.

Recommendation

The committee recommends that the government invest in the development of cost-effective analytical methods that will make possible a significantly larger volume of DLC testing to support risk management. As part of this effort, the committee recommends that the government review the current toxicity equivalents assessment standards to ensure that the standards accurately reflect the most current knowledge of the toxicity contribution of various DLC congeners, particularly for low-level exposures from foods.

Research to Support Removal of DLCs from Animal Feeds

Justification

The committee has recommended reducing the contamination of animal forage and feeds and reducing the recycling of DLCs that result from the use of animal fat in animal feeds as high-priority risk-management interventions. Achieving these goals may require significant adjustments in animal feeding practices and the development of alternative uses or acceptable disposal solutions for the several billion pounds of animal fat that are used annually in animal production systems.

Recommendation

The committee recommends that the government sponsor research on the economics of animal production and current animal feeding practices, with a view toward identifying economically feasible alternatives to current practices that will result in significant reductions in DLC contamination of animal forage and feeds. To complement the economic research, the committee recommends that the government sponsor research on disposal possibilities and alternative uses of animal fat, including use as a biofuel.

Expansion of Data Collection on DLC Body Burdens

Justification

In 1999, NHANES began collecting measurements of serum dioxins from a subsample of the population examined in this large, recurring survey on diet and health. These measurements, if continued over time, will enable scientists to monitor trends in DLC body burdens. When these data are combined with data from FDA's Total Diet Study and the USDA Economic Research Service's food-consumption data as a component of the continuing NHANES, it will be possible to develop a more refined understanding of DLC exposure through food, including its geographic and population-group variability and its impact on body burdens among various components of the population. Such data will be essential to sound risk-management decision-making over the long term, including the consideration of possible regulatory interventions.

Recommendations

The committee recommends that the government expand data collection of DLC body burdens to expedite the development of a reasonably representative and reliable database on DLC exposure patterns and body burdens.

DLC analyses in human subjects are expensive and invasive to the subject, which makes the collection of new body burden data difficult. Thus, to supplement the NHANES data collection on DLC body burdens, the committee recommends that the government support continued and expanded DLC assays of tissues collected from existing cohorts and control groups, such as in the Ranch Hands study (discussed in Chapter 2), which has a large sample base of subjects with background or low-level DLC exposures.

Research on the Effects of Dietary DLCs on Fetuses and Breastfeeding Infants

Justification

From a public health perspective, the committee places its highest risk-management priority on reducing the DLC exposure of fetuses and breastfeeding infants.

Recommendation

To support and further focus risk-management initiatives in this area, the committee recommends that the government sponsor a prospective cohort study (that includes monitoring breast-milk samples) to examine the health effects over time of pre- and postnatal DLC exposure of infants. The results of such a study would help public health officials and risk managers better understand the nature and severity of the risks posed by exposure to DLCs at this critical developmental phase of life, and thus help guide and set priorities for future risk-management initiatives. In addition, the committee recommends that breast-milk monitoring in sentinel populations be conducted to assess the magnitude of exposure through this source.

Behavioral Research on Achieving Dietary Change

Justification and Recommendation

To the extent that the government's risk-management strategy relies on achieving dietary changes to reduce DLC exposure, the committee recommends that it sponsor behavioral research to better understand how such changes can be brought about. It is clear from experience over the past decade that formulation and communication of a scientifically sound message about diet and health is not enough to change long-established dietary patterns. Dietary patterns are complex human behaviors that are affected by many factors and vary among individuals. Careful research is needed to better understand the phenomenon and to develop educational and other techniques for improving dietary patterns.

Predictive Modeling Studies on DLCs in the Food Supply

Justification

DLCs move through the atmospheric, terrestrial, and aquatic environments and into the human food supply in complex ways. In designing interventions to reduce DLC exposure through food, it is important to be able to model the movement of DLCs and their ultimate biomagnifications through the food chain, as well as to be able to predict how proposed interventions will affect levels in food and, in turn, human exposure and body burdens.

Recommendation

The committee recommends that the government sponsor research to develop improved predictive modeling tools and that it apply them in studies aimed at assessing the DLC-reduction effects of potential interventions.

CONCLUSION

DLCs are an undesirable contaminant in food, and there are good public health reasons for reducing DLC exposure through food, especially among highly exposed and sensitive populations. The committee recognizes there are serious limitations in the data available for managing the risks posed by DLCs in food and that, in light of these limitations and remaining uncertainties in risk assessments on DLCs, it is premature to recommend traditional food-safety regulatory remedies for the DLC problem. There are, however, a number of steps the government could take to reduce DLC exposure among the most vulnerable population groups in the short term and to significantly reduce DLC exposure within the general population in the long term. These steps have been outlined in this concluding chapter.

A

Data Tables

TABLE A-1 Dioxins and Dioxin-like Compounds and the Most Common Toxicity Equivalency Factor (TEF) Systems

Compound	I-TEF[a]	WHO 94[b]	WHO 98[c]
Dioxins			
2,3,7,8-tetrachlorodibenzo-*p*-dioxin	1	1	1.0
1,2,3,7,8-pentachlorodibenzo-*p*-dioxin	0.5	0.5	1.0
1,2,3,4,7,8-hexachlorodibenzo-*p*-dioxin	0.1	0.1	0.1
1,2,3,6,7,8-hexachlorodibenzo-*p*-dioxin	0.1	0.1	0.1
1,2,3,7,8,9-hexachlorodibenzo-*p*-dioxin	0.1	0.1	0.1
1,2,3,4,6,7,8-heptachlorodibenzo-*p*-dioxin	0.01	0.01	0.01
Octachlorodibenzo-*p*-dioxins	0.001	0.001	0.0001
Furans			
2,3,7,8-tetrachlorodibenzofuran	0.1	0.1	0.1
1,2,3,7,8-pentachlorodibenzofuran	0.05	0.05	0.05
2,3,4,7,8-pentachlorodibenzofuran	0.5	0.5	0.5
1,2,3,4,7,8-hexachlorodibenzofuran	0.1	0.1	0.1
1,2,3,7,8,9-hexachlorodibenzofuran	0.1	0.1	0.1
1,2,3,6,7,8-hexachlorodibenzofuran	0.1	0.1	0.1
2,3,4,6,7,8-hexachlorodibenzofuran	0.1	0.1	0.1
1,2,3,4,6,7,8-heptachlorodibenzofuran	0.01	0.01	0.01
1,2,3,4,7,8,9-heptachlorodibenzofuran	0.01	0.01	0.01
1,2,3,4,6,7,8,9-octachlorodibenzofuran	0.001	0.001	0.0001
Polychlorinated biphenyls (PCB)			
3,3'4,4'-tetrachlorobiphenyl (PCB-77)		0.0005	0.0001
3,4,4',5-tetrachlorobiphenyl (PCB-81)			0.0001
3,3',4,4',5-pentachlorobiphenyl (PCB-126)		0.1	0.1
3,3',4,4',5,5'-hexachlorobiphenyl (PCB-169)		0.01	0.01
2,3,3',4,4-pentachlorobiphenyl (PCB-105)		0.0001	0.0001
2,3',4,4',5-pentachlorobiphenyl (PCB-118)		0.0001	0.0001
2',3,4,4',5-pentachlorobiphenyl (PCB-123)		0.0001	0.0001
2,3,3',4,4',5-hexachlorobiphenyl (PCB-156)		0.0005	0.0005
2,3,3',4,4',5'-hexachlorobiphenyl (PCB-157)		0.0005	0.0005
2,3',4,4',5,5'-hexachlorobiphenyl (PCB-167)		0.00001	0.00001
2,3,4,4',5-pentachlorobiphenyl (PCB-114)		0.0005	0.0005
2,2',3,3',4,4',5-heptachlorobiphenyl (PCB-170)		0.0001	
2,2',3,4,4',5,5'-heptachlorobiphenyl (PCB-180)		0.00001	
2,3,3',4,4',5,5'-heptachlorobiphenyl (PCB-189)		0.0001	0.0001

[a]International TEF system. See EPA (1989).
[b]Ahlborg et al. (1994).
[c]van den Berg et al. (1998).

NOTE: **Bold** type indicates compounds with varying TEFs.

TABLE A-2 Toxicity Equivalency Factor (TEF) Systems Used in Governmental Reports on Dioxins and Dixon-like Compounds Reviewed by the Committee

Reference	CDDs/CDFs	PCBs
AEA Technology, 1999	I-TEQ unless otherwise specified; WHO (1994) also used	Not evaluated
ATSDR, 1998	All are presented, but use is not distinguished	Both are presented, but use is not distinguished
EPA, 2000	WHO 98 is preferred, but all systems are used and distinguished	WHO 98 is preferred, but all systems are used and distinguished
Fiedler et al., 2000	I-TEQ and WHO 98; most data given in I-TEQ	WHO 98; sometimes added into I-TEQ
IARC, 1997	I-TEQ	Not evaluated
Scientific Committee on Food, 2000, 2001	WHO 98 is preferred	WHO 98 is preferred

NOTE: CDD = chlorinated dibenzo-*p*-dioxin, CDF = chlorodibenzofuran, PCB = polychlorinated biphenyl, TEQ = toxicity equivalents.

TABLE A-3 Data for the Lowest Exposure Strata in Cancer Epidemiology Studies on Dioxins and Dioxin-like Compounds

Table Number	Exposure Definition	Risk Measure	Risk Estimate (95% confidence interval)
7-3	Lowest septile of estimated cumulative exposure (0 to < 19)	SMR for total cancer	1.14 (no lag)
7-3	Lowest septile of estimated cumulative exposure (0 to < 39)	SMR for total cancer	0.98 (10-y lag)
7-5	0–125.1 ng TCDD/kg blood (German chemical manufacturing workers)	RR for total cancer	1.24 (0.82–1.79)
7-6	Subcohorts (German chemical manufacturing workers) with median blood TCDD levels of 9.5 ppt and 8.4 ppt	RR for lung cancer RR for stomach cancer RR for total cancer	0.5 (0.1–1.8) 0 (0–1.7) 1.1 (0.6–1.8)
7-14	Low estimated potential for phenoxyacetic acid exposure	RR for soft-tissue sarcoma RR for non-Hodgkin's lymphoma	0.6 (0.3–1.1) 0.9 (0.6–1.3)
7-14	Low estimated potential for chlorophenol exposure	RR for soft-tissue sarcoma RR for non-Hodgkin's lymphoma	0.9 (0.5–1.6) 1.0 (0.7–1.3)
7-17	Pulp and paper mill workers (IARC considers this to entail low exposure)	RR for death from all cancer	0.9 (0.8–1.0)
7-19	Seveso Region R	RR and SMR for all cancer	Males 0.9 (0.9–1.0) Females 0.9 (0.8–1.1)

NOTE: SMR = standard mortality ratio, TCDD = tetrachlorodibenzo-p-dioxin, RR = relative risk, IARC = International Agency for Research on Cancer.
SOURCE: EPA (2000).

TABLE A-4 Body Burdens of Dioxins and Dioxin-like Compounds in Potentially Highly Exposed Populations

Reference	Value (Tissue) in the Exposed Population[a]	Value (Tissue) in the Referents	Comments
Breastfed infants			
EPA, 2000	Mean at 11 mo: 34.7 pg $TEQ_{DF\text{-}WHO98}$/g lipid (blood)	Mean at 11 mo: 2.7 pg $TEQ_{DF\text{-}WHO98}$/g lipid (blood)	1994–1995
EPA, 2000	Mean at 25 mo: 43.9 pg $TEQ_{DF\text{-}WHO98}$/g lipid (blood)	Mean at 25 mo: 3.3 pg $TEQ_{DF\text{-}WHO98}$/g lipid (blood)	1994–1995
EPA, 2000	Mean at 11 mo: 31.4 pg $TEQ_{P\text{-}WHO98}$/g lipid (blood)	Mean at 11 mo: 2.5 pg $TEQ_{P\text{-}WHO98}$/g lipid (blood)	1994–1995
Heavy consumers of fish (other than Northern Dwellers)			
EPA, 2000	Mean in high-fish consumers: 60 pg $I\text{-}TEQ_{DF}$/g lipid (blood)	Mean in non-fish consumers: 20 pg $I\text{-}TEQ_{DF}$/g lipid (blood)	Baltic Sea fish in Sweden
EPA, 2000	Mean in fish eaters: 292.6 pg $TEQ_{P\text{-}WHO94}$/g lipid (blood)	Mean in controls: 10.9 pg $TEQ_{P\text{-}WHO94}$/g lipid (blood)	Fishermen in the Gulf of the St. Lawrence; congener analysis done only in the 10/185 samples with the highest total PCB concentrations
People living near local sources of contamination			
EPA, 2000	Consumers of local beef and eggs for up to 15 y: 63.7 pg $I\text{-}TEQ_{DF}$/g (blood)	Comparison group: 17.0 pg $I\text{-}TEQ_{DF}$/g (blood)	Unclear if this is lipid basis
EPA, 2000	Near industrial emissions: 49–291 pg $I\text{-}TEQ_{DF}$/g lipid (blood)	Control samples: 16 and 26 pg $I\text{-}TEQ_{DF}$/g lipid (blood)	England
EPA, 2000	Near a solid waste incinerator: 22–463 pg $TEQ_{DFP\text{-}WHO94}$/g lipid (blood)	Expected mean: 55 pg $TEQ_{DFP\text{-}WHO94}$/g lipid (blood)	Japan
EPA, 2000	Near a PCB manufacturing plant: 16–39 pg $I\text{-}TEQ_D$/g lipid (blood) 7–131 pg $I\text{-}TEQ_F$/g lipid (blood)	19 pg $I\text{-}TEQ_D$/g lipid (blood) 8 pg $I\text{-}TEQ_F$/g lipid (blood)	Alabama

continued

TABLE A-4 Continued

Reference	Value (Tissue) in the Exposed Population	Value (Tissue) in the Referents	Comments
	34–360 pg $TEQ_{P\text{-}WHO94}$/g lipid (blood)	32 pg $TEQ_{P\text{-}WHO94}$/g lipid (blood)	
Northern Dwellers			
ATSDR, 1998	8.4 pg TCDD/g (plasma)	< 2 pg TCDD/g	Inuits vs Québécois
ATSDR, 1998	39.6 pg TEQ/g (plasma)	14.6 pg TEQ/g (plasma)	Inuits vs Québécois
ATSDR, 1998	184.2 pg TEQ/g (plasma)	26.1 pg TEQ/g (plasma)	Inuits vs Québécois

NOTE: TEQ = toxicity equivalents, PCB = polychlorinated biphenyl, TCDD = tetrachlorodibenzo-p-dioxin.

TABLE A-5 Intakes of Dioxins by Highly Exposed Populations

Reference	Population	Value	Compounds	Comments
Breastfed infants				
ATSDR, 1998	At 4 weeks	132.1 pg TEQ/kg	CDD	Cumulative intake
		257 pg TEQ/kg	CDD/CDF	
ATSDR, 1998		83.1 pg TEQ/kg/d	CDD/CDF	Older data
ATSDR, 1998	During first year	35–53 pg TEQ/kg/d	CDD/CDF	Older data
People living near local sources of contamination				
EPA, 2000	Near a waste incineration site, estimated daily intake of 165 pg I-TEQ_{DF}/d	Background intake from same food types, 43.2 pg I-TEQ_{DF}/d	CDD/CDF	Wales

NOTE: TEQ = toxicity equivalents, CDD = chlorinated dibenzo-p-dioxin, CDF = chlorodibenzofuran.

TABLE A-6 Existing Total Human Exposure Limits

Legislation/Guidelines	Country/Organization	pg TEQ/kg/d
Tolerable daily intake	WHO	1–4
	JECFA	2.3
	EC	2
	The Netherlands	1
	Japan	4
Tolerable weekly intake	JECFA	16.1
	EC	14
Proposed tolerable monthly intake	Australia	70
Provisional tolerable monthly intake	JECFA	70

NOTE: TEQ = toxicity equivalents, WHO = World Health Organization, JECFA = Joint Food and Agriculture Organization of the United Nations/WHO Expert Committee on Food Additives, EC = European Commission.

TABLE A-7 Summary of Existing U.S. Federal Environmental Regulations and Guidelines

Regulation/Guideline	Description
Human Exposure	
ATSDR Guidelines/CERCLA	In compliance with CERCLA, ATSDR has set up MRLs, which are estimates of daily human exposure to a substance that is likely to be without appreciable risk of adverse health effects
Air	
40 C.F.R. 60 Subpart AAAA	New Source Performance Standards for New Small Municipal Waste Combustion Units
40 C.F.R. 60 Subpart BBBB	Emission Guidelines for Existing Small Municipal Waste Combustion Units
40 C.F.R. 60 Subpart EB	New Source Performance Standards for Large Municipal Waste Combustors
40 C.F.R. 60 Subpart EC	New Source Performance Standards—HMIWI
40 C.F.R. 60 Subpart CE	Applies to HMIWI constructed on or before 6/20/96
40 C.F.R. 63, 261, and 430	New Source Performance Standards: Pulp, Paper, and Paperboard Category
40 C.F.R. 63, 261, and 430	National Emissions Standards for Hazardous Air Pollutants for Source Category: Pulp and Paper Production, Effluent Limitations Guidelines, Pretreatment Standards
40 C.F.R. 60	Hazardous Waste Incinerators, Cement Kilns, and Lightweight Aggregate Kilns—New Sources
40 C.F.R. 60 Subpart CCCC	New Source Performance Standards for CIWI constructed after 9/30/99
40 C.F.R. 60 Subpart DDDD	Emission Guidelines for Existing CIWI

Limits	Date Implementation Required	Date Compliance Required
Acute MRL = 0.0002 µg/kg/d Intermediate MRL = 0.00002 µg/kg/d Chronic MRL = 0.000001 µg/kg/d	Not applicable	Not applicable
13 ng/dscm for dioxins/furans on a total mass basis, using a 3-run average with a minimum of 4-h run duration	6/6/01	6/6/01
125 ng/dscm (total mass basis)	Implementation by state plan	12/6/05
13 ng/dscm for dioxins/furans on a total mass basis (mandatory), or 7 ng/dscm total mass (optional to qualify for less frequent testing)	6/19/96	6/19/96
Small HMIWI: 125 ng/dscm total CDD/CDF (55 g/10^9 dscf) or 2.3 ng/dscm TEQ (1.0 g/10^9 dscf) Medium and Large HMIWI: 25 ng/dscm total CDD/CDF (11 g/10^9 dscf) or .6 ng/dscm TEQ (0.26 g/10^9 dscf)	Effective 3/15/98	3/15/98 or within 180 days of initial start up
Small, medium, and large HMIWI: 125 ng/dscm total CDD/CDF (55 g/10^9 dscf) or 2.3 ng/dscm TEQ (1.0 g/10^9 dscf) Small rural HMIWI: 800 ng/dscm total CDD/CDF (350 g/10^9 dscf) or 15 ng/dscm TEQ (6.6 g/10^9 dscf)	Effective 9/14/00	8/15/01 or incrementally by 9/15/02
New Sources: 0.20 ng TEQ/dscm Existing Sources: 0.20 ng TEQ/dscm or 0.40 ng TEQ/dscm and temperature at inlet to the initial particulate matter control device ≤ 400°F	9/30/99	9/30/02
0.41 ng/dscm (TEQ basis)	1/30/01	Within 60 days, no later than 180 days after initial startup
0.41 ng/dscm (TEQ basis)	12/1/01	12/6/05

continued

TABLE A-7 Continued

Regulation/Guideline	Description
Water	
National Primary Drinking Water Regulations	Maximum contaminant level
Safe Drinking Water Act	Maximum contaminant level goal in public drinking water
Sludge/biosolids	
40 C.F.R. 503 (proposed)	Proposed Standards for the Use or Disposal of Sewage Sludge

NOTE: ATSDR = Agency for Toxic Substances and Disease Registry, CERCLA = Comprehensive Environmental Response, Compensation, and Liability Act, MRL = minimum risk level, dscm = dry standard cubic meter, HMIWI = hazardous material/industrial waste incinerator, CDD = chlorinated dibenzo-*p*-dioxin, CDF = chlorodibenzofuran, TEQ = toxicity equivalents, dscf = dry standard cubic feet, CIWI = commercial/industrial waste incinerator, TCDD = tetrachlorodibenzo-*p*-dioxin.

Limits	Date Implementation Required	Date Compliance Required
2,3,7,8-TCDD: 3×10^{-8} mg/L	1994	1994
2,3,7,8-TCDD: 0	Nonenforceable, voluntary health goal	Nonenforceable, voluntary health goal
0.0003 mg TEQ/kg dry sewage sludge for application to land	Proposed 12/23/99	Not applicable

TABLE A-8 Detailed Summary of U.S. State Environmental Regulations and Guidelines

State	Legislation
California	Dioxin Airborne Toxic Control Measure for Medical Waste Incinerators
	Air Toxic Control Measure
	California Department of Health Services: maximum contaminant levels for state primary drinking water standards
Illinois	Title 35, Subtitle F, Chapter I, Part 611: primary drinking water standards
	Title 35, Subtitle B, Chapter I, Part 229: hospital/medical/infectious waste incinerators
Massachusetts	310 CMR 7.08 (2): municipal waste combustors
	310 CMR 22.00L: drinking water regulations
	Air Quality Guidelines for Dioxins
	310 CMR 40.1600: reportable concentrations in groundwater and soil
	310 CMR 7.08(5)(e): emissions limits for hospital/medical/infectious waste incinerators as promulgated by EPA through the Clean Air Act
Missouri	10 CSR 60-4.040: Maximum Synthetic Organic Chemical Contaminant Levels, drinking water
	10 CSR 10-6.020: Emissions to Air from Municipal Waste Combustion
	10 CSR 25-5.262: Standards applicable to generators of hazardous waste
	All air standards are the same as the federal government; the only water regulations concerning dioxins are for drinking water
New Jersey	New Jersey Drinking Water Standards
New York	6 NYCRR Part 219 [1/1]: Air emission standards for incinerators
	6 NYCRR Part 703: Surface water and groundwater quality standards and groundwater effluent limitations
Wisconsin	NR 445.02: Control of Hazardous Pollutants
	Solid Waste Enforcement Standard
	NR 140: Public Health Groundwater Quality Standards

NOTE: TCDD = tetrachlorodibenzo-p-dioxin, dscm = dry standard cubic meter, TEQ = toxicity equivalents, dscf = dry standard cubic feet, EPA = U.S. Environmental Protection Agency.

Date In Force	Description/Limits
7/31/90	10 ng/kg of waste burned or dioxin emissions are reduced by 99% of uncontrolled emissions (applies to medical waste incinerators that incinerate more than 25 tons of waste/y)
1/1/04	Eliminates burning of household garbage and rubbish in burn barrels Maximum contaminant level for 2,3,7,8-TCDD = 0.00000003 mg/L
1/17/94	Maximum Contaminant Level for 2,3,7,8-TCDD = 0.00000003 mg/L
5/15/99	Small, medium, and large incinerator emission limits: 125 (55) ng/dscm, total dioxins/furans (grains per billion dscm) or 2.3 (1.0) ng/dscm TEQ (grains per billion dscf) Rural incinerators: 800 (350) ng/dscm, total dioxins/furans (grains per billion dscm) or 15 (6.6) ng/dscm TEQ (grains per billion dscf) Emission limits for large units = 60 ng/dscm (with electrostatic precipitator) Emission limits for large units = 30 ng/dscm (with fabric filter) Maximum containment level for 2,3,7,8-TCDD = 3×10^{-8} mg/L
1985	Maximum toxic equivalent concentration = 0.045 pg/m^3; the dioxin air guideline is set at a level at which no noncancer health effects would be expected as a result of lifetime exposure Groundwater = 3×10^{-8} mg/L, soil = 4×10^{-6} mg/L
1990	Dioxins/furans = 15 ng/dscm
	2,3,7,8-TCDD = 0.00000003 mg/L
	3.5×10^{-6} tons/y If a hazardous waste facility generates more than 1 g of dioxin/y, then it must be registered with the state and complete reporting procedures must be followed; there is no limit in the state on the levels of emissions
Not applicable	Not applicable
	0.00000003 mg/L Municipal and private solid waste incinerator emission limits for dioxins = 0.02 ng/dscm
1967 (last amended in August 1999)	Maximum allowable concentration = 7×10^{-7} µg/L
1988	0.0001 lb/y of hazardous air contaminants without acceptable ambient concentrations (2,3,7,8-TCDD) 0.000003 µg/L 0.00003 µg/L

TABLE A-9 Summary of Existing European Commission Environmental Regulations

Directive	Description
Waste incineration	
89/429/EEC	Existing municipal waste incinerators
89/369/EEC	New municipal waste incinerators
94/67/EC	Incineration of hazardous waste
2000/76/EC	Waste incineration directive
Water and aquatic environment	
76/464/EEC	Pollution caused by discharge into the aquatic environment
86/280/EEC	Limit values and quality objectives for discharges of certain dangerous substances included in list 1 of the annex to 76/464/EEC
80/68/EEC	Protection of groundwater against pollution
2000/60/EC	Framework for community action in the field of water policy
Risk management	
85/467/EEC	Restrictions on the marketing and use of dangerous substances (Amendment no. 6)
91/173/EEC	Restrictions on the marketing and use of certain substances and preparations (Amendment no. 9)
82/501/EEC	The Seveso Directive
96/82/EC	The Seveso II Directive

NOTE: TEQ = toxicity equivalents, PCP = pentachlorophenol, PCB = polychlorinated biphenyl, TCDD = tetrachlorodibenzo-p-dioxin, HCDD = hexachlorodibenzo-p-dioxin.

Limits	Date Implementation Required	Date Compliance Required
Air emissions: operating conditions specified	12/1/90	12/1/95; 12/1/00 to same conditions as 89/369/EEC
Air emissions: operating conditions specified	12/1/90	12/1/90
Air emissions: 0.1 ng I-TEQ/m^3 Release to water: to be agreed, Directive 80/68/EEC applies (see below)	12/31/96	Existing plant: within 3 years of implementation; new plant: on implementation
Air emissions: 0.1 ng I-TEQ/m^3 Release to water: 0.5 ng I-TEQ/L, 150 ng I-TEQ/ton of waste	12/28/02	Existing plant: within 5 years of implementation; new plant: on implementation
Organohalogen discharge to water prohibited	No date specified	No date specified
Water quality values specified for PCP content	1/1/88	1/1/88
Organohalogen discharge to water prohibited	12/16/81	12/16/81
Provides for the progressive reduction or cessation of discharges, emissions and losses of pollutants to water	10/00	
Prohibition on all use of PCBs	6/30/86	6/30/86
Use of PCP limited to 0.1% of total content	7/1/92	7/1/92
Storage: sites qualify if 2,3,7,8-TCDD stored reaches 1 kg of HCDD quantity of 100 kg	1/8/84	1/8/84
Storage: sites qualify as Major Accident Hazard if total dioxin stored reaches 1 kg	2/3/99	2/3/99

TABLE A-10 Detailed Summary of European Commission Countries' Environmental Regulations and Guidelines

Country	Legislation
Austria	Clean Air Ordinance for steam boilers (BGBl Nr 19/1989) Limit value for incinerators with steam-boiler installations Limit value for emission from steam-boiler installations fired by wood fuel (Amendment: BGBI 785/1994) Ordinance on air pollution from sintering plant for iron ore production (BGB1. II, Nr. 163/1997, 20.01.1997) Ordinance on air pollution from installations for the production or iron and steel (BGB1. II, Nr. 160/1997, 17.06.1997) Ordinance concerning the ban of PCP (BGB1. Nr. 58/1991) Ordinance concerning fertilizers, soil additives, and culture substrates (BGB1. Nr. 1007/1994) Ordinance in the province of Upper Austria on the application of sewage sludge, municipal solid waste compost and compost from sewage sludge (LGBl. Nr. 21/1993) Ordinance in the province of Lower Austria on sewage sludge (LGBl. 6160/2-0, 80/94) Guidelines
Belgium	Flemish regulation on environmental permits and conditions (Vlarem, Decree of 6/1/95) Flemish Decree of 1/19/99, Vlarem Regulation on environmental permits and conditions (BS/MB 31.03.99) regulating following sectors Municipal waste incineration Sewage sludge incineration Oil refineries, FCC catalyst regeneration

Date In Force	Description
1989	
	Limit of 0.1 ng I-TEQ/m^3 applicable to plant with more than 750 kg/h
	Limit of 0.1 ng I-TEQ/m^3 applicable to plant with more than 10 MW
1997	Limit of 0.4 ng I-TEQ/m^3 set for new sintering plant built after 1/1/04
1997	Iron and steel plant: limit value set at 0.25 ng I-TEQ/m^3 until 12/31/05; from 1/1/06 the limit value is 0.1 ng I-TEQ/m^3; plant already approved 0.4 ng I-TEQ/m^3
	Electrical arc furnaces, induction furnaces, and ladle metallurgic installations: limit value set at 0.4 ng I-TEQ/m^3; existing plants to comply within 5 years
6/2/91	Prohibits the production, use, and marketing of PCP
1/4/93	Limit value: 50 ng I-TEQ/kg in fertilizers, soil additives, or culture substrate; products containing 20–50 ng TEQ/kg have to be labeled with a warning sign "Attention contains dioxins/furans" (forbidden for use on children playgrounds and for vegetable cultivation); culture substrates are not allowed to contain more than 20 ng I-TEQ/kg
1/4/93	Dioxin limit value of 100 ng I-TEQ/kg dm
7/28/94	Dioxin limit value in sewage sludge of 100 ng I-TEQ/kg dm
	Officially, there are no guidelines for dioxin concentrations in the various environmental media or food; in the past, there were recommendations for maximum dioxin concentrations in milk, and a limit concentration of 3 ng I-TEQ/kg dm in grass fed to dairy cows
	Based on present information, it is unlikely that guide levels will be developed
	The recommended maximum human daily intake of dioxins is 10 pg 2,3,7,8-TCDD/kg body weight and if this is exceeded action should be taken to reduce exposure; a target value of 1 pg 2,3,7,8-TCDD/kg body weight should be achieved
1/8/95	New municipal solid waste incinerators: emission limit 0.1 ng I-TEQ/m^3
	Existing municipal solid waste incinerators: yearly measurement; emission limit 0.1 ng I-TEQ/m^3 from 1/1/97
1/5/99 (unless other date specified)	
	Continuous dioxin sampling required; analysis at least every 2 weeks
1/1/00	Emission limit 0.1 ng I-TEQ/m^3; continuous sampling required
1/1/02	New plant: emission limits 0.5 ng I-TEQ/m^3 (guide value: 0.1 ng I-TEQ/m^3)
	Existing plant: emission limits 2.5 ng I-TEQ/m^3 (guide value: 0.4 ng I-TEQ/m^3)

continued

TABLE A-10 Continued

Country	Legislation
	Metallurgical industries (iron and noniron)
	Sintering plant
	Crematoria Wood combustion (> 1 ton/h) Guidelines
Denmark	Danish Executive Order, Danish Environment Protection Agency Decision Number 41 on waste incinerators
	Legislation on the prohibition of the use of PCP of 7/25/95 Guidelines
Finland	Council of State Decision 23 June 1994/626 on the prevention of air pollution by municipal waste incineration
	Guidelines
France	Ministerial Decision on urban waste incineration plant from 2/24/97 Guidelines

Date In Force	Description
1/1/03	New plant: emission limits 0.5 ng I-TEQ/m^3 (guide value: 0.1 ng I-TEQ/m^3); yearly measurements required
	Existing plant: emission limit 1 ng I-TEQ/m^3 (guide value: 0.4 ng I-TEQ/m^3)
1/1/02	New plant: emission limit 0.5 ng I-TEQ/m^3 at 16% O_2 (guide value: 0.1 ng I-TEQ/m^3)
	Existing plant: emission limit 2.5 ng I-TEQ/m^3 at 16% O_2 (guide level: 0.4 ng I-TEQ/m^3)
	Emission limit 0.1 ng I-TEQ/m^3
2/2/97	Emission limit 0.1 ng I-TEQ/m^3 at 16% O_2
	There is a proposal for a recommended limit value for atmospheric deposition in the Flanders Region of 10 pg I-TEQ/m^2/d (as a yearly average); there is currently no maximum tolerable daily intake recommended in Belgium
1/14/97	Target emission concentration set at 0.1 ng I-TEQ/m^3; requires a residence time of the flue gases of > 2 sec, at a temperature of > 850°C and combustion at > 6% oxygen; also covers sewage sludge incineration
	PCP use prohibited
	Maximum tolerable daily intake (recommended by the Nordic countries) of 5 pg N-TEQ/kg body weight
1995	Old plant: limit value of 1.0 ng I-TEQ/m^3
	New plant (from 1/12/90): limit value of 1.0 ng I-TEQ/m^3 and target value of 0.1 ng I-TEQ/m^3
	Target value of 0.1 ng/m^3 for stack emissions from all municipal waste incinerators
	The Ministry of the Environment, Department for Environmental Protection has proposed a guideline of 2 ng I-TEQ/kg and a limit value of 500 ng I-TEQ/kg for contaminated soils
	Finland applies the maximum tolerable daily intake as adopted by the Nordic countries of 5 pg N-TEQ/kg body weight
2/24/97	New incinerators to meet the emission limits set in the EC Directive on Hazardous Waste incineration 0.1 ng I-TEQ/m^3
	A national recommendation from 2/97 set a guideline value for dioxin emissions from existing municipal solid waste incinerators of 0.1 ng I-TEQ/m^3; the regional authorities decide on a case-by-case basis the legal limit value for such plant; the government's intention is that, in the future, stack emissions from all types of municipal solid waste incinerators should not exceed 0.1 ng/m^3
	The recommended guide level for emissions from metal processing and paper processing is 1 ng I-TEQ/y; if measurements reveal that concentrations exceed the guideline, then abatement action is required, as well as a program of milk analysis within a radius of 3 km from the plant
	A guide level of 0.1 ng I-TEQ/m^3 is recommended for incineration in the cement and lime manufacturing industry

continued

TABLE A-10 Continued

Country	Legislation
Germany	Ordinance on bans and restrictions on the placing on the market of dangerous substances, preparations and products pursuant to the Chemicals Act (ChemVerbotsV)
	Ordinance on waste incineration plant (17 BImSchV)
	Ordinance on scavengers (19. BImSchV) Ordinance on sewage sludge (AbfKlärV 1992)
	Ordinance on crematoria (27 BImSchV 1997) Guidelines
Ireland	None
	Guidelines
Italy	Ministerial Decree (DM) No. 503
	Ministerial Decree on the water quality and characteristics of the purification of the Venice Lagoon Guidelines

Date In Force	Description
1989	Ordinance on the ban of PCP
1989	Ordinance on the ban of PCB
1990	Limit values for dioxins in substances, preparations and articles set as 1, 5, or 100 μg/kg of the chemical compounds depending on the dioxin type
1990	Limit value of 0.1 ng I-TEQ/m^3 for dioxin emissions from waste incinerators
1992	Ban on the addition of scavengers to leaded gasoline
1992	Limit value of 100 ng I-TEQ/kg dried residue for dioxins in sewage sludge used as fertilizer in agriculture, horticulture, or forestry
1997	Limit value of 0.1 ng I-TEQ/m for dioxin emission from crematoria
	In 1995, the Federal Environmental Agency issued the report *Determination of Requirements to Limit Emissions of Dioxins and Furans*; the report described measures that could reduce the emission of dioxins and furans from industrial installations; a target value of 0.1 ng I-TEQ/m^3 was recommended
	The Conference of the Ministers for the Environment adopted the report and asked the competent authorities to implement the measures
	A recommended limit value of 17 ng I-TEQ/kg dm exists for the use of compost; in the State of Baden-Württemberg, this is set as a legal limit
	Guideline measures were established for children's playgrounds and residential areas; in playgrounds, replacement of contaminated soil is required if the soil contains more than 100 ng I-TEQ/kg; in residential areas, such action is required if the soil is contaminated with more than 1,000 ng I-TEQ/kg; in industrial areas, the limit value was set to 10,000 ng I-TEQ/kg.
	No national legislation; however, any company which is licensable under the Irish Environment Protection Act, and which is seen as having "dioxin emission potential," may be required to undertake dioxin emission measurements
	The Irish EPA BATNEEC Guidance Note for the Waste Sector includes a section on dioxin emissions, which states "The aim should be to achieve a guide TEQ value of 0.1 ng I-TEQ/m^3. For hazardous waste incineration, subject to the EC establishing harmonized measurement methods by 01.07.96, this guide level becomes an emission level from 01.01.97"; where other issues arise, they tend to be guided by limits set or recommended by other EU countries (e.g., United Kingdom or Germany)
11/19/97	Regulates new municipal solid waste incinerator plant at the limit of 0.1 ng I-TEQ/m^3 in the exhaust gas; existing incineration plant must be technically upgraded to meet the above emission limit
4/23/98	Implements a zero emission concentration of dioxins and other organic compounds into the Venice Lagoon
	National Toxicology Commission proposed, in 1985, PCDD and PCDF reference technical limits for land rehabilitation; for farmable land the reference value was 750 ng/m^2 and for nonfarmable land 5,000 ng/m^2;

continued

TABLE A-10 Continued

Country	Legislation
Luxembourg	Regulation of the Grand Duke, Memorial A No 89, p 1897, 12/30/91
	Guidelines: Ministry Memorandum of 5/27/94, requiring the application of the best available technology by defining recommended limits for emissions into the air caused by industrial and crafts plants
Netherlands	Guideline on incineration of municipal solid waste and related processes (1989)
	Order on emissions from waste incineration Plant (Stcrt no. 15, 1992) and regulation on measurement methods for emissions from waste incineration plant
	Incineration Decree (Sb 36)
	Ministerial Order of the Ministry of Housing, Land Use, Planning and Environmental Protection (Stb Nos 176 to 182, 4/28/92)
	Pesticides Act 1989
	Guidelines

Date In Force	Description
	these were adopted, but recently were shown as inappropriate levels; a request made by the Tuscany Regional Authority that they be reviewed. A maximum tolerable daily intake 10 pg I-TEQ/kg body weight (excluding PCBs) was adopted as a guideline value in 1989
1991	Adopts Directive 89/429/EEC and 89/369/EEC. Emission limit of 0.1 ng I-TEQ/m^3 applies to new and existing municipal solid waste incinerators
1994	Recommends emission limit of 0.1 ng I-TEQ/m^3 for industrial and crafts plants; if no suitable technology is available in the market in order to meet this limit, then the process operator of existing industrial and crafts plants may apply for a derogation, of not more than 5 years duration, to 1 ng I-TEQ/m^3
8/89	Guideline on incineration of municipal solid waste and related processes (e.g., incineration of chemical waste, hospital waste, and sludge). Atmospheric emission standard of 0.1 ng I-TEQ/m^3 was recommended for new incinerators; existing incinerators had to meet the standard by 1/95
10/92	Adopted to implement Directive 89/369/EEC Emission standard concentrations set as 0.1 ng I-TEQ/m^3
1/95	The guideline of 1989 is transformed into law by this Decree
4/92	Limits for discharges of PCP (Stb 178) and hexachlorobenzene (Stb 181) into controlled waters
1991	Production and use of PCP and NaPCP prohibited
	For sintering plant for iron ore production, best available technology has been introduced, resulting in a recommended emission level of 0.4 ng I-TEQ/m^3
	For the application of sewage sludge a standard of 190 ng I-TEQ/kg dm has been proposed; the maximum permissible application of sewage sludge on arable land is 2,000 kg dm/ha and on pasture land 1,000 kg dm/ha
	A standard of 63 ng I-TEQ/kg dm of dioxins in compost was been proposed in July 1994; the maximum permissible application of compost on arable land is 6,000 kg dm/ha and on pasture land 3,000 kg dm/ha.
	No legal standards have been set for dioxin concentrations in soil, but in 1987 guidance levels were proposed for soil pollution in residential areas and agricultural areas of 1,000 ng I-TEQ/kg dm, for aquatic sediments 100 ng I-TEQ/kg dm and for dairy farming 10 ng I-TEQ/kg dm
	Rehabilitation of dioxin-contaminated areas, such as production sites, waste disposal sites and harbor sediments, is a topical issue in the Netherlands
	The recommended maximum tolerable daily intake is currently 10 pg I-TEQ/kg body weight; the government is striving for a maximum daily intake of 1 pg I-TEQ/kg body weight

continued

TABLE A-10 Continued

Country	Legislation
Spain	Guidelines
Sweden	SEPA Regulation on emissions to air from plant for incinerating municipal waste with a nominal capacity less than 6 tons per hour with a permit according to the Environment Protection Act (SNFS 1969:387) issued before 1/1/94 SNFS 1993:13.
	SEPA Regulation on emissions to air from plant for incinerating municipal waste with a permit according to the Environment Protection Act (1969:387) later than 1/1/94 and plant with a nominal capacity equal to or larger than 6 tons per hour with a permit issued according to the same law before 1/1/94, SNFS 1993:14
	SEPA regulation on discharges of industrial wastewater containing certain substances, SNFS 1995:7
	Guidelines
United Kingdom (U.K.)	Guidelines

NOTE: TEQ = toxicity equivalents, PCP = pentachlorophenol, dm = dry matter, TCDD = tetrachlorodibenzo-p-dioxin, PCB = polychlorinated biphenyl, PCDD = polychlorinated dibenzo-p-dioxin, PCDF = polychlorinated dibenzofuran.

Date In Force	Description
	In Catalonia, the Regional Environmental Protection Agency has recommended that all municipal waste incinerators meet a stack emission limit of 0.1 ng/m^3 The Catalonian EPA is currently developing a guideline concentration for dioxin in soils.
6/12/93	Rules and regulations from 1987 have required that all new municipal solid waste incinerators emit less than 0.1 ng TCDD/m^3; older plant also had to reduce their emissions towards this limit; for existing incinerators, limit values are in the range of 0.1–2 ng TCDD/m^3
6/30/95	Use of chlorine in Sweden's pulp and paper industry stopped; the production and use of PCP has been banned within Sweden for at least 10 years Tolerable daily intake = 5 pg TEQ/kg body weight The government provides guidance on concentrations of dioxin emissions achievable for various industrial processes in the form of the IPC Guidance Notes, issued by the U.K. Environment Agency; these guiding levels are used to assist in setting limits in individual plant authorizations; once an authorization is agreed this becomes a legally binding limit for the plant The recommended emission limits set for various incineration processes (including municipal, clinical, chemical, sewage sludge, animal carcasses, crematoria, and recovered oil) is 1.0 ng/m^3 with an objective of achieving 0.1 ng TEQ/m^3 For combustion processes (including large boilers and furnaces, combustion of fuel, reheat and heat treatment furnaces, coke manufacturers, compression ignition engines, and cement and lime manufacturers) the recommended limit value is 0.1 ng TEQ/m^3 For various metal processes (including integrated iron and steel works, ferrous foundry processes, production of zinc, lead, copper, and aluminum) and papermaking the recommended limit value is 1.0 ng TEQ/m^3 The maximum tolerable daily intake of 10 pg TEQ/kg body weight (including PCBs) was endorsed by the independent Committee on the Toxicity of Chemicals in Food, Consumer Products and the Environment

TABLE A-11 Summary of Existing Environmental Regulations and Limits by European Countries

			Country			
Regulated Area	Applicable Directive	Directive Limits	Austria	Belgium	Denmark	Finland
Existing municipal waste incineration plants (ng I-TEQ/m^3)^a						
Air emissions	89/429/EEC	None set	**0.1**	**0.1**	0.1	**0.1**
New municipal waste incineration plants (ng I-TEQ/m^3)^a						
Air emissions	89/369/EEC	None set	**0.1**	**0.1**	0.1	**0.1**
Incineration of hazardous waste (ng I-TEQ/m^3)^a						
Air emissions	94/67/EC	0.1	C	C	C	C
Releases to water	94/67/EC	None set	C	C	C	C
Air pollution from industrial processes (ng I-TEQ/m^3)^c						
Metal production and processing^d	NA^e	NA	**0.1**	**0.5**		
Sintering plant for iron ore production	NA	NA	**0.4**	**0.5**		
Combustion plant emission^f	NA	NA	**0.1**	**0.1**		
Papermaking processes	NA	NA				
Coke manufacturing	NA	NA				
Cement and lime manufacturing	NA	NA				
Water and aquatic environment						
Protection of ground water	80/68/EEC	Organohalogens prohibited	C	C	C	C
Discharge into aquatic environment	76/464/EEC	Content of PCP; Organohalogens prohibited	C	C	C	C
Marketing and use of chemicals						
PCBs	85/467/EEC	P^g	C	C	C	C
PCPs	91/173/EEC	0.1%	**P**	C	**P**	C
Major accident hazards						
The Seveso Directive	82/501/EEC	1kg of 2,3,7,8-TCDD	C	C	C	C
Sewage sludge (ng I-TEQ/kg dm)						
Application	NA	NA	**100**			
Compost use	NA	NA	**100**			

France	Germany	Greece	Ireland	Italy	Luxembourg	Netherlands	Portugal	Spain	Sweden	U.K.
0.1	**0.1**	C[b]	0.1	**0.1**	**0.1**	**0.1**	C	0.1	0.1	0.1
0.1	**0.1**	C	0.1	**0.1**	**0.1**	**0.1**	0.1	0.1	0.1	0.1
C	C	C	C	C	C	C	C	C	C	C
C	C	C	C	C	C	C	C	C	C	C
1.0	0.1				0.1					1.0
	0.1				0.1	0.4				
	0.1				0.1					0.1
1.0	0.1				0.1					1.0
	0.1				0.1					0.1
	0.1				0.1					0.1
C	C	C	C	C	C	C	C	C	C	C
C	C	C	C	C	C	C	C	C	C	C
C	**P**	C	C	C	C	C	C	C	C	C
C	**P**	C	C	C	C	**P**	C	C	**P**	C
C	C	C	C	C	C	C	C	C	C	C
	100					190				

continued

TABLE A-11 Continued

| | | | Country | | | |
| | Applicable | Directive | | | | |
Regulated Area	Directive	Limits	Austria	Belgium	Denmark	Finland
Soils and terrestrial environment (ng I-TEQ/kg dm)						
Soil: residential	NA	NA				500
Soil: agricultural	NA	NA				500
Soil: dairy farming	NA	NA				
Children's playground	NA	NA				
Industrial areas	NA	NA				
Fertilizer/soil additives	NA	NA	**50**			
Human exposure (pg I-TEQ/kg body weight/d)						
	NA	NA	10		5^h	5^h

[a]Measured at 11% O_2, 0°C, 101.3 kPa.
[b]Assume compliance with Directive.
[c]Measured at 16% O_2, dry gases, 0°C, 101.3 kPa.
[d]Includes iron and steel plants.
[e]None applicable.
[f]Includes boilers and/or crematoria.
[g]Prohibited production, marketing, and use. [h]pg N-TEQ/kg body weight/d.

NOTE: Figures in **bold** are legislative limits, others are guidelines. TEQ = toxicity equivalents, PCB = polychlorinated biphenyl, PCP = pentachlorophenol, TCDD = tetrachlorodibenzo-*p*-dioxin.

France	Germany	Greece	Ireland	Italy	Luxembourg	Netherlands	Portugal	Spain	Sweden	U.K.
	1,000					1,000			10	
	40					1,000			10	
						10			10	
	100								10	
	10,000								250	
	17								10	
1	10			10		10			5^h	10

TABLE A-12 Dioxin Environmental Regulations and Guidelines in Other Countries and International Organizations

Country/Organization	Regulated Area	Legislation/Guideline	Date In Force	Description/Limits
World Health Organization	Human exposure	Tolerable daily intake	1998	1–4 pg WHO-toxicity equivalents (TEQ)/kg body weight
Joint Expert Committee on Food Additives	Human exposure	Provisional tolerable monthly intake	2001	70 pg TEQ/kg body weight/mo; equivalent to 2.3 pg TEQ/kg body weight/d and 16.3 pg TEQ/kg body weight/wk
Australia	Human exposure	Proposed tolerable monthly intake	Proposed in 2002	70 pg TEQ/kg body weight/mo
Japan	All media and human exposure	Law concerning special measures against dioxins (dioxin law)	1/00	This law set a tolerable daily intake at 4 pg TEQ/kg of body weight
	Air quality	Dioxin law and environmental quality standard	1/00	Annual (air) average shall not exceed 0.6 pg TEQ/m^3
	Water quality	Dioxin law and environmental quality standard	1/00	Annual (water) average shall not exceed 1 pg TEQ/L
	Soil quality	Dioxin law and environmental quality standard	1/00	Soil level shall not exceed 1,000 pg TEQ/g
	Waste incinerator	Dioxin law and emission standard	1/01–11/02	New facility: 0.1–5 ng TEQ/m^3 N Existing facility: 80 ng TEQ/m^3 N
			12/02	New facility: 0.1–5 ng TEQ/m^3 N Existing facility: 1–10 ng TEQ/m^3 N
	Electric steel-making furnaces	Dioxin law and emission standard	1/01–11/02	New facility: 0.5 ng TEQ/m^3 N Existing facility: 20 ng TEQ/m^3 N
			12/02	New facility: 0.5 ng TEQ/m^3 N

continued

Facility	Regulation	Date	Standard
Sintering facilities for steel industry	Dioxin law and emission standard	1/01–12/02	New facility: 0.1 ng TEQ/m³ N Existing facility: 2 ng TEQ/m³ N
		12/02	New facility: 0.1 ng TEQ/m³ N Existing facility: 1 ng TEQ/m³ N
Facilities for collecting zinc	Dioxin law and emission standard	1/01–12/02	New facility: 1 ng TEQ/m³ N Existing facility: 40 ng TEQ/m³ N
		12/02	New facility: 1 ng TEQ/m³ N Existing facility: 10 ng TEQ/m³ N
Facilities for manufacturing aluminum base alloy	Dioxin law and emission standard	1/01–12/02	New facility: 1 ng TEQ/m³ N Existing facility: 20 ng TEQ/m³ N
		12/02	New facility: 1 ng TEQ/m³ N Existing facility: 5 ng TEQ/m³ N
Bleaching facilities Resolving facilities for waste of PCB Cleansing facility for PCB contaminated matter	Dioxin law and emission standard	1/01	New facility: 10 pg TEQ/L Existing facility: 10 pg TEQ/L
Cleansing facilities for waste gas and wet dust	Dioxin law and emission standard	1/01–1/03	New facility: 10 pg TEQ/L Existing facility: 20 pg TEQ/L
Cleansing facilities for dichloroethane in manufacturing of vinyl chloride monomer		1/03	New facility: 10 pg TEQ/L Existing facility: 10 pg TEQ/L
Waste gas washing facilities, wet dust collectors or ash landfill facilities for discharging	Dioxin law and emission standard	1/01–1/03	New facility: 10 pg TEQ/L Existing facility: 50 pg TEQ/L
		1/03	New facility: 10 pg TEQ/L Existing facility: 10 pg TEQ/L

TABLE A-12 Continued

Country/Organization	Regulated Area	Legislation/Guideline	Date In Force	Description/Limits
	polluted effluent of municipal waste incinerator			
	Sewage treatment plants that treat effluent from the facilities above	Dioxin law and emission standard	1/01	New facility: 10 pg TEQ/L Existing facility: 10 pg TEQ/L
	Effluent spillage from final waste disposal sites	Standards for maintenance and management of final waste disposal sites	1/01	10 pg TEQ/L

TABLE A-13 Summary of Existing U.S. Federal Food and Feed Regulations and Guidelines

Regulation/Guideline	Description	Limits	Date Implementation Required	Date Compliance Required
There are no regulatory levels for dioxins in food or feed				
40 C.F.R. 180.302	Tolerance and exemptions from tolerances for pesticide chemicals in or on raw agricultural commodities limit of 2,3,7,8-TCDD in technical grade hexachlorophene	0.1 ppm	1971	
FDA Guidelines	FDA guidelines for levels of 2,3,7,8-TCDD in fish	< 25 ppt no serious health effects, > 50 ppt, fish should not be consumed		
21 C.F.R. 109.30 and 509.30	FDA has also established an action level for the presence of PCBs in red meat	3 ppm (lipid basis)	Guidelines—not enforceable	Guidelines—not enforceable

NOTE: TCDD = tetrachlorodibenzo-*p*-dioxin, FDA = U.S. Food and Drug Administration, PCB = polychlorinated biphenyl.

TABLE A-14 Summary of Existing by European Commission Countries' Food and Feed Regulations and Limits

Directive	Description
Agricultural/food	
98/60/EC	Citrus pulp pellets as feedstuffs (amendment to 74/63/EEC)
1999/29/EC	Undesirable Substances and Products in Animal Nutrition (Maximum limits for animal feeds) (amended in 2001)
Commission Regulation No 2439/1999	Binders, anti-caking agents and coagulants in feeding stuffs
Commission Regulation EC/446/2001	Maximum limits for dioxins in food (from proposed regulations)
Nonbinding European Commission recommendations	Dioxin action limits in feedstuffs
Human exposure	
The Scientific Committee on Food of the European Union	Established a group TWI for dioxins and dioxin-like PCBs

NOTE: TEQ = toxicity equivalents, TWI = tolerable weekly intake, PCB = polychlorinated biphenyl.

Limits	Implementation Date	Compliance Date
500 pg I-TEQ/kg upper bound detection limit	7/31/98	7/31/98
Plant origin feed: 0.75 ng TEQ/kg Minerals: 1.0 ng TEQ/kg Animal lipid: 2.0 ng TEQ/kg Other land animal products: 0.75 ng TEQ/kg Fish oil: 6 ng TEQ/kg Nonoil fish products: 1.25 ng TEQ/kg Compound feeding stuffs: 0.75 ng TEQ/kg Feeding stuffs for fish: 2.25 ng TEQ/kg	7/1/02	
Provides conditions for the authorization of additives to binders, anti-caking agents and coagulants in feeding stuffs	11/17/99	
Ruminant meat products: 3 pg TEQ/g lipid Poultry meat products: 2 pg TEQ/g lipid Pig meat products: 1 pg TEQ/g lipid Liver: 6 pg TEQ/g lipid Muscle meat of fish: 4 pg TEQ/g fresh wt Milk and milk products: 3 pg TEQ/g lipid Hen eggs and egg products: 3 pg TEQ/g lipid Ruminant lipid: 3 pg TEQ/g lipid Poultry lipid: 2 pg TEQ/g lipid Pig lipid: 1 pg TEQ/g lipid Mixed lipid: 2 pg TEQ/g lipid Vegetable oil: 0.75 pg TEQ/g lipid Fish oil: 2 pg TEQ/g lipid	7/1/02	
Fish oil (feed) = 4.5 ng TEQ/kg Fish and pet feed = 1.5 ng TEQ/kg Animal lipid = 1.2 ng TEQ/kg Nonoil fish products = 1.0 ng TEQ/kg Plant origin feed = 0.5 ng TEQ/kg Mineral and binders = 0.5 ng TEQ/kg Other land animal products = 0.5 ng TEQ/kg Compound feeding stuffs = 0.4 ng TEQ/kg	2002	
TWI = 14 pg TEQ/kg body weight/wk, equivalent to 2 pg TEQ/kg body weight/d and 60 pg TEQ/kg body weight/mo	2001	

TABLE A-15 Detailed Summary of European Commission Countries' Food and Feed Regulations and Guidelines

Country	National Legislation	In Force
Belgium	Royal Decree of 23 April 1998 on maximum levels of dioxin in food	12/6/98
	Royal Decree of 23 June 1998 for the effective withdrawal from the market of milk products	10/7/98
France	Guidelines	
Germany	Guidelines	
Netherlands	Commodities Act 1991	1991
United Kingdom	Guidelines	

NOTE: TCDD = tetrachlorodibenzo-*p*-dioxin, TEQ = toxicity equivalents, TDI = tolerable daily intake, PCB = polychlorinated biphenyl.

Description

Maximum concentrations set for milk and dairy products: 5 pg 2,3,7,8-TCDD/g milk lipid for foods with > 2% lipid, 100 pg 2,3,7,8-TCDD/g food for foods with ≤ 2% lipid

Withdrawal of milk and milk products from the market when maximum dioxin concentration is exceeded

The French Ministry of Agriculture recommended a maximum limit value of 5 pg I-TEQ/g lipid for milk and milk products; at this level products are removed from the market

An objective has been established of achieving less than 1 pg I-TEQ/g lipid

A maximum tolerable daily intake of 1 pg I-TEQ/kg body weight is recommended by Conseil Supérieur d'Hygiène Publique of France

In 1992 and 1993, a Joint Working Group of the Federal and Lander Ministers of the Environment on dioxins established recommendations and reference values for dioxin concentrations in soils and milk; for preventative reasons, and as a long-term objective, the dioxin concentrations in soil used for agricultural purposes should be reduced to below 5 ng I-TEQ/kg; cultivation of foodstuffs is not restricted if the soil contains 5–40 ng I-TEQ/kg, although it is recommended that critical land uses, for example grazing, should be avoided if increased dioxin levels are found in foodstuffs grown in such soils; the cultivation of certain feedstuffs and foodstuffs is restricted if the dioxin contamination is above 40 ng I-TEQ/kg soil; unlimited cultivation is allowed for plant with minimum dioxin transfer (e.g., corn).

The recommended maximum dioxin concentration in milk should not exceed 5.0 pg I-TEQ/g milk lipid; thus, milk and dairy products should not be on the market if the contamination exceeds this value

To reduce the human impact via consumption of dairy products, a limit value of 3 pg I-TEQ/g lipid was set

The target concentration of 0.9 pg I-TEQ/g milk lipid was set as an objective to be achieved; if the TDI exceeds 10 pg I-TEQ/kg body weight (excluding PCBs) action should be taken to reduce the daily dioxin intake; a target value of 1 pg I-TEQ/kg body weight should be achieved

Based on the TDI and on the inventory of food consumption, a standard of 6 pg I-TEQ/g lipid for milk and milk products was derived originally in 1989; the decision was taken that the standard should remain unaltered in 1991

The Ministry for Agriculture, Fisheries and Foods established a recommended maximum tolerable concentration of dioxins, furans, and dioxin-like PCBs in cows' milk; this is currently set at 0.66 ng TEQ/kg of whole milk (approximately 16.6 ng TEQ/kg of milk lipid)

TABLE A-16 Summary of Existing Food and Feed Regulations and Limits by European Countries

Regulated Area	Applicable Directive	Directive Limits	Country			
			Austria	Belgium	Denmark	Finland
Food (pg I-TEQ/g lipid)[a]						
Milk and dairy products > 2% lipid	NA[b]	NA		5		
Milk and dairy products ≤ 2% lipid	NA	NA		100[c]		
Milk and dairy products	NA	NA	3			
Beef	NA	NA	6			
Pork	NA	NA	2	3		
Poultry and eggs	NA	NA	5			
Animal nutrition (pg I-TEQ/g)[a]						
Citrus pulp pellets as feedstuffs	98/60/EC	500	C[d]	C	C	C

[a]Commission Regulation EC/446/2001 goes into effect July 2002; therefore, limits are not listed on this table.

[b]NA = none applicable.

[c]pg I-TEQ/g food.

[d]Assume compliance with Directive.

NOTE: Figures in bold are legislative limits, others are guidelines. TEQ = toxicity equivalents.

TABLE A-17 U.S. Federal Monitoring Programs

Environmental Media	Program
Ambient air/deposition	U.S. Environmental Protection Agency (EPA), National Dioxin Air Monitoring Network (1998): stations established nationally to periodically measure dioxins in ambient air EPA, Urban Transect Study (Phase 1 in 2000): monitors for dioxins within and surrounding urban centers to show how sources of dioxin compounds influence air quality
Sediments	EPA and U.S. Department of Energy, Sediment Core Study: measured for dioxin-like compounds in 11 lakes in the United States
Animal feed	EPA, National Study on Animal Feeds: characterizes feeds for cattle, poultry, and swine nationally; samples of feed components taken to determine which components contribute the bulk of dioxin to terrestrial food animals

France	Germany	Greece	Ireland	Italy	Luxembourg	Netherlands	Portugal	Spain	Sweden	U.K.
5	5, 3			3 6 2 5	**6**			5		16.6
C	C	C	C	C	C	C	C	C	C	C

TABLE A-17 Continued

Environmental Media	Program
	Phase 1 with U.S. Department of Agriculture (USDA): lactating cows
	Phase 2 with U.S. Food and Drug Administration (FDA), Preliminary National Survey of Dioxin-like Compounds in Animal Lipids, Animal Meals, Oilseed Deodorizer Distillates, and Molasses (2000): determined background levels of dioxin-like compounds in lipidity and other feed ingredients commonly used in animal feeds; 48 samples collected
	Phase 2 with FDA, Preliminary National Survey of Dioxin-like Compounds in Oilseed Meals, Lipid-soluble Vitamins, Complete Feeds, Milk Products, Minerals, and Wood Products: appears to be underway; determine background levels of dioxin-like compounds in feed ingredients and complete animal feeds

continued

TABLE A-17 Continued

Environmental Media	Program
	Phase 3: follow-on animal feed study in planning stage
Food	EPA, National Milk Study on PBTs [persistent, bioaccumultive, and toxic compounds] (1996–1997): national milk sampling network to measure the concentrations of dioxins; 48 samples were collected, another round of sampling is currently in the planning stage
	FDA, Total Diet Study (TDS) (annually): using samples obtained for the TDS, FDA analyzes between 200 and 300 food samples that are either likely to contain animal lipids or that have not been analyzed for dioxins previously; dioxin analyses have been ongoing for four years
	FDA, targeted sampling (annually): 500 and 1,000 samples collected per year; targets foods that tend to have variation (e.g., fish species, vegetable oils, supplements)
	FDA, Sampling and Testing Programs for Catfish (1997): sampling is designed to ensure that catfish that might contain dioxin in amounts greater than 1 ppt do not enter the commercial channels
	USDA Food Safety and Inspection Service, (1994) sampled beef from 13 states for dioxins and resurveyed beef, pork, and chicken to understand background levels in U.S. food supply
	USDA/EPA: (1) USDA and EPA conducted a joint program of three surveys for dioxin in beef, pork, and poultry using 60–80 samples in each survey taken from federally inspected slaughterhouses in the United States; not repeated or continuous studies, but rather one-time events; 63 beef samples were collected in May and June 1994 and examined for chlorinated dibenzo-p-dioxins (CDDs) and chlorodibenzofurans (CDFs); sampling for the pork survey took place in August and September 1995 and yielded 78 final samples; sampling was conducted in September/October 1996 for poultry with a final sample size of 80, (2) used samples from four lipid reservoirs of cattle to evaluate distribution of dioxins in beef lipid matrices
	FDA, Food Compliance Program (annual): uses surveillance and sampling of foods to ensure compliance with regulations; if there is a potential issue with dioxin contamination, analysis could be requested

Country/State	Ambient Air/Deposition	Water	Soils, Sediments, Animal Feed	Food
California	California Ambient Dioxin Air Monitoring Program: Nine site, two-year dioxin air monitoring program; first comprehensive monitoring for dioxins in an urban setting; Air Resource Board will also conduct stationary and mobile source testing on such sources as medical waste incinerators, catalytic oxidizers, refineries, drum reconditioners, landfills, and secondary metal recovery facilities; also looking into approaches to determine how to estimate dioxin contribution from motor vehicles	Monitoring and Reporting for US Navy, Naval Air Facility in El Centro: monitor and report on dioxin in effluent	No information	California does not conduct dioxin monitoring in meat and poultry; also does not look for dioxins in any of its food monitoring programs
Illinois	No information	Groundwater monitoring	No information	Department of Food, Drugs, and Dairy: no monitoring programs in place in Illinois to track dioxins in food or agriculture
New York	No information	New York/New Jersey Harbor Containment Assessment and Reduction Project: study of sediment contaminants, the identification and elimination of the sources of contamination of harbor sediments, the remediation of contaminated areas, and the pursuit and sanction of polluting entities	No information	Assistant director of Department of Agriculture states that New York state complies with the federal regulations
Wisconsin	No information	Groundwater monitoring	No information	

TABLE A-19 European Commission Member Countries' Dioxin Monitoring Programs (as of 1999)

Country	Ambient Air/Deposition	Soils	Sediments
Austria	Federal Environmental Agency, 1997–2001: eight locations (industrial, urban, rural, background) of air samples; monitoring to track trend concentrations	Federal Environmental Agency, 1999: analysis of 30 samples to obtain overview of background soil contamination from unmanaged soils	Federal Environmental Agency, 1996: Spruce Needle Biomonitoring program in city of Linz for analysis of spatial distribution in an industrial urban region
Belgium	Flemish Environment Agency: monitors air close to potential emission sources; monitors deposition in vicinity of incinerators and known industrial emitters General Directorate for the Environment of theWalloon Region: intends to implement monitoring program		
Denmark			
Finland		National Public Health Institute: monitors soil contamination every five years	
France	Ministry of Environment: developing nationwide air monitoring program to be run by the French Environment Agency		
Germany	Federal Environmental Agency and Agency for Consumer Protection and Veterinary Medicine, 1993: Dioxin Reference Program monitors particulate deposition, grass, foodstuffs and fodder, soil, milk, human blood, breast milk and indicator matrices (pine needles and sediment) collected from urban, suburban and rural regions; sampling frequency depends on matrix; biannual evaluations of the data planned		

Vegetation	Food	Human	Sewage Sludge
	Ministry of Health and Ministry of Agriculture: monitors milk and milk products over three regions (Brussels, Flanders, and Wallony) twice a year		
			Danish Environment Protection Agency: monitors samples from various waste water treatment plants
French Environment Agency: developing vegetation monitoring program	Ministry of Agriculture, 1994–1997: monitored milk and milk products; measurements will continue to be taken by the Veterinary Services of each region	Ministry of Environment, 1998: nationwide monitoring program	

continued

TABLE A-19 Continued

Country	Ambient Air/Deposition	Soils	Sediments
Luxembourg	Environment Agency, 1995: biomonitoring around industrial plants, which measured levels in moss and cabbage; evaluated levels in air, deposition, accumulation in soil, sediment, and vegetables ARBED Steel Company (funded by Department of the Environment): measurement programs on electric arc furnaces		
Netherlands			Institute for Quality Control of Agricultural Products (funded by the Ministry of Agriculture, Nature Management and Fisheries): monitors levels in primary food products National Institute of Public Health and Environmental Protection (funded by Ministry of Public Health, Welfare and Sports) 1997–1998: monitored milk
Portugal	No national programs Instituto do Ambiente e Desenvolvimento: monitors levels in ambient air, soil, sediment, cow milk, vegetable matter, breast milk, and human blood in Porto; also monitored levels in ambient air before commissioning of new waste incinerator plant in Lisbon		
Sweden			National Dioxin Survey (funded by Swedish Environmental Protection Agency): monitors fish and guillemots in Baltic Sea
United Kingdom (U.K.)	Department of the Environment, Transport and the Regions (DETR) Hazardous Air Pollutants: monitors levels in three urban and three rural sites from samples collected twice a year The Environment Agency: monitors various industrial processes for stack emissions		Central Science Laboratory Ministry of Agriculture, Fisheries, and Food (MAFF): monitored levels in eggs, freshwater and marine fish and fish products, shellfish, lipid oils, cow's milk, infant formula, and samples collected as part of the 1997 Total Diet Study (TDS); cow's milk surveys are carried out in the vicinity of various industrial installations; monitors dioxins in TDS every five years

Vegetation	Food	Human	Sewage Sludge
	Every 5 years the body burden of dioxins are monitored by analyzing levels in breast milk		
	National Environmental Program (funded by Swedish Environmental Protection Agency): monitors breast milk		
	MAFF: conducted several studies on human milk and the analysis of pooled samples of human milk will continue on a regular basis		Laboratory of the Government Chemist, on behalf of the Department of Trade and Industry: monitors sewage sludge Lancaster University, on behalf of DETR, Environment Agency and UK WIR: monitors organics in sewage sludge

TABLE A-20 Other Countries' and Organizations' Dioxin Monitoring Program

Country/ Other	Ambient Air/ Deposition	Soils	Sediments	Vegetation
Australia				
Japan				Ministry of the Environment Survey on the State of Dioxin Accumulation in Wildlife: findings of the Fiscal 1999 Survey
World Health Organization				

Food	Human	Water	Sewage Sludge
Department of Health and Aging, the Australia New Zealand Food Authority, and the Australian Government Analytical Laboratories: dioxins in foods; results will be released by mid-2002			
Ministry of the Environment: The State of Dioxin Accumulation in the Human Body, Blood, Wildlife, and Food: Findings of the Fiscal 1998 Survey Detailed Study of Dioxin Exposure: Findings of the Fiscal 1999 Survey			
Global Environmental Monitoring System: Food Monitoring and Assessment Program	Several monitoring studies on dioxins in mother's milk and food		

TABLE A-21 Dioxin-Furan Results for Fishmeal, Fish Feed, and Fish Oil from the Canadian Food Inspection Agency 1989–1999 Preliminary Survey

Country of Origin (# samples)	Mean Dioxins and Furans Toxicity Equivalents (ppt) (minimum-maximum values)	
	Fish Meal and Fish Feed	Fish Oil
Canada (14)	1 (0.11–3.73)	
Canada (2)		9.9 (7.06–12.67)
United States (7)	1.1 (0.47–1.71)	
Iceland (1)	0.23	7.92
Peru (1)	0	
Russia (1)	0.22	
South America (4)		3.7 (0.15–11.47)
Mexico (2)		9.8 (8.83–10.81)

TABLE A-22 Selected U.S. Federal Research Programs

Environmental Media	Program
Ambient air/ deposition	U.S. Environmental Protection Agency (EPA), National Modeling of Sources (currently): all known anthropogenic sources of dioxin air emissions in the United States are being put into EPA's long-range transport and deposition model to study the relationship between emissions and depositions
	EPA, Incinerator Impacts (1994): evaluated local impacts from emissions of dioxins from municipal solid waste incinerator known to have been emitting large amounts of dioxins
	EPA and Harvard School of Public Health, Modeling Incinerator Impacts (1999): using incinerator data from above paper, tested EPA's ISCST3 air dispersion/deposition model
	EPA, Exposure Estimation in EPA Regulations (1998 and 2000): papers regarding how exposure modeling is used in regulation of dioxin sources in EPA
	EPA, Emissions from Uncontrolled Burning of Domestic Waste (1999 and 2000): measured dioxin emissions from simulated backyard burning conditions
	EPA, Release of Dioxins from Pentachlorophenol-Treated Utility Poles (1999): investigated potential for dioxins to be released into the environment
	EPA, Diesel Emissions (1997): measured emissions of dioxins from diesel trucks under normal road conditions
Soils	EPA, Nocturnal-Diurnal Releases from Soil (1999): investigated possibility that soils may be reservoir source of release into atmosphere
Food	U.S. Department of Agriculture (USDA): currently (1) identifying routes of exposure and distribution in animals being raised for food and effects on food supply, (2) exploring methods to minimize burden of dioxin compounds that persist in animals' bodies
	EPA, Effect of Cooking (1997): paper describing the impact of cooking various meats and fish on concentrations and masses of dioxins
	EPA, Air-to-plant-to-animal modeling (1994, 1995, 1998): papers describing derivation and validation of model predicting beef concentrations starting from air concentrations
	EPA, Analytical Methods (1996, 1997, 1999): papers describing development, testing and verification of analytical methods to evaluate dioxins in a variety of matrices
	EPA/USDA, Mass Balance of Dioxins in Lactating Cows (2000): study on mass balance of dioxins and furans in lactating cows
	EPA, Historical Meat and Milk Sampling Study (1998): sampled canned meats and dried milk from decades in past to determine temporal variability of dioxins in animal lipids

TABLE A-22 Continued

Environmental Media	Program
Human	EPA, Dioxins in Ceramics and Potter Products (2000): results of sampling of pottery and ceramic products made from ball clay
	EPA, Modeling Past Exposures to Dioxins (1998): presents derivation and calibration of a model to predict past exposures to 2,3,7,8-tetrachlorodibenzo-p-dioxin
	National Institute of Environmental Health Sciences: currently researching dioxin and cancer, endometriosis and immune function, and in connection with Agent Orange; also conducting research into the human response to dioxin

TABLE A-23 European Commission Member Countries' Dioxin Research Programs (as of 1999)

Country	Ambient Air/Deposition	Sediment	Vegetation
Belgium			
Denmark			
Finland	Finnish Environment Institute: 1996 SIPS program assessed atmospheric emissions from all emission sources; developed national air emission database		
France	ADEME: study the emissions from six small incinerators; study emissions during start-up of plant; evaluate deposition and lipid around a source		ADEME: study soils around industrial plant
Italy	Research, 1998: analysis of sources according to type Study dioxin contamination in urban air		
Luxembourg	Study stack emissions		
Spain	Central Government and Catalonia Environment Protection Agency has		

Food	Human	Sewage Sludge
Ministry of Health and Ministry of Agriculture: results from monitoring program used in European Commission's Scientific Cooperation Project to estimate exposure to consumers		
		National Environmental Research Institute: researched dioxin levels in waste water and sewage sludge
	ADEME: study levels in breast milk	
Research dioxin levels in cow's milk; study of various foodstuffs Instituto Superiore di Sanita: study dioxin levels in edible mollusks and fishes	Instituto Superiore di Sanita supported by the Ministry of the Environment: evaluate human exposure and risk from microcontaminants in urban areas and Venice Lagoon	

funded projects over the past years: identify representative levels throughout Spain

continued

TABLE A-23 Continued

Country	Ambient Air/Deposition	Sediment	Vegetation
Sweden		University of Uppsala, National Food Administration, and Karolinska: study sediments, food, and human exposure	
United Kingdom	Lancaster University: conducted a project to model dioxins and persistent organic pollutants (POPs) and aimed to match sources to the environmental distribution of POPs		Lancaster University: studied air to herbage transfer of POPs

TABLE A-24 Trace Analytical Methods and Cost Estimates

		Cost per Sample ($)	
Matrix	Method[a]	Minimum	Maximum
Food, soil/sediment	HRGC-HRMS	900[b]	1,040
Human serum	HRGC-HRMS	500[b]	1,000
Soil/sediment	HRGC-HRMS		2,000
Soil/sediment	HRGC-HRMS	900	1,800

[a]HRGC-HRMS = high resolution gas chromatography and mass spectophotometry.
[b]Lowest cost with economy of scale.

Food	Human	Sewage Sludge
University of Uppsala, National Food Administration, and Karolinska: study sediments, food, and human exposure	University of Uppsala, National Food Administration, and Karolinska: study sediments, food, and human exposure	
Central Science Laboratory, Norwich, University of East Anglia and Milk Marque: study effects of dioxins in sediment deposited on pasture by flooding on the concentration of these compounds in cow's milk	MAFF, Department of Health, Department of the Environment, Transport and the Regions, Health and Safety Executive: plan to establish an archive of human milk samples from individual nursing mothers. University of Birmingham; study of the bioavailability of dioxins by analyzing accumulation in the body	
Several commercial organizations on behalf of Ministry of Agriculture, Fisheries, and Food (MAFF): develop improved analytical techniques to dioxins in foodstuffs	MAFF: conducted several studies on human milk and the analysis of pooled samples of human milk; will continue on a regular basis	
	University of Lancaster: model human exposure to dioxins	

Source

Alta Analytical Perspectives (Y. Tondeur, personal communication)
National Center for Health Statistics (J. Perkle, personal communication)
US Army Corps of Engineers R&D Center (http:www.wes.army.mil/el/resbrief/rapidsc.html)
Environmental Technology Commercialization Center (http://www.etc2.org/pr3.html)

TABLE A-25 Concentrations of DLCs in Foods

Food	ATSDR (1998)[a]	Fiedler et al. (2000)[b]	AEA Technology (1999)[c]
Beef	National average: 0.89 pg I-TEQ/g lipid New York: 0.04–1.5 pg I-TEQ/g ww	Europe: all meats: 0.09–20 pg I-TEQ/g lipid	Denmark: meat, 2.6 pg I-TEQ/g lipid Finland: < 0.2 pg N-TEQ/g lipid Germany: 1.44–3.5 pg I-TEQ/g lipid Netherlands: 1.25–1.8 pg TEQ/g lipid Spain: 1.76 pg I-TEQ/g lipid Sweden: range 0.4–1.5 pg N-TEQ/g lipid
Pork	National average: 1.30 pg I-TEQ/g lipid New York: 0.3 pg I-TEQ/g ww	See beef	Finland: < 0.1 pg N-TEQ/g lipid Germany: 0.13–0.5 pg I-TEQ/g lipid Netherlands: 0.25–0.43 pg TEQ/g lipid Spain: 0.90 pg I-TEQ/g lipid Sweden: range 0.06–1.2 pg N-TEQ/g lipid
Chicken	National average: 0.40–0.98 pg I-TEQ/g lipid New York: 0.03 pg I-TEQ/g ww	See milk	Denmark: 2.2 pg I-TEQ/g lipid Germany: 0.70–2.3 pg I-TEQ/g lipid Netherlands: 0.66–1.6 pg TEQ/g lipid Spain: 1.15 pg I-TEQ/g lipid Sweden, range: 0.42–1.1 pg N-TEQ/g lipid United Kingdom: 0.13 pg I-TEQ/g fw
Other meat	New York: 0.12–0.4 pg I-TEQ/g ww	See beef	Germany: veal, 0.70–7.4 pg I-TEQ/g lipid Germany: lamb, 2.0 pg I-TEQ/g lipid Netherlands: liver, 5.7–42 pg TEQ/g lipid Spain: lamb, 1.76 pg I-TEQ g/lipid Sweden: mutton, range 0.55–1.3 pg N-TEQ/g lipid United Kingdom: duck, 0.4 pg I-TEQ/g fw United Kingdom: carcass, 0.13 pg I-TEQ/g fw

Scientific Committee on Food (2000, 2001)[d]	IARC (1997)[e]	EPA (2000)[f]
Beef and veal: PCDD/PCDF, 0.681 (0.38–1.1) pg I-TEQ/g lipid; PCB, 0.914 (0.86–1.08) pg TEQ$_{WHO94}$/g lipid	Mean of all data for all meats: 6.5 pg I-TEQ/g lipid Europe: 1.77 pg I-TEQ/g lipid United States: 0.89 pg I-TEQ/g lipid	Preferred value: 1.06 pg TEQ$_{DF\text{-}WHO98}$/g lipid 0.49 pg TEQ$_{P\text{-}WHO98}$/g lipid Other: 0.152–0.254 pg I-TEQ$_{DF}$/g ww 0.04–1.5 pg I-TEQ$_{DF}$/g ww 0.29 pg I-TEQ$_{DF}$/g ww Canada: 0.10–0.39 pg TEQ$_{DFP\text{-}WHO94}$/g Germany: 0.71–0.95 pg I-TEQ$_{DF}$/g lipid Netherlands: 1.75 pg I-TEQ$_{DF}$/g lipid Spain: 1.76 pg I-TEQ$_{DF}$/g lipid
PCDD/PCDF, 0.258 (0.13–3.8) pg I-TEQ/g lipid	Europe: 0.42 pg I-TEQ/g lipid	Preferred value: 1.48 pg TEQ$_{DF\text{-}WHO98}$/g lipid 0.06 pg TEQ$_{P\text{-}WHO98}$/g lipid Other: 0.029–0.26 pg I-TEQ$_{DF}$/g ww Canada: 0.049–0.053 pg TEQ$_{DFP\text{-}WHO94}$/g Germany: 0.39 pg I-TEQ$_{DF}$/g lipid Netherlands: 0.43 pg I-TEQ$_{DF}$/g lipid Spain: 0.90 pg I-TEQ$_{DF}$/g lipid
PCDD/PCDF, 0.524 (0.37–1.4) pg I-TEQ/g lipid	Europe: 1.62–1.68 pg I-TEQ/g lipid	Preferred value: 0.77 pg TEQ$_{DF\text{-}WHO98}$/g lipid 0.29 pg TEQ$_{P\text{-}WHO98}$/g lipid Other: 0.043-0.085 pg I-TEQ$_{DF}$/g ww 0.03 pg I-TEQ$_{DF}$/g ww Canada: 0.062–0.076 pg TEQ$_{DFP\text{-}WHO94}$/g Germany: 0.62 pg I-TEQ$_{DF}$/g lipid Netherlands: 1.65 pg I-TEQ$_{DF}$/g lipid Spain: 1.15 g I-TEQ$_{DF}$/g lipid
Game: PCDD/PCDF, 1.81 (0.97–1.97) pg I-TEQ/g lipid; PCB, 3.15 pg TEQ$_{WHO94}$/g lipid	Europe: 0.4–61.2 pg I-TEQ/g lipid	No preferred value

continued

TABLE A-25 Continued

Food	ATSDR (1998)[a]	Fiedler et al. (2000)[b]	AEA Technology (1999)[c]
Cow milk	National average: 0.82 pg I-TEQ/g lipid	Europe: 0.2–3 pg I-TEQ/g lipid Europe: milk, milk products, poultry, eggs, midpoints in range of 0.75–1.7 pg I-TEQ/g lipid	Denmark: 2.6 pg I-TEQ/g lipid Finland: 0.83–1.17 pg N-TEQ/g lipid France: 1.81 pg I-TEQ/g lipid Germany: 0.71–0.87 pg I-TEQ/g lipid Ireland: 0.08–0.51 pg I-TEQ/g lipid Netherlands: 0.38–1.6 pg TEQ/g lipid Spain: 1.2–2.0 pg I-TEQ/g lipid Sweden: 0.93–2 pg N-TEQ/g lipid United Kingdom: 1.01 pg I-TEQ/g lipid
Other dairy	New York: 0.04–0.7 pg I-TEQ/g ww	See milk Europe: 0.5–4 pg I-TEQ/g lipid	Denmark: cheese, 2.2 pg I-TEQ/g lipid Finland: 0.83 pg N-TEQ/g lipid France: butter, cheese, cream, 1.01–1.34 pg I-TEQ/g lipid Germany: milk/dairy products, 0.75–1.02 pg I-TEQ/g lipid Italy: butter, 8.4 pg I-TEQ/g fw Netherlands: butter, cheese, 1.4–1.8 pg TEQ/g lipid Spain: 1.25 pg I-TEQ/g lipid Sweden: butter, range 0.35–0.5 pg N-TEQ/g lipid

Scientific Committee on Food (2000, 2001)[d]	IARC (1997)[e]	EPA (2000)[f]
PCDD/PCDF, 0.972 (0.26–3.57) pg I-TEQ/g lipid PCB, 1.25 (0.23–1.8) pg TEQ_{WHO94}/g lipid	Mean of all data: 2.3 pg I-TEQ/g lipid Median: 1.7 pg I-TEQ/g lipid 95th percentile: 4.5 pg I-TEQ/g lipid Europe: 1.31–3.94 pg I-TEQ/g lipid United States: 0.99 pg I-TEQ/g lipid	Preferred value: 0.98 pg $TEQ_{DF\text{-}WHO98}$/g lipid 0.49 pg $TEQ_{P\text{-}WHO98}$/g lipid Other: 0.061 pg TEQDF/g 0.012–0.026 pg I-TEQ_{DF}/g ww 0.1–1.2 pg I-TEQ_{DF}/g lipid Canada: 0.025–0.072 pg $TEQ_{DFP\text{-}WHO94}$/g France: 1.91 pg I-TEQ_{DF}/g lipid Germany: 0.71 pg I-TEQ_{DF}/g lipid Spain: 1.02–1.20 pg I-TEQ_{DF}/g lipid United Kingdom: 0.06–0.08 pg I-TEQ_{DF}/g
PCDD/PCDF, 0.612 (0.30–1.5) pg I-TEQ/g lipid PCB, 0.564 (0.38–0.78) pg TEQ_{WHO94}/g lipid	Mean of all data: 2.4 pg I-TEQ/g lipid Europe: 0.75–2.3 pg I-TEQ/g lipid	Preferred value: assumed to be the same, on a lipid basis, as for milk Other: 0.082–0.38 pg I-TEQ_{DF}/g 0.254–0.770 pg I-TEQ_{DF}/g ww 0.04–0.73 pg I-TEQ_{DF}/g ww Canada: 0.138–0.93 pg $TEQ_{DFP\text{-}WHO94}$/g France: 1.01–1.34 pg I-TEQ_{DF}/g lipid Germany: 0.64–0.66 pg I-TEQ_{DF}/g lipid Netherlands: 1.4–1.8 pg I-TEQ_{DF}/g lipid Spain: 1.25 pg I-TEQ_{DF}/g lipid

continued

TABLE A-25 Continued

Food	ATSDR (1998)[a]	Fiedler et al. (2000)[b]	AEA Technology (1999)[c]
Fish	New York: 0.02–0.13 pg I-TEQ/g ww New York/ New Jersey: wild crab: 78.2 pg TEQ/g ww	Europe: 2–300 pg I-TEQ/g lipid	Sweden: 0.35–6.3 pg N-TEQ/g fw Sweden: herring, 9.1–420 pg N-TEQ/g lipid United Kingdom: 16–700 pg I-TEQ/g lipid Denmark: 50 pg I-TEQ/g lipid Finland: herring, 30.2 pg N-TEQ/g lipid Finland: farmed trout, 4.2–33.4 pg N-TEQ/g lipid Germany: 3.3–22.3 pg I-TEQ/g lipid Italy: 8.7–14.6 pg TEQ/g lipid Italy: other, 5.9–57.5 pg TEQ/g lipid Netherlands: 2.4–121.5 pg TEQ/g lipid Netherlands: other, 66.8–76.5 pg TEQ/g lipid Spain: 2.57–7.90 pg I-TEQ/g lipid Spain: seafood, 10.59 pg I-TEQ/g lipid Sweden: 0.35–42.8 pg N-TEQ/g fw United Kingdom: farmed, 5.1 pg I-TEQ/g lipid United Kingdom: farmed, including PCBs, 24 pg I-TEQ/g lipid United Kingdom: wild, 1.9–34 pg I-TEQ/g lipid
Eggs	No data	See milk Europe: 1–4 pg I-TEQ/g lipid	Sweden: 0.08 pg N-TEQ/g fw Denmark: 1.5 pg I-TEQ/g lipid Finland: 1.2 pg N-TEQ/g lipid Germany: 1.36–4.58 pg I-TEQ/g lipid Netherlands: 2 pg TEQ/g lipid Spain: 1.22 pg I-TEQ/g lipid Sweden: range 0.89–1.3 pg N-TEQ/g lipid United Kingdom: 0.6–1.2 pg I-TEQ/g fw

Scientific Committee on Food (2000, 2001)[d]	IARC (1997)[e]	EPA (2000)[f]
Wild: PCDD/PCDF, 9.92 (0.125–2.25) pg I-TEQ/g lipid; PCB, 35.3 (1.61–168) pg TEQ$_{WHO94}$/g lipid Cultured: PCDD/PCDF, 8.84 (2.33–27.9) pg I-TEQ/g lipid; PCB: 19.6 (9.92–39.7) pg TEQ$_{WHO94}$/g lipid	Mean of all data: 25 pg I-TEQ/g lipid; median: 31 pg I-TEQ/g lipid 95th percentile: 54 pg I-TEQ/g lipid Europe: 2.72–48.6 pg I-TEQ/g lipid United States: farmed, 5.0–42.9 pg I-TEQ/g lipid	Freshwater fish, shellfish, and estuarine fish: WM 1.0 pg TEQ$_{DF-WHO98}$/g fw WM 1.2 pg TEQ$_{P-WHO98}$/g fw Marine fish and shellfish: WM 0.26 pg TEQ$_{DF-WHO98}$/g fw WM 0.25 pg TEQ$_{P-WHO98}$/g fw Farmed: 1.19–2.64 pg I-TEQ$_{DF}$/g fw Wild: 0.027-0.72 pg I-TEQ$_{DF}$/g fw 11.13–25.33 pg TEQ$_{DFP-WHO98}$/g lipid 0.22–2.0 pg TEQ$_{DF-WHO98}$/g 1.57–2.12 pg TEQ$_{P-WHO94}$/g fw 0.25–0.69 pg I-TEQ$_{DF}$/g fw 0.1–6.9 pg TEQ$_{P-WHO94}$/g mt Shellfish: 0.033–2.34 pg I-TEQ$_{DF}$/g fw 0.23–0.26 pg TEQ$_{DF-WHO98}$/g 0.03–1.99 pg TEQ$_{P-WHO94}$/g fw 0.1–5.4 pg TEQ$_{P-WHO94}$/g mt Hepatopancreas, 5.2–1,820 pg TEQ$_{P-WHO94}$/g Canada: 0.12–0.62 pg TEQ$_{DFP-WHO94}$/g France: fish, 104–523 pg TEQ $_{P-WHO94}$/g lipid Germany: 7.44–104.1 pg I-TEQ$_{DF}$/g lipid Netherlands: 2.4–48.65 pg I-TEQ$_{DF}$/g lipid Spain: 2.57–10.59 pg I-TEQ$_{DF}$/g lipid United Kingdom: farmed fish, 5.1 pg I-TEQ$_{DF}$/g lipid 19.0 pg TEQ$_{P-WHO94}$/g lipid; wild: 0.7–25 pg TEQ$_{DF-WHO98}$/g lipid 5.4–59 pg TEQ$_{P-WHO98}$/g lipid United Kingdom/Norway: wild, 5–18 pg TEQ$_{DF-WHO98}$/g lipid 9–25 pg TEQ$_{P-WHO98}$/g lipid
PCDD/PCDF, 1.19 (0.46–7.32) pg I-TEQ/g lipid	Europe: 1.26–2.02 pg I-TEQ/g lipid	Preferred value: 0.081 pg TEQ$_{DF-WHO98}$/g ww 0.1 pg TEQ$_{P-WHO98}$/g ww Other: 0.032 pg TEQ$_{DF-WHO98}$/g ww 0.17 pg TEQ$_{DF}$/g 0.019–0.038 pg I-TEQ$_{DF}$/g ww Germany: 2.10 pg I-TEQ$_{DF}$/g lipid Netherlands: 2.0 pg I-TEQ$_{DF}$/g lipid Spain: 1.22 pg I-TEQ$_{DF}$/g lipid

continued

TABLE A-25 Continued

Food	ATSDR (1998)[a]	Fiedler et al. (2000)[b]	AEA Technology (1999)[c]
Lipids and oils	No data	Europe: 0.2–2 pg I-TEQ/g lipid Europe: midpoint, 1.2 pg I-TEQ/g lipid	Denmark: 0.5 pg I-TEQ/g lipid Germany: lard, 0.8 pg I-TEQ/g lipid Netherlands: 0.03–0.17 pg TEQ/g lipid Netherlands: fish oil, 0.99 pg TEQ/g lipid Spain: 0.49–0.64 pg I-TEQ/g lipid United Kingdom: fish oils, 0.25–9.2 pg TEQ/g oil United Kingdom: 0.2 pg I-TEQ/g fw
Vegetables	No data	Europe: 0.002–0.3 pg I-TEQ/g fw Europe: fruits and vegetables, midpoint, 0.1 pg I-TEQ/g fw	Denmark: 0.015 pg I-TEQ/g lipid Finland: < 0.00005–0.04 pg I-TEQ/g fw Germany: 0.0064–0.017 pg I-TEQ/g fw Netherlands: 0.13 pg TEQ/g dw Spain: 0.14–0.19 pg I-TEQ/g fw United Kingdom: 0.3–0.4 pg I-TEQ/g fw
Fruits	No data	See vegetables	Denmark: 0.015 pg I-TEQ/g lipid Finland: < 0.00005 pg I-TEQ/g fw Germany: < 0.01–0.015 pg I-TEQ/g dm Spain: 0.09 pg I-TEQ/g fw United Kingdom: 0.3 pg I-TEQ/g fw
Bread and grains	No data	Europe: 0.1–2 pg I-TEQ/g fw Europe: midpoint, 1.2 pg I-TEQ/g fw	Denmark: 0.1 pg I-TEQ/g lipid Finland: 0.00048–0.0014 pg I-TEQ/g ww Netherlands: 0.4–0.85 pg TEQ/g lipid Spain: 0.25 pg I-TEQ/g fw United Kingdom: 0.03–0.17 pg I-TEQ/g fw

[a]PCDD/PCDF. Nondetected congeners were assumed present at ½ the detection limit.

[b]Ranges are estimated from a log-scale graphic.

[c]From Task 4 Annex 1. PCDD/PCDF unless otherwise indicated. Most data are mean values.

[d]Estimated mean (range) from Table 2.

[e]From Appendix 1. PCDD/PCDF only. Nondetected congeners were assumed present at detection limits. All data: irrespective of country or sampling year.

[f]From part I, volume 3, chapter 3. In most cases, nondetected congeners were assumed present at ½ the detection limit. For preferred values on a whole-food basis, see chapter 3, tables 3-59 and 3-60.

NOTE: TEQ = toxicity equivalents, fw = fresh weight, ww = wet weight, dw = dry weight, dm = dry matter, mt = muscle tissue, WM = weighted mean (U.S. Environmental Protection Agency [EPA] averaged dioxin and dioxin-like compound concentrations for each fish/shellfish species by the average consumption rate and summed for a category mean; WM values are EPA's preferred values).

Scientific Committee on Food (2000, 2001)[d]	IARC (1997)[e]	EPA (2000)[f]
No data	Europe: 0.03–2.24 pg I-TEQ/g lipid	Preferred value: 0.056 pg TEQ $_{DF-WHO98}$/g 0.037 TEQ$_{WHO98}$/g Canada: 0.31–0.44 pg TEQ$_{DFP-WHO94}$/g Netherlands: 0.006–2.2 pg I-TEQ$_{DF}$/g lipid Spain: 0.49–0.64 pg I-TEQ$_{DF}$/g ww
Fruit and vegetables, PCDD/PCDF, 0.029 (0.004–0.090) pg I-TEQ/g fw	No data	No preferred value Germany: 4.3–16.9 pg I-TEQ$_{DF}$/g lipid Spain: 0.09–0.25 pg I-TEQ$_{DF}$/g ww
See vegetables	No data	No data See vegetables
PCDD/PCDF, 0.019 (0.01–0.02) pg I-TEQ/g fw PCB, 0.11 pg TEQ$_{WHO94}$/g fw	Europe: 1.34–2.66 pg I-TEQ/g lipid	No data See vegetables

TABLE A-26 U.S. Mean Polychlorinated Dibenzo-*p*-Dioxins (PCDDs), Polychlorinated Dibenzofurans (PCDFs), and Polychlorinated Biphenyls (PCB) Toxicity Equivalents (TEQ) Concentrations in Beef, Pork, Young Chickens, and Milk

Description	Beef	Pork	Young Chickens	Milk
1993–Present PCDD/PCDF, pg TEQ/g lipid	0.89 (0.35)	1.30 (0.46)	0.64 (0.41)	0.84 (0.84)
1993–Present PCB, pg TEQ/g lipid	0.51 (0.51)	0.06 (0.06)	0.28 (0.28)	0.43 (0.43)

NOTE: Results assume nondetects = ½ limit of detection; results calculated at nondetects = 0 shown in parentheses.

TABLE A-27 Polychlorinated Dibenzo-*p*-Dioxins (PCDDs), Polychlorinated Dibenzofurans (PCDFs), and Polychlorinated Biphenyls (PCB) Toxicity Equivalents (TEQ) Concentrations and Percent Differences from Current Levels

Description	PCDD/PCDF TEQ pg/g Lipid	PCB TEQ pg/g Lipid	Percent Difference from Current PCDD/PCDF Levels	Percent Difference from Current PCB Levels
1908 Beef ration	0.34 (0.15)	0.07 (0.07)	38 (42)	15 (15)
1945 Beef and pork	0.98 (0.75)	0.36 (0.36)	89 (197)	140 (146)
1957 Dried cream	2.05 (0.81)	3.56 (3.54)	244 (96)	827 (824)
1968 Bacon bar	3.01 (2.94)	1.05 (1.05)	231 (638)	1,747 (2,620)
1968 Deviled ham	3.73 (3.71)	0.61 (0.61)	287 (805)	1,019 (1,529)
1971 Beef	1.36 (0.02)	2.48 (1.98)	153 (7)	540 (540)
1971 Bacon wafer	1.75 (1.62)	1.98 (1.98)	135 (352)	3,301 (4,952)
1977 Raw chicken	1.29 (1.18)	2.72 (2.72)	202 (287)	970 (970)
1977 Cooked chicken	1.33 (1.20)	2.83 (2.83)	209 (292)	1,009 (1,009)
1979 Pork slices	1.46 (1.20)	0.04 (0.04)	112 (262)	72 (105)
1980 Beef steak	0.94 (0.73)	0.93 (0.93)	106 (207)	203 (203)
1982 Ham slice	1.36 (1.04)	0.07 (0.07)	105 (227)	119 (178)
1983 Beef in barbeque	0.50 (0.03)	0.79 (0.79)	56 (8)	171 (171)
1983 Turkey with gravy	0.55 (0.23)	0.32 (0.31)	85 (57)	113 (113)

NOTE: Results assume nondetects = ½ limit of detection; results calculated at nondetects = 0 shown in parentheses.

TABLE A-28 Sources of Dioxins in the United States, May 2000

Source	1987 Emissions (g TEQ$_{DF-WHO98}$/y)	1995 Emissions (g TEQ$_{DF-WHO98}$/y)	2002/2004 Emissions (g TEQ$_{DF-WHO98}$/y)
Municipal solid waste incineration, air	8,877.0	1,250.0	12.0
Backyard barrel burning, air	604.0	628.0	628.0
Medical waste incineration, air	2,590.0	488.0	7.0
Secondary copper smelting, air	983.0	271.0	5.0
Cement kilns (hazardous waste), air	117.8	156.1	7.7
Sewage sludge/land applied, land	76.6	76.6	76.6
Residential wood burning, air	89.6	62.8	62.8
Coal-fired utilities, air	50.8	60.1	60.1
Diesel trucks, air	27.8	35.5	35.5
Secondary aluminum smelting, air	16.3	29.1	29.1
2,4-Dichlorophenoxyacetic acid, land	33.4	28.9	28.9
Iron ore sintering, air	32.7	28.0	28.0
Industrial wood burning, air	26.4	27.6	27.6
Bleached pulp and paper mills, water	356.0	19.5	12.0
Cement kilns (nonhazardous waste), air	13.7	17.8	17.8
Sewage sludge incineration, air	6.1	14.8	14.8
Endocrine disrupting chemicals/ vinyl chloride, air	NA	11.2	11.2
Oil-fired utilities, air	17.8	10.7	10.7
Crematoria, air	5.5	9.1	9.1
Unleaded gasoline, air	3.6	5.9	5.9
Hazardous waste incineration, air	5.0	5.8	3.5
Lightweight ag kilns, hazardous waste, air	2.4	3.3	0.4
Kraft black liquor boilers, air	2.0	2.3	2.3
Petrol refine catalyst reg., air	2.2	2.2	2.2
Leaded gasoline, air	37.5	2.0	2.0
Secondary lead smelting, air	1.2	1.7	1.7
Paper mill sludge, land	14.1	1.4	1.4
Cigarette smoke, air	1.0	0.8	0.8
Endocrine disrupting chemicals/ vinyl chloride, land	NA	0.7	0.7
Endocrine disrupting chemicals/ vinyl chloride, water	NA	0.4	0.4
Boilers/industrial furnaces, air	0.8	0.4	0.4
Tire combustion, air	0.1	0.1	0.1
Drum reclamation, air	0.1	0.1	0.1
Totals	13,995	3,252	1,106
Reduction from 1987		77%	92%

NOTE: TEQ = toxicity equivalents.

REFERENCES

AEA Technology. 1999. *Compilation of EU Dioxin Exposure and Health Data.* Prepared for the European Commission DG Environment. Oxfordshire, England: AEA Technology.

Ahlborg V, Becking G, Birnbaum L, Brower A, Derks HJGM, Feeley M, Golor G, Hanberg A, Larsen JC, Liem AKD, Safe SH, Schlatter C, Waern F, Younes M, Yrjanheikki E. 1994. Toxic equivalency factors for dioxin-like PCBs. *Chemosphere* 28:1049–1067.

ATSDR. 1998. *Toxicological Profile for Chlorinated Dibenzo-p-dioxins.* Atlanta, GA: ATSDR.

EPA (U.S. Environmental Protection Agency). 1989. *Interim Procedures for Estimating Risks Associated with Exposures to Mixtures of Chlorinated Dibenzo-p-dioxins and -dibenzo-furans (CDDs and CDFs) and 1989 Update.* EPA/625/3-89.016. Washington, DC: EPA.

EPA. 2000. *Exposure and Human Health Reassessment of 2,3,7,8-Tetrachlorodibenzo-p-Dioxin (TCDD) and Related Compounds.* Draft Final Report. Washington, DC: EPA.

Fiedler H, Hutzinger O, Welsch-Pausch K, Schmiedinger A. 2000. *Evaluation of the Occurrence of PCDD/PCDF and POPs in Wastes and Their Potential to Enter the Foodchain.* Prepared for the European Commission DG Environment. Bayreuth, Germany: University of Bayreuth.

IARC (International Agency for Research on Cancer). 1997. *IARC Monographs on the Evaluation of Carcinogenic Risks to Humans. Volume 69: Polychlorinated Dibenzo-para-Dioxins and Polychlorinated Dibenzofurans.* Lyon, France: World Health Organization.

Scientific Committee on Food. 2000. *Opinion of the Scientific Committee on Food on the Risk Assessment of Dioxins and Dioxin-like PCBs in Food.* European Commission, Health and Consumer Protection Directorate-General. SCF/CS/CNTM/DIOXIN/8 Final. Brussels: European Commission.

Scientific Committee on Food. 2001. *Opinion of the Scientific Committee on Food on the Risk Assessment of Dioxins and Dioxin-like PCBs in Food. Update.* European Commission, Health and Consumer Protection Directorate-General. CS/CNTM/DIOXIN/20 Final. Brussels: European Commission.

van den Berg M, Birnbaum L, Bosveld AT, Brunstrom B, Cook P, Feeley M, Giesy JP, Hanberg A, Hasegawa R, Kennedy SW, Kubiak T, Larsen JC, van Leeuwen FX, Liem AK, Nolt C, Peterson RE, Poellinger L, Safe S, Schrenk D, Tillitt D, Tysklind M, Younes M, Waern F, Zacharewski T. 1998. Toxic equivalency factors (TEFs) for PCBs, PCDDs, PCDFs for humans and wildlife. *Environ Health Perspect* 106:775–792.

B

Total Diet Study Report: Dioxin Concentrations in Foods

Data on dioxin concentrations (as toxicity equivalents [TEQs]) in foods were obtained from the U.S. Food and Drug Administration's Total Diet Study (TDS) program. Although polychlorinated biphenyls account for approximately 50 percent of the TEQs in foods, this class of compounds was not included in the analysis. This study involves periodic retail purchase, preparation, and analysis of a variety of foods representative of the U.S. food supply. The TDS was initiated in 1961 in response to public concerns regarding the radionuclide contamination of foods. The program has been expanded to include analyses of concentrations of specific nutrients, pesticide residues, heavy metals, and other contaminants in TDS foods. Foods collected in the TDS "market baskets" remained fairly constant from 1992 (when the food list underwent a major revision) through 2002. The food list was significantly revised in 2002 to reflect current dietary patterns. Foods on the revised list will be collected beginning with the first market basket of 2003.

Data generated on constituent concentrations in TDS foods are used in estimating total dietary intakes by the total U.S. population and by specific age and sex groups. Population-specific consumption factors are applied to constituent levels in specific TDS foods and the weighted intakes for each food are summed to estimate total dietary intakes for each population group.

Three TEQ values were generated for each TDS sample food, reflecting the assignment of zero, half the limit of detection (LOD), or LOD values to congener nondetects. Mean TEQ values for TDS foods based on each nondetect scenario are listed in Table B-2.

The 1994–1996 Continuing Survey of Food Intakes by Individuals (CSFII) was conducted between January 1994 and January 1997 with noninstitutionalized individuals in the United States. In each of the three survey years, 24-hour recall data were collected through two in-person interviews (conducted approximately one week apart) from a nationally representative sample of individuals of all ages. The 1998 CSFII was designed as a supplement to the 1994–1996 CSFII, using the same survey methodology to increase the sample size for children from birth through age 9 years. In the merged surveys (1994–1996 and 1998 CSFIIs), 21,662 individuals provided intake data for the first survey day; 20,607 provided intake data for a second day.

The committee conducted its assessment of exposure to dioxins and dioxin-like compounds through foods using data from the TDS analysis coupled with food consumption data and the intake scenarios. The committee gratefully acknowledges the work of Judith Douglass and Mary Murphy for conducting the analysis.

TABLE B-1 Dioxin Congeners Included in the Total Diet Study Food Analyses

Congener Description
1,2,3,4,6,7,8-heptachlorodibenzo-*p*-dioxin
1,2,3,4,6,7,8-heptachlorodibenzofuran
1,2,3,4,7,8-hexachlorodibenzo-*p*-dioxin
1,2,3,4,7,8-hexachlorodibenzofuran
1,2,3,4,7,8,9-heptachlorodibenzofuran
1,2,3,6,7,8-hexachlorodibenzo-*p*-dioxin
1,2,3,6,7,8-hexachlorodibenzofuran
1,2,3,7,8-pentachlorodibenzo-*p*-dioxin
1,2,3,7,8-pentachlorodibenzofuran
1,2,3,7,8,9-hexachlorodibenzo-*p*-dioxin
1,2,3,7,8,9-hexachlorodibenzofuran
2,3,4,6,7,8-hexachlorodibenzofuran
2,3,4,7,8-pentachlorodibenzofuran
2,3,7,8-tetrachlorodibenzo-*p*-dioxin
2,3,7,8-tetrachlorodibenzofuran
Octachlorodibenzo-*p*-dioxin
Octachlorodibenzofuran

TABLE B-2 Mean Dioxin Toxicity Equivalents (TEQ)[a] for Total Diet Study[b] Foods

Food Category	Total Diet Study Food Description	Mean TEQ (parts per trillion)		
		Nondetects[c] = 0	Nondetects = 0.5 (LOD[d])	Nondetects = LOD
Dairy	Baby food, custard pudding, strained/junior	—[e]	0.0241	0.0481
	Cheese, American	0.0276	0.0601	0.0925
	Cheese, Cheddar	0.2426	0.2534	0.2643
	Cheese, cottage, 4% fat	0.0021	0.0261	0.0500
	Cheese, cream	0.0678	0.0835	0.0993
	Cheese, Swiss	0.0833	0.1345	0.1857
	Cream substitute, frozen	0.0002	0.0171	0.0341
	Cream, half & half	0.0497	0.0502	0.0507
	Cream, sour	0.0618	0.0622	0.0625
	Ice cream, vanilla	0.0242	0.0401	0.0560
	Ice cream, vanilla, light	0.0001	0.0195	0.0388
	Infant formula, milk-based, high iron	0.0001	0.0105	0.0208
	Infant formula, milk-based, low iron	—	0.0122	0.0243
	Infant formula, soy-based	0.0003	0.0039	0.0075
	Milk shake, chocolate, fast-food	—	0.0300	0.0599
	Milk, chocolate, fluid	0.0006	0.0148	0.0290
	Milk, whole, fluid	0.0105	0.0110	0.0115
	Pudding, chocolate, instant[f]	0.0034	0.0071	0.0108
	Yogurt, low-fat, fruit	0.0003	0.0035	0.0067
	Yogurt, low-fat, plain	0.0013	0.0121	0.0230
Meat	Baby food, beef, strained/junior	0.0029	0.0257	0.0484
	Baby food, vegetables and beef, strained/junior	0.0004	0.0073	0.0143
	Beef chuck roast, baked	0.4592	0.4595	0.4598
	Beef steak, pan-cooked	0.0119	0.0347	0.0575
	Beef stew	0.0423	0.0497	0.0570
	Beef stroganoff	0.0076	0.0226	0.0375
	Bologna, sliced	0.1558	0.1558	0.1558
	Brown gravy, homemade	0.2054	0.2055	0.2056
	Cheeseburger, fast-food	0.0670	0.0739	0.0808
	Chili con carne w/beans	0.0127	0.0196	0.0265
	Frankfurter, beef, boiled	0.1983	0.1983	0.1983
	Frankfurter, fast-food	0.1060	0.1065	0.1070
	Green peppers, stuffed	0.0046	0.0637	0.1228
	Ground beef, pan-cooked	0.1210	0.1290	0.1369
	Ham luncheon meat	0.0080	0.0432	0.0785
	Ham, baked	0.0435	0.0463	0.0492
	Hamburger, fast-food	0.0675	0.0678	0.0680
	Lamb chop, pan-cooked	0.0857	0.0888	0.0919
	Liver, beef, fried	0.1099	0.1198	0.1298
	Meatloaf, homemade	0.0384	0.0857	0.1330
	Pork bacon, pan-cooked	0.2114	0.2116	0.2118
	Pork chop, pan-cooked	0.0161	0.0198	0.0234

continued

TABLE B-2 Continued

Food Category	Total Diet Study Food Description	Mean TEQ (parts per trillion)		
		Nondetects[c] = 0	Nondetects = 0.5 (LOD[d])	Nondetects = LOD
	Pork roast, baked	0.0888	0.0963	0.1039
	Pork sausage, pan-cooked	0.1959	0.1962	0.1965
	Salami, sliced[g]	0.0200	0.1476	0.2751
	Salisbury steak, frozen meal	0.0052	0.0136	0.0219
	Veal cutlet, pan-cooked	0.0181	0.0255	0.0330
Poultry	Baby food, chicken noodle dinner, strained/junior	0.0005	0.0068	0.0131
	Baby food, chicken, strained/junior	0.0008	0.0094	0.0180
	Baby food, turkey and rice, strained/junior	0.0005	0.0062	0.0119
	Chicken breast, roasted	0.0083	0.0166	0.0248
	Chicken nuggets, fast-food	0.0041	0.0118	0.0196
	Chicken potpie, frozen	0.0100	0.0196	0.0292
	Chicken, fried, fast-food	0.0392	0.0580	0.0768
	Chicken, fried, homemade	0.0556	0.0559	0.0562
	Turkey breast, roasted	0.0481	0.0538	0.0595
	Turkey, frozen meal	0.0050	0.0112	0.0174
Fish	Fish sandwich, fast-food	0.0124	0.0138	0.0152
	Fish sticks, frozen, heated	0.0246	0.0335	0.0424
	Salmon	0.3254	0.3257	0.3260
	Shrimp, boiled	0.0597	0.0597	0.0597
	Soup, clam chowder, canned	0.0026	0.0054	0.0082
	Tuna noodle casserole	0.0257	0.0334	0.0412
	Tuna, canned in oil[g]	—	0.0838	0.1675
	White sauce, homemade	—	0.0204	0.0408
Eggs	Egg/cheese/ham/muffin, fast-food	0.0215	0.0373	0.0530
	Eggs, boiled	0.0107	0.0313	0.0519
	Eggs, scrambled	0.0498	0.0498	0.0498
Fruits/ vegetables	Apple juice, bottled	0.0002	0.0029	0.0055
	Apple, red, raw	0.0003	0.0065	0.0128
	Applesauce, bottled	—	0.0055	0.0110
	Apricot, raw	—	0.0022	0.0044
	Asparagus, boiled	0.0003	0.0032	0.0060
	Avocado, raw	0.0034	0.0073	0.0113
	Baby food, apple juice, strained	0.0008	0.0171	0.0333
	Baby food, apples/applesauce w/apricots, strained/junior	0.0033	0.0119	0.0206
	Baby food, applesauce, strained/junior	—	0.0022	0.0043
	Baby food, apricots w/tapioca, strained/junior	0.0161	0.0339	0.0517
	Baby food, banana/pineapple, strained/junior	0.0010	0.0108	0.0205
	Baby food, bananas w/tapioca, strained/junior	0.0006	0.0075	0.0145
	Baby food, carrots, strained/junior	—	0.0160	0.0320

TABLE B-2 Continued

		Mean TEQ (parts per trillion)		
Food Category	Total Diet Study Food Description	Nondetects[c] = 0	Nondetects = 0.5 (LOD[d])	Nondetects = LOD
	Baby food, creamed corn, strained/junior	—	0.0033	0.0066
	Baby food, creamed spinach, strained/junior	—	0.0147	0.0294
	Baby food, dessert, banana, apple, strained/junior	0.0044	0.0180	0.0316
	Baby food, dessert, peach cobbler, strained/junior	0.0066	0.0281	0.0496
	Baby food, dessert, yogurt w/fruit, strained/junior	0.0003	0.0061	0.0120
	Baby food, Dutch apple/apple betty, strained/junior	0.0021	0.0168	0.0315
	Baby food, fruit dessert or pudding, strained/junior	—	0.0079	0.0157
	Baby food, green beans, strained/junior	—	0.0134	0.0267
	Baby food, juice, apple-banana, strained	—	0.0053	0.0107
	Baby food, juice, apple-cherry, strained	—	0.0017	0.0033
	Baby food, juice, apple-grape, strained	0.0001	0.0005	0.0010
	Baby food, juice, grape, strained	0.0004	0.0025	0.0047
	Baby food, juice, mixed fruit, strained	—	0.0006	0.0011
	Baby food, juice, pear, strained	—	0.0007	0.0014
	Baby food, mixed vegetables, strained/junior	—	0.0101	0.0202
	Baby food, orange juice, strained	0.0004	0.0020	0.0036
	Baby food, peaches, strained/junior	0.0004	0.0073	0.0142
	Baby food, pears & pineapple, strained/junior	0.0296	0.0325	0.0355
	Baby food, pears, strained/junior	—	0.0416	0.0833
	Baby food, peas, strained/junior	—	0.0184	0.0368
	Baby food, plums/tapioca, strained/junior	0.0003	0.0055	0.0107
	Baby food, squash, strained/junior	0.0002	0.0094	0.0186
	Baby food, sweet potatoes, strained/junior	—	0.0076	0.0153
	Baby food, vegetables and beef, strained/junior	0.0004	0.0073	0.0143
	Baby food, vegetables and chicken, strained/junior	—	0.0053	0.0107
	Baby food, vegetables and ham, strained/junior	0.0012	0.0249	0.0487
	Banana, raw	0.0494	0.0517	0.0539

continued

TABLE B-2 Continued

Food Category	Total Diet Study Food Description	Mean TEQ (parts per trillion)		
		Nondetects[c] = 0	Nondetects = 0.5 (LOD[d])	Nondetects = LOD
	Beans, green, boiled	0.0098	0.0216	0.0333
	Beets, boiled	0.0001	0.0017	0.0034
	Broccoli, boiled	—	0.0035	0.0071
	Brussels sprouts, boiled	0.0018	0.0038	0.0058
	Cabbage, fresh, boiled	0.0005	0.0036	0.0067
	Cantaloupe, raw	—	0.0072	0.0145
	Carrot, fresh, boiled	0.0001	0.0035	0.0069
	Cauliflower, boiled	0.0005	0.0019	0.0034
	Celery, raw	0.0002	0.0023	0.0045
	Cherries, sweet raw	0.0051	0.0089	0.0127
	Coleslaw w/dressing	—	0.0103	0.0206
	Collards, boiled	0.0004	0.0061	0.0118
	Corn, boiled	0.0038	0.0051	0.0065
	Corn, cream style, canned	0.0010	0.0182	0.0353
	Cucumber, raw[g]	—	0.1173	0.2345
	Eggplant, fresh, boiled	0.0012	0.0031	0.0050
	Fruit cocktail, canned	0.0005	0.0078	0.0150
	Grape juice, from concentrate	—	0.0017	0.0034
	Grapefruit juice, from concentrate	—	0.0043	0.0086
	Grapefruit, raw	0.0006	0.0073	0.0141
	Grapes, seedless, raw	0.0017	0.0154	0.0291
	Lettuce, iceberg, raw	—	0.0043	0.0086
	Lima beans, boiled	0.0002	0.0027	0.0053
	Mixed vegetables, boiled	—	0.0061	0.0122
	Mushrooms, raw	—	0.0028	0.0056
	Mustard, yellow	—	0.0079	0.0157
	Okra, boiled	0.0014	0.0028	0.0042
	Olives, black	—	0.0154	0.0308
	Onion, mature, raw	—	0.0045	0.0090
	Orange juice, from concentrate	—	0.0302	0.0604
	Orange, raw	—	0.0073	0.0146
	Peach, canned	—	0.0085	0.0169
	Peach, raw	0.0021	0.0096	0.0172
	Pear, canned in light syrup	—	0.0055	0.0109
	Pear, raw	0.0006	0.0043	0.0080
	Peas, green, boiled	0.0004	0.0020	0.0035
	Pepper, green, raw	0.0027	0.0039	0.0051
	Pickles, cucumber, dill	0.0001	0.0018	0.0034
	Pickles, cucumber, sweet	—	0.0044	0.0088
	Pineapple juice, from concentrate	—	0.0023	0.0047
	Pineapple, canned in juice	—	0.0209	0.0419
	Plums, raw	0.0042	0.0066	0.0091
	Potato chips	0.0017	0.0263	0.0510
	Potato, white, baked, skin	0.0282	0.0301	0.0321
	Potato, white, boiled, no skin	0.0001	0.0184	0.0368
	Potatoes, French fries, fast-food	0.0038	0.0147	0.0256

TABLE B-2 Continued

		Mean TEQ (parts per trillion)		
Food Category	Total Diet Study Food Description	Nondetects[c] = 0	Nondetects = 0.5 (LOD[d])	Nondetects = LOD
	Potatoes, French fries, frozen, heated	0.0492	0.0614	0.0736
	Potatoes, mashed, from flakes	0.0073	0.0105	0.0138
	Potatoes, scalloped	0.0904	0.1188	0.1471
	Prune juice, bottled	0.0001	0.0018	0.0035
	Prunes, dried	0.0135	0.0165	0.0194
	Radish, raw	—	0.0043	0.0085
	Raisins, dried	—	0.0136	0.0271
	Sauerkraut, canned	0.0005	0.0044	0.0082
	Sherbet, fruit flavor	—	0.0034	0.0067
	Soup, mushroom, canned	0.0010	0.0034	0.0057
	Soup, tomato, canned	—	0.0042	0.0085
	Soup, vegetable beef, canned	0.0012	0.0024	0.0035
	Spinach, boiled	—	0.0081	0.0163
	Squash, summer, boiled	0.0174	0.0228	0.0282
	Squash, winter, baked	0.0003	0.0019	0.0035
	Strawberries, raw	—	0.0236	0.0472
	Sweet potato, fresh, baked	0.0127	0.0340	0.0553
	Tomato catsup	0.0001	0.0019	0.0037
	Tomato juice, bottled	0.0001	0.0017	0.0034
	Tomato sauce, bottled	—	0.0017	0.0034
	Tomato, red, raw	—	0.0407	0.0815
	Tomato, stewed, canned	—	0.0101	0.0202
	Turnip, boiled	—	0.0104	0.0208
	Watermelon, raw	0.0032	0.0046	0.0059
Fats/oils	Butter, regular (salted)	0.2200	0.2202	0.2205
	Margarine, stick (salted)	0.0044	0.0270	0.0496
	Mayonnaise	0.0096	0.0136	0.0177
	Oil, olive or safflower	0.0565	0.0641	0.0716
	Salad dressing, French	0.0065	0.0101	0.0136
	Salad dressing, Italian, low-calorie	0.0019	0.0095	0.0170
Other foods	Baby food, arrowroot cookies	0.0168	0.0220	0.0272
	Baby food, cereal, prepared from dry, barley	0.0006	0.0046	0.0086
	Baby food, macaroni, tomatoes, beef, strained/junior	—	0.0106	0.0211
	Baby food, rice infant cereal, instant, w/whole milk	0.0010	0.0097	0.0183
	Baby food, rice/apple cereal, strained/junior	0.0102	0.0441	0.0780
	Baby food, teething biscuits	0.0040	0.0175	0.0310
	Baby food, zwieback toast	0.0159	0.0205	0.0251
	Bagel, plain	—	0.0191	0.0383
	Beef chow mein, carry-out	0.0021	0.0068	0.0115
	Biscuit, baked	0.0094	0.0147	0.0200

continued

TABLE B-2 Continued

Food Category	Total Diet Study Food Description	Mean TEQ (parts per trillion)		
		Nondetects[c] = 0	Nondetects = 0.5 (LOD[d])	Nondetects = LOD
	Bread, cracked wheat	0.0013	0.0047	0.0081
	Bread, rye	0.0017	0.0058	0.0099
	Bread, white	0.0029	0.0071	0.0112
	Bread, white roll	—	0.0113	0.0225
	Bread, whole wheat	0.0048	0.0068	0.0088
	Brownies, commercial	0.0082	0.0151	0.0220
	Cake, chocolate snack	0.0109	0.0176	0.0242
	Cake, chocolate with icing	0.0188	0.0244	0.0299
	Cake, yellow, w/white icing	0.0344	0.0379	0.0414
	Candy, bar, milk chocolate	0.0219	0.0222	0.0225
	Candy, caramel	0.0030	0.0071	0.0113
	Candy, suckers, any flavor	0.0020	0.0049	0.0079
	Cereal, corn flakes	—	0.0106	0.0211
	Cereal, corn grits, regular, cooked	0.0100	0.0185	0.0270
	Cereal, crisped rice	—	0.0157	0.0314
	Cereal, fruit flavored	—	0.0150	0.0300
	Cereal, granola	0.0011	0.0042	0.0072
	Cereal, oat ring	0.0029	0.0224	0.0420
	Cereal, oatmeal, quick, cooked	0.0013	0.0063	0.0114
	Cereal, raisin bran	0.0007	0.0067	0.0126
	Cereal, shredded wheat	0.0053	0.0114	0.0175
	Cereal, wheat farina, quick, cooked	—	0.0169	0.0337
	Coffee, decaffeinated, instant	0.0001	0.0031	0.0062
	Coffee, from ground	0.0018	0.0042	0.0067
	Cola, carbonated	—	0.0183	0.0365
	Cola, carbonated, low-calorie	—	0.0020	0.0041
	Cookies, chocolate chip	0.0014	0.0206	0.0399
	Cookies, sandwich, creme-filled	0.0013	0.0125	0.0236
	Cookies, sugar, commercial	0.0106	0.0166	0.0227
	Corn chips	0.0076	0.0110	0.0144
	Cornbread, homemade	0.0460	0.0504	0.0548
	Crackers, butter-type	0.0010	0.0082	0.0154
	Crackers, graham	0.0020	0.0118	0.0216
	Crackers, saltine	0.0161	0.0267	0.0374
	Doughnuts, cake type, w/icing	0.0059	0.0124	0.0188
	Egg noodles, boiled	—	0.0271	0.0542
	English muffin, toasted	0.0020	0.0100	0.0179
	Fruit drink, canned	—	0.0019	0.0038
	Fruit drink, from powder	—	0.0033	0.0066
	Fruit drink, lemonade, from concentrate	—	0.0043	0.0086
	Fruit flavor sherbet	—	0.0034	0.0067
	Fruit-flavored drink, carbonated	0.0029	0.0043	0.0057
	Gelatin dessert, any flavor	—	0.0087	0.0173
	Honey	—	0.0249	0.0498
	Jelly, any flavor	0.0036	0.0074	0.0112
	Kidney beans, dry, boiled	0.0013	0.0029	0.0046

TABLE B-2 Continued

Food Category	Total Diet Study Food Description	Mean TEQ (parts per trillion)		
		Nondetects[c] = 0	Nondetects = 0.5 (LOD[d])	Nondetects = LOD
	Lasagna w/meat	0.0377	0.0408	0.0439
	Macaroni and cheese, box	0.0172	0.0242	0.0312
	Macaroni, boiled	0.0001	0.0056	0.0110
	Muffin, blueberry	0.0005	0.0142	0.0279
	Nuts, mixed, no peanuts	0.0032	0.0155	0.0277
	Pancake from mix	0.0102	0.0124	0.0145
	Pancake syrup	0.0005	0.0043	0.0080
	Peanut butter, smooth	0.0039	0.0068	0.0098
	Peanuts, dry roasted	0.0056	0.0098	0.0140
	Peas, mature, dry, boiled	—	0.0381	0.0762
	Pie, apple	0.0201	0.0327	0.0453
	Pie, pumpkin	0.0023	0.0391	0.0760
	Pinto beans, dry, boiled	—	0.0046	0.0092
	Pizza, cheese, carry-out	0.0108	0.0185	0.0262
	Pizza, pepperoni, carry-out	0.0071	0.0205	0.0339
	Popcorn, popped in oil	0.0337	0.0411	0.0486
	Popsicle, any flavor	—	0.0053	0.0106
	Pork and beans, canned	0.0024	0.0053	0.0082
	Pretzels, hard, salted	0.0076	0.0131	0.0186
	Rice, white, cooked	0.0009	0.0041	0.0072
	Soup, chicken noodle, canned	0.0001	0.0054	0.0108
	Soup, bean, canned	0.0015	0.0032	0.0049
	Spaghetti and meatballs	0.0024	0.0109	0.0193
	Spaghetti, canned	—	0.0062	0.0124
	Sweet roll or Danish	0.0139	0.0184	0.0230
	Syrup, chocolate	0.0006	0.0033	0.0060
	Taco or tostada, carry-out	0.0472	0.0619	0.0765
	Tea, from tea bag	—	0.0171	0.0342
	Tortilla, flour	0.0023	0.0051	0.0078
	White sugar, granulated	0.0003	0.0050	0.0098

[a]Sum of TEQs for dioxin congeners.

[b]Total Diet Study, administered by the U.S. Food and Drug Administration. Data are for samples collected in 2001, Market Basket 2, and analyzed for dioxin congener concentrations by gas chromatograph/high-resolution mass spectrometer (HRMS), except as noted otherwise.

[c]Samples with no amount of a dioxin congener detected.

[d]Limit of detection (LOD) for relevant congener.

[e]TEQ < 0.00005 ppt.

[f]Sample collected in 2000, Market Basket 2, analyzed using HRMS.

[g]Sample collected in 2000, Market Basket 2, analyzed using gas chromatograph-ion trap instrumentation.

TABLE B-3 Estimated Intake of Dioxins from Food by Boys and Girls, 1–5 Years Old ($n = 6,409$)

Food Categories	Percent Consuming	Food Intake (g/kg body weight/d)
All foods	100.0	90.5
If skim milk consumed[c]	100.0	90.5
Dairy foods and mixtures	98.3	29.5
If skim milk consumed[c]	98.3	29.5
Meat and mixtures	80.3	4.7
Poultry and mixtures	61.1	3.6
Fish and mixtures	17.0	2.8
Eggs and mixtures	33.2	2.7
Fruits, vegetables, and mixtures	98.9	21.9
Fats, oils, and mixtures	59.8	0.4
Other foods and mixtures[e]	99.9	32.3

[a]Reflects treatment of samples for which no dioxin congener was detected.
[b]LOD = limit of detection.
[c]Dioxin TEQ intake from milk assuming that all plain milk consumed is skim milk rather than whole milk, as was otherwise assumed due to lack of analytical data on milks other than whole milk. Dioxin TEQ for skim milk was estimated assuming that all dioxin congeners concentrate in milk fat, that whole milk contains 3.34% fat, and that skim milk contains 0.18% fat. (Nutrient Data Laboratory. 2002. *USDA National Nutrient Database for Standard Reference, Release 15.* Online. ARS, USDA. Available at http://www.nal.usda.gov/fnic/foodcomp. Accessed December 9, 2002.)
[d]< 0.0005 pg/kg body weight/d. [e]Grains and mixtures, legumes and mixtures, beverages (other than milk and juice), candy.

NOTE: Data represent 2-day averages. Breastfeeding children are excluded.
DATA SOURCE: Dioxin concentrations: U.S. Food and Drug Administration Total Diet Study (1999–2001); food consumption: 1994–1996, 1998 U.S. Department of Agriculture Continuing Survey of Food Intakes by Individuals.

Consumers' Dioxin Toxic Equivalents (TEQ) Intake (pg/kg body weight/d)

Nondetects[a] = 0				Nondetects = 0.5 (LOD)[b]				Nondetects = LOD			
Mean	Percentile			Mean	Percentile			Mean	Percentile		
	10	50	90		10	50	90		10	50	90
1.09	0.43	0.93	1.89	1.67	0.85	1.50	2.63	2.24	1.18	2.04	3.51
0.85	0.25	0.69	1.61	1.42	0.64	1.24	2.34	1.98	0.93	1.78	3.21
0.34	0.08	0.27	0.67	0.42	0.12	0.34	0.81	0.50	0.14	0.41	0.94
0.10	0.01	0.04	0.24	0.16	0.01	0.09	0.40	0.23	0.01	0.13	0.56
0.46	0.03	0.30	1.02	0.51	0.06	0.35	1.09	0.56	0.08	0.39	1.19
0.11	0.01	0.05	0.31	0.13	0.02	0.08	0.33	0.16	0.03	0.10	0.36
0.28	—[d]	0.05	0.89	0.33	0.03	0.12	0.90	0.38	0.04	0.18	0.93
0.12	0.02	0.10	0.24	0.13	0.04	0.10	0.24	0.13	0.05	0.11	0.25
0.11	—	0.05	0.31	0.33	0.05	0.25	0.70	0.54	0.09	0.40	1.16
0.02	—	—	0.06	0.02	—	0.01	0.06	0.03	—	0.01	0.07
0.11	0.02	0.08	0.25	0.33	0.11	0.27	0.62	0.55	0.18	0.45	1.05

TABLE B-4 Estimated Intake of Dioxins from Food by Boys and Girls, 6–11 Years Old ($n = 1,913$)

Food Categories	Percent Consuming	Food Intake (g/kg body weight/d)
All foods	100.0	53.4
If skim milk consumed[c]	100.0	53.4
Dairy foods and mixtures	98.1	14.4
If skim milk consumed[c]	98.1	14.4
Meat and mixtures	84.8	3.4
Poultry and mixtures	57.1	2.4
Fish and mixtures	15.6	2.2
Eggs and mixtures	23.9	1.6
Fruits, vegetables, and mixtures	98.4	10.1
Fats, oils, and mixtures	66.8	0.4
Other foods and mixtures[e]	100.0	24.0

[a]Reflects treatment of samples for which no dioxin congener was detected.
[b]LOD = limit of detection.
[c]Dioxin TEQ intake from milk assuming that all plain milk consumed is skim milk rather than whole milk, as was otherwise assumed due to lack of analytical data on milks other than whole milk. Dioxin TEQ for skim milk was estimated assuming that all dioxin congeners concentrate in milk fat, that whole milk contains 3.34% fat, and that skim milk contains 0.18% fat. (Nutrient Data Laboratory. 2002. *USDA National Nutrient Database for Standard Reference, Release 15.* Online. ARS, USDA. Available at http://www.nal.usda.gov/fnic/foodcomp. Accessed December 9, 2002.)
[d]< 0.0005 pg/kg body weight/d.
[e]Grains and mixtures, legumes and mixtures, beverages (other than milk and juice), candy.

NOTE: Data represent 2-day averages generated using U.S. Department of Agriculture (USDA) sample weights.
DATA SOURCE: Dioxin concentrations: U.S. Food and Drug Administration Total Diet Study (1999–2001); food consumption: 1994–1996, 1998 USDA Continuing Survey of Food Intakes by Individuals.

Consumers' Dioxin Toxic Equivalents (TEQ) Intake (pg/kg body weight/d)

Nondetects[a] = 0				Nondetects = 0.5 (LOD)[b]				Nondetects = LOD			
	Percentile				Percentile				Percentile		
Mean	10	50	90	Mean	10	50	90	Mean	10	50	90
0.69	0.24	0.57	1.24	1.10	0.53	0.99	1.76	1.51	0.76	1.38	2.43
0.58	0.17	0.46	1.11	0.99	0.44	0.88	1.64	1.40	0.66	1.27	2.31
0.17	0.04	0.13	0.33	0.23	0.07	0.20	0.43	0.29	0.08	0.25	0.58
0.06	—[d]	0.03	0.15	0.12	0.01	0.08	0.28	0.18	0.01	0.13	0.41
0.32	0.02	0.21	0.65	0.36	0.05	0.25	0.72	0.40	0.07	0.29	0.82
0.08	—	0.03	0.20	0.09	0.01	0.05	0.21	0.11	0.02	0.07	0.25
0.25	—	0.04	0.83	0.29	0.02	0.08	0.83	0.33	0.03	0.13	0.87
0.07	0.01	0.05	0.14	0.07	0.02	0.06	0.14	0.08	0.03	0.06	0.15
0.06	—	0.02	0.17	0.18	0.02	0.12	0.41	0.30	0.04	0.20	0.69
0.01	—	—	0.04	0.02	—	0.01	0.04	0.02	—	0.01	0.04
0.09	0.02	0.06	0.18	0.27	0.10	0.23	0.48	0.45	0.16	0.38	0.80

TABLE B-5 Estimated Intake of Dioxins from Food by Adolescent Males, 12–19 Years Old (n = 696)

Food Categories	Percent Consuming	Food Intake (g/kg body weight/d)
All foods	100.0	38.6
Dairy foods and mixtures	92.3	7.2
Meat and mixtures	88.4	2.9
Poultry and mixtures	57.3	1.9
Fish and mixtures	17.5	1.9
Eggs and mixtures	25.5	1.1
Fruits, vegetables, and mixtures	97.2	5.7
Fats, oils, and mixtures	62.8	0.3
Other foods and mixtures[d]	100.0	22.0

[a]Reflects treatment of samples for which no dioxin congener was detected.

[b]LOD = limit of detection.

[c]< 0.0005 pg/kg body weight/d.

[d]Grains and mixtures, legumes and mixtures, beverages (other than milk and juice), candy.

NOTE: Data represent 2-day averages generated using U.S. Department of Agriculture (USDA) sample weights.

DATA SOURCE: Dioxin concentrations: U.S. Food and Drug Administration Total Diet Study (1999–2001); food consumption: 1994–1996, 1998 USDA Continuing Survey of Food Intakes by Individuals.

Consumers' Dioxin Toxic Equivalents (TEQ) Intake (pg/kg body weight/d)

Nondetects[a] = 0				Nondetects = 0.5 (LOD)[b]				Nondetects = LOD			
	Percentile				Percentile				Percentile		
Mean	10	50	90	Mean	10	50	90	Mean	10	50	90
0.52	0.17	0.40	0.92	0.87	0.39	0.74	1.39	1.23	0.55	1.08	1.98
0.09	0.01	0.06	0.19	0.12	0.03	0.09	0.25	0.16	0.03	0.12	0.33
0.27	0.02	0.16	0.56	0.31	0.04	0.20	0.62	0.34	0.06	0.22	0.67
0.07	—[c]	0.04	0.17	0.08	0.01	0.05	0.18	0.09	0.02	0.06	0.20
0.14	—	0.04	0.50	0.17	0.02	0.08	0.50	0.20	0.02	0.11	0.51
0.05	0.01	0.04	0.10	0.05	0.02	0.04	0.10	0.05	0.02	0.04	0.10
0.05	—	0.01	0.12	0.13	0.01	0.08	0.30	0.22	0.02	0.12	0.50
0.01	—	—	0.02	0.01	—	—	0.02	0.01	—	0.01	0.03
0.07	0.01	0.05	0.16	0.27	0.09	0.23	0.51	0.47	0.14	0.40	0.89

TABLE B-6 Estimated Intake of Dioxins from Food by Men 20+ Years Old ($n = 4,751$)

Food Categories	Percent Consuming	Food Intake (g/kg body weight/d)
All foods	100.0	32.3
Dairy foods and mixtures	88.9	3.4
Meat and mixtures	87.6	2.4
Poultry and mixtures	57.2	1.6
Fish and mixtures	26.7	1.3
Eggs and mixtures	34.3	0.8
Fruits, vegetables, and mixtures	98.1	5.1
Fats, oils, and mixtures	73.1	0.3
Other foods and mixtures[d]	100.0	20.4

[a]Reflects treatment of samples for which no dioxin congener was detected.
[b]LOD = limit of detection.
[c]< 0.0005 pg/kg body weight/d.
[d]Grains and mixtures, legumes and mixtures, beverages (other than milk and juice), candy.

NOTE: Data represent 2-day averages generated using U.S. Department of Agriculture (USDA) sample weights.
DATA SOURCE: Dioxin concentrations: U.S. Food and Drug Administration Total Diet Study (1999–2001); food consumption: 1994–1996, 1998 USDA Continuing Survey of Food Intakes by Individuals.

Consumers' Dioxin Toxic Equivalents (TEQ) Intake (pg/kg body weight/d)

Nondetects[a] = 0				Nondetects = 0.5 (LOD)[b]				Nondetects = LOD			
	Percentile				Percentile				Percentile		
Mean	10	50	90	Mean	10	50	90	Mean	10	50	90
0.41	0.13	0.32	0.80	0.68	0.30	0.58	1.17	0.95	0.44	0.83	1.58
0.05	0.01	0.04	0.11	0.07	0.01	0.05	0.15	0.09	0.01	0.07	0.19
0.22	0.01	0.13	0.51	0.25	0.03	0.16	0.55	0.28	0.04	0.19	0.61
0.05	—[c]	0.03	0.12	0.06	0.01	0.04	0.14	0.07	0.01	0.05	0.16
0.15	—	0.04	0.41	0.17	0.01	0.07	0.43	0.18	0.02	0.10	0.44
0.03	0.01	0.03	0.07	0.04	0.01	0.03	0.07	0.04	0.01	0.03	0.08
0.04	—	0.02	0.11	0.10	0.01	0.07	0.24	0.17	0.02	0.12	0.38
0.01	—	—	0.02	0.01	—	—	0.03	0.01	—	0.01	0.03
0.05	0.01	0.03	0.11	0.19	0.06	0.16	0.37	0.34	0.10	0.27	0.66

TABLE B-7 Estimated Intake of Dioxins from Food by Adolescent Females, 12–19 Years Old, not Pregnant or Lactating ($n = 692$)

Food Categories	Percent Consuming	Food Intake (g/kg body weight/d)
All foods	100.0	30.6
If skim milk consumed[c]	100.0	30.6
Dairy foods and mixtures	88.7	5.3
If skim milk consumed[c]	88.7	5.3
Meat and mixtures	78.6	2.2
Poultry and mixtures	53.1	1.7
Fish and mixtures	17.4	1.4
Eggs and mixtures	23.4	0.9
Fruits, vegetables, and mixtures	96.8	5.4
Fats, oils, and mixtures	60.1	0.3
Other foods and mixtures[e]	100.0	17.4

[a]Reflects treatment of samples for which no dioxin congener was detected.

[b]LOD = limit of detection.

[c]Dioxin TEQ intake from milk assuming that all plain milk consumed is skim milk rather than whole milk, as was otherwise assumed due to lack of analytical data on milks other than whole milk. Dioxin TEQ for skim milk was estimated assuming that all dioxin congeners concentrate in milk fat, that whole milk contains 3.34% fat, and that skim milk contains 0.18% fat. (Nutrient Data Laboratory. 2002. *USDA National Nutrient Database for Standard Reference, Release 15.* Online. ARS, USDA. Available at http://www.nal.usda.gov/fnic/foodcomp. Accessed December 9, 2002.)

[d]< 0.0005 pg/kg body weight/d.

[e]Grains and mixtures, legumes and mixtures, beverages (other than milk and juice), candy.

NOTE: Data represent 2-day averages generated using U.S. Department of Agriculture (USDA) sample weights.

DATA SOURCE: Dioxin concentrations: U.S. Food and Drug Administration Total Diet Study (1999–2001); food consumption: 1994–1996, 1998 USDA Continuing Survey of Food Intakes by Individuals.

Consumers' Dioxin Toxic Equivalents (TEQ) Intake (pg/kg body weight/d)

Nondetects[a] = 0				Nondetects = 0.5 (LOD)[b]				Nondetects = LOD			
	Percentile				Percentile				Percentile		
Mean	10	50	90	Mean	10	50	90	Mean	10	50	90
0.37	0.11	0.29	0.66	0.66	0.28	0.58	1.08	0.96	0.40	0.85	1.67
0.33	0.09	0.26	0.66	0.63	0.25	0.54	1.07	0.92	0.92	0.81	1.63
0.07	0.01	0.05	0.15	0.10	0.02	0.07	0.20	0.13	0.02	0.09	0.27
0.03	0.00	0.01	0.08	0.06	0.00	0.03	0.14	0.09	0.09	0.05	0.21
0.21	0.01	0.13	0.46	0.24	0.03	0.15	0.52	0.26	0.04	0.17	0.61
0.05	—[d]	0.03	0.12	0.06	0.01	0.04	0.13	0.07	0.01	0.05	0.15
0.10	—	0.01	0.31	0.14	0.01	0.06	0.31	0.17	0.02	0.10	0.34
0.04	—	0.03	0.07	0.04	0.01	0.04	0.07	0.04	0.02	0.04	0.07
0.03	—	0.01	0.10	0.11	0.01	0.07	0.25	0.18	0.02	0.11	0.45
0.01	—	—	0.02	0.01	—	—	0.02	0.01	—	0.01	0.02
0.06	0.01	0.04	0.13	0.22	0.07	0.18	0.43	0.38	0.11	0.31	0.74

TABLE B-8 Estimated Intake of Dioxins from Food by Women, 20+ Years Old, not Pregnant or Lactating (*n* = 4,470)

Food Categories	Percent Consuming	Food Intake (g/kg body weight/d)
All foods	100.0	29.3
Dairy foods and mixtures	90.9	3.2
Meat and mixtures	78.5	1.7
Poultry and mixtures	58.3	1.5
Fish and mixtures	25.6	1.3
Eggs and mixtures	30.5	0.7
Fruits, vegetables, and mixtures	98.0	5.4
Fats, oils, and mixtures	73.9	0.3
Other foods and mixtures[d]	99.9	18.1

[a]Reflects treatment of samples for which no dioxin congener was detected.
[b]LOD = limit of detection.
[c]< 0.0005 pg/kg body weight/d.
[d]Grains and mixtures, legumes and mixtures, beverages (other than milk and juice), candy.

NOTE: Data represent 2-day averages generated using U.S. Department of Agriculture (USDA) sample weights.
DATA SOURCE: Dioxin concentrations: U.S. Food and Drug Administration Total Diet Study (1999–2001); food consumption: 1994–1996, 1998 USDA Continuing Survey of Food Intakes by Individuals.

Consumers' Dioxin Toxic Equivalents (TEQ) Intake (pg/kg body weight/d)

Nondetects[a] = 0				Nondetects = 0.5 (LOD)[b]				Nondetects = LOD			
	Percentile				Percentile				Percentile		
Mean	10	50	90	Mean	10	50	90	Mean	10	50	90
0.33	0.10	0.25	0.60	0.57	0.25	0.50	0.96	0.82	0.37	0.74	1.36
0.04	—[c]	0.03	0.10	0.06	0.01	0.05	0.14	0.08	0.01	0.06	0.18
0.16	0.01	0.09	0.38	0.18	0.02	0.11	0.42	0.21	0.03	0.13	0.45
0.04	—	0.02	0.10	0.05	0.01	0.03	0.11	0.06	0.01	0.04	0.13
0.15	—	0.03	0.42	0.17	0.01	0.07	0.44	0.19	0.02	0.12	0.46
0.03	—	0.02	0.06	0.03	0.01	0.03	0.06	0.04	0.01	0.03	0.07
0.04	—	0.02	0.10	0.11	0.01	0.08	0.23	0.17	0.02	0.13	0.38
0.01	—	—	0.02	0.01	—	—	0.02	0.01	—	0.01	0.03
0.04	0.01	0.03	0.10	0.18	0.05	0.15	0.34	0.31	0.09	0.25	0.60

TABLE B-9 Estimated Intake of Dioxins from Food by Females, 12+ Years Old, Pregnant and/or Lactating ($n = 112$)

Food Categories	Percent Consuming	Food Intake (g/kg body weight/d)
All foods	100.0	33.3
Dairy foods and mixtures	97.9	6.1
Meat and mixtures	82.8	2.1
Poultry and mixtures	56.1	1.5
Fish and mixtures	25.0	1.0
Eggs and mixtures	36.1	1.0
Fruits, vegetables, and mixtures	99.1	6.7
Fats, oils, and mixtures	71.7	0.3
Other foods and mixtures[d]	100.0	17.4

[a]Reflects treatment of samples for which no dioxin congener was detected.
[b]LOD = limit of detection.
[c]< 0.0005 pg/kg body weight/d.
[d]Grains and mixtures, legumes and mixtures, beverages (other than milk and juice), candy.

NOTE: Data represent 2-day averages generated using U.S. Department of Agriculture (USDA) sample weights.
DATA SOURCE: Dioxin concentrations: U.S. Food and Drug Administration Total Diet Study (1999–2001); food consumption: 1994–1996, 1998 USDA Continuing Survey of Food Intakes by Individuals.

Consumers' Dioxin Toxic Equivalents (TEQ) Intake (pg/kg body weight/d)

Nondetects[a] = 0				Nondetects = 0.5 (LOD)[b]				Nondetects = LOD			
	Percentile				Percentile				Percentile		
Mean	10	50	90	Mean	10	50	90	Mean	10	50	90
0.38	0.15	0.31	0.64	0.65	0.32	0.57	1.06	0.91	0.47	0.82	1.51
0.08	0.01	0.06	0.16	0.11	0.02	0.09	0.22	0.13	0.02	0.10	0.30
0.17	0.02	0.09	0.39	0.19	0.03	0.11	0.41	0.22	0.04	0.13	0.41
0.04	—[c]	0.02	0.11	0.05	0.01	0.03	0.12	0.06	0.01	0.04	0.15
0.05	—	0.02	0.14	0.08	0.01	0.06	0.15	0.10	0.02	0.08	0.16
0.04	0.01	0.04	0.07	0.04	0.01	0.04	0.09	0.05	0.01	0.04	0.09
0.04	—	0.01	0.11	0.12	0.02	0.09	0.28	0.20	0.03	0.14	0.46
0.01	—	—	0.02	0.01	—	—	0.02	0.01	—	0.01	0.03
0.06	0.01	0.04	0.13	0.19	0.04	0.17	0.36	0.31	0.07	0.28	0.59

TABLE B-10 Estimated Intake of Dioxins from Food by Males and Females, 1+ Years Old (*n* = 19,043)

Food Categories	Percent Consuming	Food Intake (g/kg body weight/d)
All foods	100.0	38.0
Dairy foods and mixtures	91.5	7.0
Meat and mixtures	83.0	2.5
Poultry and mixtures	57.7	1.8
Fish and mixtures	23.4	1.5
Eggs and mixtures	30.8	1.0
Fruits, vegetables, and mixtures	98.0	7.1
Fats, oils, and mixtures	70.4	0.3
Other foods and mixtures[d]	100.0	20.7

[a]Reflects treatment of samples for which no dioxin congener was detected.
[b]LOD = limit of detection.
[c]< 0.0005 pg/kg body weight/d.
[d]Grains and mixtures, legumes and mixtures, beverages (other than milk and juice), candy.

NOTE: Data represent 2-day averages generated using U.S. Department of Agriculture (USDA) sample weights. Breastfeeding children are excluded. Pregnant and/or lactating females are included.
DATA SOURCE: Dioxin concentrations: U.S. Food and Drug Administration Total Diet Study (1999–2001); food consumption: 1994–1996, 1998 USDA Continuing Survey of Food Intakes by Individuals.

Consumers' Dioxin Toxic Equivalents (TEQ) Intake (pg/kg body weight/d)

Nondetects[a] = 0				Nondetects = 0.5 (LOD)[b]				Nondetects = LOD			
	Percentile				Percentile				Percentile		
Mean	10	50	90	Mean	10	50	90	Mean	10	50	90
0.46	0.12	0.33	0.96	0.77	0.30	0.62	1.41	1.07	0.43	0.88	1.91
0.09	0.01	0.04	0.21	0.12	0.01	0.07	0.28	0.15	0.02	0.09	0.35
0.23	0.01	0.13	0.53	0.26	0.03	0.16	0.58	0.29	0.04	0.18	0.64
0.06	—[c]	0.03	0.13	0.07	0.01	0.04	0.15	0.08	0.01	0.05	0.17
0.16	—	0.04	0.45	0.18	0.01	0.08	0.46	0.21	0.02	0.11	0.49
0.04	0.01	0.03	0.09	0.05	0.01	0.03	0.09	0.05	0.01	0.04	0.10
0.05	—	0.02	0.12	0.13	0.01	0.08	0.29	0.22	0.03	0.14	0.48
0.01	—	—	0.02	0.01	—	—	0.03	0.01	—	0.01	0.03
0.06	0.01	0.04	0.13	0.21	0.06	0.17	0.41	0.36	0.10	0.29	0.72

TABLE B-11 Estimated Intake of Dioxin-like Compounds from Food by Consumers of High versus Low Combined Amounts of Meat, Poultry, and Fish

Population	Consumption of Meat, Poultry, and Fish	*n*
Males and females, 1–5 y, not breastfeeding	High	5,306
	Low	1,035
Males and females, 6–11 y	High	1,735
	Low	166
Males, 12–19 y	High	675
	Low	18
Males, 20+ y	High	4,608
	Low	110
Females, 12–19 y, not pregnant or lactating	High	615
	Low	71
Females, 20+ y, not pregnant or lactating	High	4,128
	Low	281
Females, pregnant or lactating	High	108
	Low	3
Males and females, 1+ y, includes pregnant and/or lactating women	High	17,175
	Low	1,684

[a]Reflects treatment of samples for which no dioxin congener was detected.
[b]LOD = limit of detection.

NOTE: Data represent 2-day averages generated using U.S. Department of Agriculture sample weights. Low combined meat, poultry, and fish consumption defined as 2-day average intakes less than 1 oz (29 g).

Dioxin Toxic Equivalents Intake (pg/kg body weight/d)			Dietary Comparison Measures		
Nondetects[a] = 0	Nondetects = 0.5 (LOD[b])	Nondetects = LOD	Energy (kcal/d)	Protein (g/d)	Total Fat (g/d)
1.17	1.76	2.35	1,554	56.6	57.1
0.76	1.26	1.77	1,241	39.7	42.3
0.72	1.14	1.55	1,906	67.3	69.6
0.39	0.77	1.15	1,576	45.9	51.2
0.53	0.89	1.25	2,737	99.4	101.9
0.18	0.47	0.77	2,119	63.2	69.3
0.42	0.69	0.96	2,422	95.5	92.1
0.19	0.40	0.61	1,772	51.4	56.0
0.39	0.69	1.00	1,884	66.7	68.9
0.17	0.41	0.66	1,435	40.2	45.6
0.34	0.59	0.83	1,620	63.8	60.0
0.16	0.38	0.60	1,303	37.7	40.2
0.38	0.65	0.91	2,092	78.7	76.7
0.29	0.54	0.79	2,014	62.5	69.7
0.47	0.78	1.08	2,015	77.5	75.4
0.33	0.64	0.95	1,420	42.0	45.5

C

Open Session and Workshop Agendas

OPEN SESSION

Implications of Dioxin in the Food Supply

December 19, 2001
National Academy of Sciences
Washington, DC

1:00 p.m. Environmental Protection Agency
William Farland

Agency for Toxic Substances and Disease Registry
Christopher DeRosa

European Union
Antoine Liem

2:15 Break

2:30 Interagency Working Group on Dioxin
Clifford Gabriel

U.S. Department of Health and Human Services
William Raub

U.S. Department of Agriculture
James Schaub

5:00 Adjourn

WORKSHOP #1

Implications of Dioxin in the Food Supply

February 19, 2002
National Academy of Sciences
Washington, DC

8:30 a.m.   Welcome on Behalf of the Food and Nutrition Board
Ann Yaktine, Food and Nutrition Board

8:35   Introductory Comments on Behalf of the Committee on the
Implications of Dioxin in the Food Supply
Robert Lawrence, Committee Chair

8:45   Dioxins in Human Foods and Animal Feeds: Effects of Cooking
Methods on Dioxins in Foods
Janice Huwe, U.S. Department of Agriculture

9:25   Intake Data from Consumption of Animal Products and Future
Directions of the National Health and Nutrition Examination
Survey
Clifford Johnson, Centers for Disease Control and Prevention

10:05   Break

10:25   Assays for Dioxin in Human Tissues: Update on Dioxin Assays in
NHANES Participants
Don Patterson, Centers for Disease Control and Prevention

11:20   Health Effects of Dioxins
Linda Birnbaum, U.S. Environmental Protection Agency

12:00 p.m. Lunch

1:00   Introductory Comments
Moderator: Robert Lawrence, Committee Chair

Populations at Risk from Potential Exposure to Dioxins in Food,
Tribal Council of Arizona and Tribal Women, Infants and
Children Program
Tamera Dawes, Inter Tribal Council of Arizona
Ken Jock, St. Regis Mohawk Tribe

2:00   Break

2:15   Issues with Dioxins in Foods and the Environment
        Michael F. Jacobson, Center for Science in the Public Interest
        Stephen Lester, Center for Health and Environmental Justice
        Gina Solomon, Natural Resources Defense Council
        Karen Perry, Physicians for Social Responsibility

3:15   Break

3:30   Perspectives from Industry
        Hilary Shallo, Egg Nutrition Center
        Sean Hays, Chlorine Chemistry Council

5:15   Open Forum

5:45   Adjourn

WORKSHOP #2

Implications of Dioxin in the Food Supply

April 2, 2002
National Academy of Sciences
Washington, DC

9:00 a.m.   Welcome on Behalf of the Food and Nutrition Board
            Ann Yaktine, Food and Nutrition Board

            Introductory Comments
            Robert Lawrence, Committee Chair

Methodological Analysis of Dioxins in Food and Feed

9:10   Economic Impact Studies of Food Safety Regulations
        John Eyraud, Eastern Research Group

10:00   Break

10:30   Food Monitoring for Dioxins
        Richard Canady, Food and Drug Administration

11:30   Group Discussion of Total Diet Study
        Richard Canady, Food and Drug Administration
        Elke Jensen, Food and Drug Administration
        P. Michael Bolger, Food and Drug Administration
        Karen Hulebak, U.S. Department of Agriculture

12:00 p.m. Lunch

Accumulation and Elimination of Dioxins in Humans

1:00	Dioxin Intake from Food: Current State of Knowledge and Uncertainties in Assumptions *Dwain Winters, U.S. Environmental Protection Agency*
2:00	Kinetics of Dioxin Elimination in the Ranch Hand Cohort *Joel Michalek, Brooks Air Force Base*
2:50	Break
3:00	Open Forum
4:00	Adjourn

D

Committee Member Biographical Sketches

ROBERT S. LAWRENCE, M.D. *(chair)*, is associate dean for Professional Education Programs, the Edyth Schoenrich Professor of Preventive Medicine, and professor of health policy in the Johns Hopkins Bloomberg School of Public Health. His expertise and research interests include community and social medicine, preventive medicine, international health, and the use of evidence-based decision rules to develop policy for clinical preventive services and community health services. Dr. Lawrence is a member of the Institute of Medicine (IOM) and has previously served the IOM as chair of several committees addressing pertinent issues in public health, including Exposure of American People to I-131 from Nevada Atomic Bomb Test: Implications for Public Health. Dr. Lawrence is a master of the American College of Physicians, a fellow of the American College of Preventive Medicine, and holds membership in the American Public Health Association, the Association of Teachers of Preventive Medicine, and Physicians for Human Rights.

DENNIS M. BIER, M.D., is professor of pediatrics, director of the U.S. Department of Agriculture (USDA) Children's Nutrition Research Center, and program director of the National Institutes of Health (NIH) General Clinical Research Center at the Baylor College of Medicine. He is a member of the IOM and currently serves as associate editor of the *Annual Review of Nutrition*, president of the NIH General Clinical Research Centers Programs Directors Association, and as member of the Expert Advisory Panel on Nutrition and Electrolytes of The United States Pharmacopeial Convention. Previously, Dr. Bier was professor of pediatrics and internal medicine at Washington University School of Medicine where he was codirector of the Division of Pediatric Endocrinology and Metabolism, director of the NIH Mass Spectrometry Resource, and program director of the NIH General Clinical Research Center at St. Louis Children's Hospital. He has been president of the American Society of Clinical Nutrition, editor-in-chief of *Pediatric Research*, chairman of the USDA Human Studies Review, councilor of the American Pediatric Society, chairman of the NIH Nutrition Study Section, chairman of the NIH General Clinical Research Centers Committee, chairman of the National Institute of Child Health and Human Development's Expert Panel Five-year Plan for Nutrition Research and Training, and chairman of the Kellogg Grain Nutrition Board. He has also served as a member of various other scientific advisory panels, including the HHS/USDA Dietary Guidelines Advisory Committee, the IOM Food and Nutrition Board, the Food and Drug Administration (FDA) Food and Advisory Committee, the Medical Science Advisory Board of the Juvenile Diabetes Foundation, the Task Force and Steering Committee of the Pediatric Scientist Development Program, and the Advisory Board of the National Stable Isotopes Resource at Los Alamos National Laboratory. Dr. Bier has authored more than 230 scientific publications and, for his research work, he has received the E. V. Cullum award from the American Institute of Nutrition, the Grace A. Goldsmith award from the American College of Nutrition, and the General Clinical Research Centers award for Excellence in Clinical Research from NIH.

ROBERT E. BROYLES retired from Purina Mills, Inc., after 34 years in regulatory, quality, environmental, and safety activities. In that time, he served as director of the Regulatory, Quality, and Safety Department, director of Regulatory Affairs, and manager of Regulatory Affairs and Quality Assurance of the Health Industries Division. He currently serves as a consultant to Purina Mills. Mr. Broyles is past chair of the Animal Health Institute's Regulatory Committee, Animal Drug Section, and of the American Feed Industry Association's Feed Control Committee. He is a member of the National Grain and Feed Association's Feed Industry Committee and serves as a faculty member for its Feed Quality Assurance Workshops. Mr. Broyles was awarded the American Feed Industry Association's Member of the Year Award for 1992–1993, and a Distinguished Service Award in 1995 by the Association of American Feed Control Officials for his contributions in regulatory and quality service.

DOROTHY R. CALDWELL, M.S., is the coordinator of the North Carolina Initiative for Healthy Weight in Children and Youth in the North Carolina Division of Public Health. She has extensive experience in public policy and nutrition, including food and nutrition assistance programs at local, state, and federal levels, and has been involved in many policy initiatives in national professional associations. Mrs. Caldwell has previously held the post of deputy administrator, Special Nutrition Programs, Food and Nutrition Services, USDA, where she was responsible for child nutrition programs, the Special Supplemental Program for Women, Infants and Children, and commodity distribution programs.

DAVID O. CARPENTER, M.D., is a professor of environmental health and toxicology, and the director of the Institute for Health and the Environment, School of Public Health at the State University of New York, Albany. He is principal investigator for a Fogarty International Center training grant in environmental and occupational health for fellows from Eastern Europe, Russia, Uzbekistan, and Mongolia. Dr. Carpenter serves on the Board of Directors of Healthy Schools Network, Inc. and is treasurer of the Pacific Basin Consortium for Hazardous Waste, Health, and Environment. He is a member of the New York State Public Health Association and the American Public Health Association; chair of the Board of Directors of Albany-Tula, Inc., an alliance between the capital district of New York and Tula, Russia; and cochair of the Workgroup on Ecosystem Health of the Science Advisory Board of the International Joint Commission. Dr. Carpenter is a long-time spokesperson for environmentally related health concerns of the Mohawk tribe of Akwesasne, New York. In 1999, Dr. Carpenter was awarded the Homer N. Calver Award from the American Public Health Association for studies in environmental health.

JULIE A. CASWELL, PH.D., is a professor of resource economics and adjunct professor of food science at the University of Massachusetts. Her research interests include the operation of domestic and international food systems, analyzing food system efficiency, and evaluating government policy as it affects systems operation and performance, with particular interest in the economics of food quality, safety, and nutrition. Her edited book publications include *Economics of Food Safety*, *Valuing Food Safety and Nutrition*, and *Global Food Trade and Consumer Demand for Quality*. Dr. Caswell has provided her expertise to the U.N. Food and Agriculture Organization and the Organization for Economic Cooperation and Development on food safety issues. From 1989–2002 she chaired Regional Research Project NE-165, an international group of over 100 economists who analyzed the operation and performance of the food system. She has held numerous senior positions with the American Agricultural Economics Association and the Northeastern Agricultural and Resource Economics Association.

KEITH R. COOPER, PH.D., is a professor and dean of research and graduate programs and senior associate director of the New Jersey Agricultural Experiment Station, Cook College, Rutgers—The State University of New York. Dr. Cooper is also the associate director of the Joint Graduate Program in Toxicology NIEHS Training Grant, and former chair of the Department of Biochemistry and Microbiology at Rutgers University. Dr. Cooper is a member of the Center for Marine and Coastal Research, the Environmental/Science Graduate Program, and the Joint Graduate Program in Toxicology. His research interest is xenobiotic metabolism in aquatic animals, including endocrine disrupting compounds, particularly dioxins, dibenzofurans, and phthalates, on finfish and bivalve mollusks. He is also developing both food web and physiological based pharmacokinetic models to better predict chemical movement both in the environment and within the organism of concern.

JAMES K. HAMMITT, PH.D., is a professor of economics and decision sciences in the Department of Health Policy and Management, the Department of Environmental Health, and the Center for Risk Analysis, and is director of the Program in Environmental Science and Risk Management at the Harvard University School of Public Health. His research interests include the development and application of quantitative methods of decision and risk analysis, health-risk management and benefit-cost analysis, and mathematical modeling to health and environmental policy. Dr. Hammitt is currently researching management of long-term environmental issues with important scientific uncertainties such as global climate change and stratospheric-ozone depletion, the evaluation of ancillary benefits and countervailing risks associated with risk-control measures, and the characterization of social preferences over health and environmental risks using revealed-preference and contingent-valuation methods. Dr. Hammitt is a member of the U.S. Environmental Protection Agency's Science Advisory Board and holds professional memberships in the Association of Environmental and Resource Economists and the Society for Risk Analysis.

GAIL G. HARRISON, PH.D., is a professor and chair of the Department of Community Health Sciences, UCLA School of Public Health. She also serves as associate director for Public Health Programs of the UCLA Center for Human Nutrition and assistant director for the Program for Healthy and At-Risk Populations in the Division of Cancer Prevention and Control, UCLA/Jonsson Comprehensive Cancer Center. Her research interests include dietary and nutritional assessment, international health and nutrition, and pediatric and maternal nutrition. She is a former member of the IOM Food and Nutrition Board and has served on several of its committees including the Committee on International Nutrition Programs.

JAMES T. HEIMBACH, PH.D., is president of JHeimbach LLC, which specializes in food and nutrition consulting. He has national and international experience with a broad range of issues regarding food and nutrition policy; food consumption behavior; assessment of dietary intakes of nutrients, food additives, and contaminants; safety evaluation of food and dietary supplement ingredients; and food regulation. Dr. Heimbach was formerly the chief operating officer of Technical Assessment Systems and a principal of ENVIRON International Corporation, following public service at FDA and as associate administrator and acting administrator of the Human Nutrition Information Service of USDA. Dr.

Heimbach is a fellow of the American College of Nutrition; a member of the Foods/Dietary Supplement Oversight committee of the Food and Drug Law Institute; and councilor, past division chair, and past section chair of the Institute of Food Technologists.

BARBARA A. KNUTH, PH.D., is a professor and chair of the Department of Natural Resources at Cornell University. She also serves as leader of the Human Dimensions Research Unit. Her research interests include assessment of the need for and effectiveness of risk communication, particularly with regard to contaminated fisheries. Dr. Knuth is first vice president of the American Fisheries Society and a member of the Fish and Wildlife Executive Committee of the National Association of State Universities and Land Grant Colleges and a previous member of the Great Lakes Science Advisory Board of the International Joint Commission, and the Board of Technical Experts of the Great Lakes Fishery Commission. She has previously served on the Committee on Improving the Collection and Use of Fisheries Data.

JAMES D. MCKEAN, D.V.M., J.D., is an extension veterinarian and professor in the Department of Veterinary Diagnostic and Production Animal Medicine at Iowa State University's College of Veterinary Medicine. His research interests include swine medicine and extension, food safety, and food law, and the assessment of chemical and drug residues in feed and food animals. Dr. McKean also serves as associate director of the Iowa Pork Industry Center, and has previously served on several national committees for governmental policy development, including the Swine Futures Team, the Taskforce on the Future of FSIS Veterinarians, and as chair of the AASV Pork Safety Committee.

PIETER J.J. SAUER, M.D., is a professor and chair of the Department of Pediatrics, University Hospital at the University of Groningen, the Netherlands. His research interests include nutrition, growth, and development in infants, specifically with respect to the effects of environmental contaminants like polychlorinated biphenyls on health and development in children. Dr. Sauer is a member of the National Health Council (the Netherlands) and the Neonatology Group of the Dutch Society of Pediatrics. He has been European chief editor of the journal Pediatric Research and president of the International Pediatric Research Foundation. He has served as the Dutch delegate at the Conference of European Pediatric Specialists and is a member of the Ethical Committee.

ROBERT E. SMITH, PH.D., is president of R. E. Smith Consulting, Inc., as well as adjunct professor in the Department of Food Science at the University of Illinois. Prior to consulting to the food industry, Dr. Smith spent almost 30 years as head of corporate research and development at Nabisco, Del Monte, Swift, and The Quaker Oats companies. He has national and international experience with the development of a broad range of food and feed products. His technical expertise includes food technology and nutrition, food engineering, packaging, food safety, quality assurance and food regulation. Dr. Smith is a former member of the Food and Nutrition Board and is a former president of the Institute of Food Technologists.

MICHAEL R. TAYLOR, J.D., is a senior fellow and director of the Risk, Resource and Environmental Management Division of Resources for the Future (RFF). Mr. Taylor also leads the Food System Program at RFF which addresses policy and institutional issues affecting the success of the global food and agricultural system in the areas of food security in developing countries, food safety, and the natural resource and environmental sustainability of agriculture. He is a former administrator of the USDA's Food Safety and Inspection Service and a former deputy commissioner for policy at FDA. In addition to his public service, Mr. Taylor has been a partner in the law firm of King & Spalding and vice president for Public Policy at Monsanto Company. He is currently cochair of the National Academies' Committee on Use of Third Party Toxicity Research with Human Participants and a member of the Subcommittee on Defining Science-based Concerns Associated with Products of Animal Biotechnology, and has previously served on the Committee on Scientific and Regulatory Issues Underlying Pesticide Use Patterns and Agricultural Innovation (IOM).

KATHERINE L. TUCKER, PH.D., is director of the Dietary Assessment and Epidemiology Research Program at the Jean Mayer USDA Human Nutrition Research Center on Aging. In addition, she is an associate professor and director of the Nutritional Epidemiology program in the School of Nutrition Science and Policy and an adjunct associate professor for the Department of Family Medicine and Community Health in the School of Medicine at Tufts University. Dr. Tucker's research interests include diet and health, dietary assessment methodology, and the nutritional status of high-risk populations. She is on the editorial board of the *Journal of Nutrition* and the *Ecology of Food and Nutrition*, and has previously chaired the Nutritional Epidemiology Research Interest Section of the American Society for Nutritional Sciences.